Household Demography and Household Modeling

The Plenum Series on Demographic Methods and Population Analysis

Series Editor: Kenneth C. Land, *Duke University, Durham, North Carolina*

ADVANCED TECHNIQUES OF POPULATION ANALYSIS
Shiva S. Halli and K. Vaninadha Rao

CURBING POPULATION GROWTH: An Insider's Perspective on the Population Movement
Oscar Harkavy

THE DEMOGRAPHY OF HEALTH AND HEALTH CARE
Louis G. Pol and Richard K. Thomas

FORMAL DEMOGRAPHY
David P. Smith

HOUSEHOLD COMPOSITION IN LATIN AMERICA
Susan M. De Vos

HOUSEHOLD DEMOGRAPHY AND HOUSEHOLD MODELING
Edited by Evert van Imhoff, Anton Kuijsten, Pieter Hooimeijer, and Leo van Wissen

MODELING MULTIGROUP POPULATIONS
Robert Schoen

THE POPULATION OF MODERN CHINA
Edited by Dudley L. Poston, Jr. and David Yaukey

A Continuation Order Plan is available for this series. A continuation order will bring delivery of each new volume immediately upon publication. Volumes are billed only upon actual shipment. For further information please contact the publisher.

Household Demography and Household Modeling

Edited by

Evert van Imhoff

Netherlands Interdisciplinary Demographic Institute
The Hague, The Netherlands

Anton Kuijsten

University of Amsterdam
Amsterdam, The Netherlands

Pieter Hooimeijer

Utrecht University
Utrecht, The Netherlands

and

Leo van Wissen

Netherlands Interdisciplinary Demographic Institute
The Hague, The Netherlands

Plenum Press • New York and London

Library of Congress Cataloging-in-Publication Data

On file

ISBN 0-306-45187-5

A Division of Plenum Publishing Corporation
233 Spring Street, New York, N. Y. 10013

10 9 8 7 6 5 4 3 2 1

Printed in the United States of America

CONTRIBUTORS

Thomas K. Burch • Population Studies Centre, University of Western Ontario, London, Ontario N6A 5C2, Canada

Daniel Courgeau • Institute National d'Etudes Démographiques, 27 rue du Commandeur, F-75675 Paris Cedex 14, France

Joop de Beer • Statistics Netherlands, Department of Population, P.O. Box 400, 2270 JM Voorburg, The Netherlands

Jenny de Jong Gierveld • Netherlands Interdisciplinary Demographic Institute (NIDI), P.O. Box 11650, 2502 AR The Hague, The Netherlands

Heinz P. Galler • Department of Econometrics and Statistics, Martin-Luther University of Halle-Wittenberg, Grosse Steinstrasse 73, D-06099 Halle, Germany

Hans Heida • FOCUS BV, Jacoba van Beierenlaan 81-83, 2613 JC Delft, The Netherlands

Pieter Hooimeijer • Faculty of Geographical Sciences, Utrecht University, P.O. Box 80115, 3508 TC Utrecht, The Netherlands

Nico W. Keilman • Statistics Norway, Division for Social and Demographic Research, P.O. Box 8131 Dep., N-0033 Oslo, Norway

Anton Kuijsten • Department of Planning and Demography, University of Amsterdam, Nieuwe Prinsengracht 130, 1018 VZ Amsterdam, The Netherlands

Jan H.M. Nelissen • Faculty of Social Sciences, Tilburg University, P.O. Box 90153, 5000 LE Tilburg, The Netherlands

Åke Nilsson • Population Unit, Statistics Sweden, S-70189 Örebro, Sweden

Notburga Ott • Johann Wolfgang Goethe University, Department of Economics, Senckenberganlage 31, D-60054 Frankfurt am Main, Germany

Christopher Prinz • European Centre for Social Welfare Policy and Research, Berggasse 17, A-1090 Vienna, Austria

Hakan Sellerfors • Population Unit, Statistics Sweden, S-70189 Örebro, Sweden

Evert van Imhoff • Netherlands Interdisciplinary Demographic Institute (NIDI), P.O. Box 11650, 2502 AR The Hague, The Netherlands

Leo van Wissen • Netherlands Interdisciplinary Demographic Institute (NIDI), P.O. Box 11650, 2502 AR The Hague, The Netherlands

Richard Wall • Cambridge Group for the History of Population and Social Structure, 27 Trumpington Street, Cambridge CB2 1QA, England

PREFACE

In 1992, a summer course 'Demographic Perspectives on Living Arrangements' as well as a one-day workshop 'Recent Issues in Household Modelling' were held in Wassenaar, The Netherlands. This volume is based on the lectures delivered during the summer course, as well as on the presentations made in the workshop. As such, the present volume combines the two elements of transfer of knowledge, on the one hand, and updating the state-of-the-art in the field of household demography, especially in household modelling, on the other hand. In organizing the contents and structure of this volume, we have aimed at creating a book that covers the field of household demography and household modelling in a certain logical and comprehensive way.

The purpose of this book is to offer a comprehensive treatment of recent developments in various aspects of the growing field of household demography. Since these recent developments have particularly occurred in household analysis and modelling, these topics will receive special emphasis. The book was written for demographers, social scientists, and planners who are involved in the study and projection of population in general, and of households in particular.

Apart from the introductory first chapter and the concluding final chapter, the book has been organized in three parts. Part I, 'Trends and Theories', puts household demography in perspective, both from a historical and a theoretical point of view. Part II is devoted to issues of 'Data and Analysis'. Recently, a range of household models have been developed and are being applied for simulation and projection purposes. In Part III, 'Models', five contributions in this area are brought together.

The summer course was organized with financial assistance from PDOD (Netherlands Graduate School of Research in Demography), NIDI (Netherlands Interdisciplinary Demographic Institute), NIAS (Netherlands Institute for Advanced Study in the Humanities and Social Sciences), and the Praemium Erasmianum Fund, under the auspices of the European Association for Population Studies (EAPS). The workshop, which was inserted into the summer course, was organized and financed by the Northern Centre for Population Studies (CEPOP/N) of the Netherlands Organization for Scientific Research (NWO).

In finalizing this publication, the editors were assisted by smoothly cooperating contributing authors and the support of Plenum Publishing Corporation. Angie Pleit-Kuiper of NIDI did the English editing. Joan Vrind of NIDI served as technical editor. Jolande Siebenga of NIDI prepared the subject index. The help of all these persons is gratefully acknowledged.

The Hague, June 1995

CONTENTS

Part III: Models

1. INTRODUCTION

Evert van Imhoff[1], Anton Kuijsten[2], and Leo van Wissen[1]

[1]Netherlands Interdisciplinary Demographic Institute (NIDI)
P.O. Box 11650
2502 AR The Hague
The Netherlands
[2]Department of Planning and Demography
University of Amsterdam
Nieuwe Prinsengracht 130
1018 VZ Amsterdam
The Netherlands

1.1. HOUSEHOLD DEMOGRAPHY: A STAR STILL RISING

In much of contemporary demography, attention is shifting from the individual to the level of living arrangements, families, and households. This shift is motivated by at least three factors.

Firstly, demographic processes are to a large degree dependent on the household situation of the individuals involved. Thus, the fertility of women living alone is quite negligible compared to women living together with a man. Mortality is typically lower for individuals living with a partner than for persons living alone or for lone parents. Migration propensities are strongly related to household composition. Many more of these examples might be given.

Secondly, in many social, economic, and cultural processes, the household rather than the individual is the relevant unit of analysis. For a wide range of human behaviour in general, the household rather than the individual is the relevant unit of decision-making. Therefore, developments in the number and composition of households are crucial in understanding many societal trends and phenomena. Examples include budget expenditure, housing demand, transportation, labour force participation, demand for public services such as social security and care, attitudes, and so on.

Household Demography and Household Modeling
Edited by E. van Imhoff *et al.*, Plenum Press, New York, 1995

Thirdly, these developments as such in the number and composition of households in many countries, especially in Europe, over the past three decades have been impressive indeed. In many countries, the share of families in the total number of households has decreased steadily over the last decades, while various new forms of co-residence (the basic conceptual criterion in household demography) have emerged: one-parent households, consensual unions, living-apart-together relationships, to mention only a few. These trends have caught the attention not only of demographers, other social scientists, and policymakers, but also of much wider audiences. Dating and mating behaviour, intimate relationships, living arrangements, and parenthood are booming topics that have invaded popular magazines and television shows. Paradoxically, while issues such as intimate relationships, living arrangements, and reproductive behaviour have become regarded as purely private matters of individualized decision makers, at the same time people seem to be more than ever before inclined to put their private matters on public display, at the verge of media exploitation.

Household and family demography, broadly defined, has not needed this growing public interest in order to flourish during the past decades. As documented by Burch in Chapter 4 of this volume, especially since the mid-1980s, a growing number of review articles and collections of papers have delineated the field, developed the methods of description, analysis, and modelling, and presented results that enriched the discipline with a better, albeit still far from complete, understanding of the current processes of household dynamics.

It is always difficult to indicate where and when such new developments started precisely, if only because publications often lag behind the emergence of new ideas for several years. Defining our topic in a rather broad sense, and emphasizing the aspect of international cooperation and comparative research as much as that of analysis and modelling of household and family structures, the OECD-commissioned publication of Le Bras (1979), which focusses on the problems of international comparison of statistical data on the position of children in households and families in modern-industrialized countries, might be looked upon as the first substantial contribution to the field, and without any doubt was an eye-opener for all who ever might have thought that the later defined core problems of family and household demography were easy to solve. This publication was soon followed by a number of review articles (Burch 1979 and 1982; Bongaarts 1983; Brass 1983; Ryder 1985; DeVos & Palloni 1989) which, when taken together, provided a concise state-of-the-art overview as well as an impressive research agenda.

What strikes most, however, is the speed with which the new topic and its research agenda was adopted by the international scientific community, particularly in the way of institutionalized network sponsoring. Pride of place in this respect must be given to the International Union for the Scientific Study of

Population (IUSSP) which mandated two subsequent so-called Scientific Committees, for a period of six years altogether, to the task of elaborating the field of family demography. As described by Höhn (1992, p. 4), the principal objective of the first committee was formal, whereas the second was asked to devote its activities to the substantive aspects of the field. The main activities of both committees were the organization of international workshops and seminars and the publication of their proceedings: two resulting from the activities of the first committee (Bongaarts, Burch & Wachter 1987; Grebenik, Höhn & Mackensen 1989), and three emerging from the activities of the second (Prioux 1990; Berquó & Xenos 1992; Xenos & Kono, forthcoming). Closely related to our topic were a seminar followed by a publication of the IUSSP Committee on Economic Consequences of Alternative Demographic Patterns (Ermisch & Ogawa 1994), and of the IUSSP Committee on Gender and Population (Mason & Jensen, forthcoming).

Other organizations rose to the occasion too. The Committee for International Coordination of National Research in Demography (CICRED) implemented a four-year Inter-Centre Co-operative Research Project on current methodological aspects and specific issues of family demography (CICRED 1980, 1984). The European Co-ordination Centre for Research and Documentation in Social Sciences (Vienna Centre) launched an international comparative research project on 'Changes in the life patterns of families in Europe' (Boh *et al.* 1989), and the University of Bielefeld co-ordinated a similar international comparative project on 'Family policies and family life in Europe' (Kaufmann *et al.*, forthcoming). The European Population Committee of the Council of Europe commissioned an Expert Committee to provide an overview of household and family trends in its Member Countries (Council of Europe 1990), and the United Nations called an international conference on 'Ageing Populations in the Context of the Family' (United Nations 1994).

Looking back at all these activities and publications, one gets the impression that during this largely institutionally-induced process of growing towards maturity of the field of family and household demography, the scientific community involved has managed quite well in keeping the balance between theory, empirical analysis, and formal modelling as was so strongly advocated by those who initiated the process (in particular, Burch 1979 and Bongaarts 1983). Concern for keeping such a balance was an important point of departure too for the editors of a book (Keilman, Kuijsten & Vossen 1988) that resulted from an international workshop on household formation and dissolution, held in December 1984 in The Hague, organized by the Netherlands Interdisciplinary Demographic Institute (NIDI), in co-operation with the University of Amsterdam, the Catholic University of Brabant, and the Netherlands Demographic Society, and under the auspices of the European Association for Population Studies (EAPS). In view of the fact that household demography was regarded as still being in its infancy, the main purpose of the workshop and of the book published afterwards was to make an inventory of existing opinions and insights.

The 1984 workshop resulted in a growing interest in family and household demography, also in the Netherlands. When in 1991 the Netherlands Organization for Scientific Research (NWO) initiated its Research Program for Population Studies, household demography was designated as one of the key areas for the newly-founded Northern Centre for Population Studies (CEPOP/N).

At the same time, stimulated by the apparent success of this 1988 volume, in 1991 the organizers of the 1984 workshop took up the idea of realizing some kind of follow-up. This initiative resulted in the NIDI/PDOD Summer Course 'Demographic Perspectives on Living Arrangements'.

1.2. DEMOGRAPHIC PERSPECTIVES ON LIVING ARRANGEMENTS

The NIDI/PDOD Summer Course 'Demographic Perspectives on Living Arrangements' was held in Wassenaar, The Netherlands, from 8-22 July 1992, at the premises of the Netherlands Institute for Advanced Study in the Humanities and Social Sciences (NIAS), and under the auspices of the European Association for Population Studies (EAPS). The course was organized with financial assistance from PDOD (Netherlands Graduate School of Research in Demography), NIDI, NIAS, and the Praemium Erasmianum Fund. Twenty-two junior researchers from eleven different countries participated in the course. All students were required to submit a paper before the start of the summer course. This paper was presented and discussed during the summer course, both in a plenary session and in meetings with individual faculty members. Each student was assigned at least one, and often two, supervising faculty members.

A total of ten persons (representing seven different countries) acted as faculty members. The task of the faculty members was twofold: to assist individual students in developing their paper and future research plans, and to deliver an extensive plenary lecture on a particular topic in household demography. The topics raised during the summer course covered a rather wide range within the field of household and family demography. In chronological order, the lectures covered the following issues: historical and current trends, explanatory theories, data sources and their properties and peculiarities, and models for analysis and projection.

The main objective of the summer course was clearly the transfer and dissemination of knowledge. On the other hand, the gathering of several international experts in the field of household demography also offered an ideal opportunity for scientific exchange and for updating the state-of-the-art. With this objective in mind, a one-day workshop on 'Recent Issues in Household Modelling' was inserted into the summer course, organized by the NWO Northern Centre for Population Studies (CEPOP/N).

In the present volume, we have tried to combine these two elements, transfer of knowledge, on the one hand, and updating the state-of-the-art in the field of household demography, especially in household modelling, on the other hand. The contributions in this volume are based on the lectures delivered during the summer course, as well as on the presentations made in the workshop. In organizing the contents and structure of this volume, we have aimed at creating a book that covers the field of household demography and household modelling in a certain logical and comprehensive way.

1.3. PURPOSE AND OUTLINE OF THIS BOOK

The purpose of this book is to offer a comprehensive treatment of recent developments in various aspects of the growing field of household demography. Since these recent developments have particularly occurred in household analysis and modelling, these topics will receive special emphasis. The book was written for demographers, social scientists, and planners who are involved in the study and projection of population in general, and of households in particular.

Apart from this introductory first chapter and the concluding final chapter, the book has been organized in three parts. Part I, 'Trends and Theories', puts household demography in perspective, both from a historical and a theoretical point of view. Part II is devoted to issues of 'Data and Analysis'. Recently, a range of household models have been developed and are being applied for simulation and projection purposes. In Part III, 'Models', five contributions in this area are brought together. In the remainder of this introductory chapter, the contents of and contributions in each part will be discussed in turn.

1.3.1. Trends and Theories

Most social scientists would agree that any scientific investigation of a particular social phenomenon should contain several elements. Typical elements include a description of the wider context of the phenomenon under study ("where does it happen"), the development of an explanatory framework ("why does it happen"), the collection and analysis of empirical data ("how does it happen"), and the construction of projection models ("what will happen"). These same elements can be found in this book, since household demography is part of the social sciences.

Fewer social scientists would agree on the order in which these various elements should be tackled by the scientific investigator. The rigid Popperians would argue that one should always start with a (testable) theory, subsequently to be confronted with empirical data. The empiricists usually start with collecting data, and then look for theoretical explanations of what the data show. Some forecasters are primarily interested in reliable forecasts and construct extrapolation models that are almost void of theoretical and contextual considerations. Historians tend to pay less attention to theory, being content with anything that historical data may show in its own right.

In organizing the various contributions in this book, the editors have tried to follow an ordering that seems to be most proper for the topic at hand, i.e. household demography. Although we tend to follow the Popperian view that all scientific inquiry should start with theory (conjectures) and end with data (refutations), we simultaneously realize that, in real life, theory in the social sciences is never independent of the social context in which the researchers find themselves. Our theoretical notions of the processes that drive household formation and dissolution behaviour are not developed in the secluded areas of our studies and libraries. Rather, our thinking is either explicitly or implicitly shaped by our empirical knowledge of household trends in general.

For this reason, Part I of this volume is devoted to both trends and theories. The two chapters on household trends sketch the social context in which the chapter on theories is embedded. Without the trends, to be described in Chapters 2 and 3, the theories described in Chapter 4 would not have been what they are.

The social context, or general household trends, contains a time dimension as well as a spatial dimension. The time dimension is concerned with the historical development of household structures, whereas the spatial dimension is concerned with national or regional differences in household structures. The two dimensions are not independent, since regional differences change over time, and historical trends differ between regions.

Both dimensions are studied in Chapters 2 and 3, although on a different time scale and with different emphases. Together, these two papers sketch a comprehensive picture of the past and present trends in household structures in Europe. In addition, both papers address conceptual issues relevant for defining household structures. Chapter 2, by Richard Wall, is of a truly historical nature: it studies household structures over a time span of several centuries. For such a long period, the actual inter-temporal analysis, in the sense of studying *changes* in household structures over time, is obviously hampered by severe data limitations. Consequently, the inter-temporal analysis is only made for England and Wales. However, this analysis is linked to a spatial analysis of European household patterns in the 17th/18th century. Wall's main conclusion is that there has been a considerable degree of continuity between household patterns in the past and present.

Chapter 3, by Anton Kuijsten, has a more modest time perspective, being limited to the last few decades. On the other hand, the spatial dimension is more prominently presented in Kuijsten's analysis. The chapter presents household and family structures for a large number of European countries at different points in time. These data permit Kuijsten to study how and when the so-called Second Demographic Transition has affected demographic behaviour in different parts of Europe. The results of this exercise are relevant for the discussion on convergence versus divergence of household structures over time.

After these two sketches of the social context, in Chapter 4 Thomas Burch treats the truly theoretical aspects of household demography. This chapter reviews existing theory, and suggests directions along which new, better theory could be developed. Burch places recent developments in household demography in the realm of the broader framework of decision-making theory. He observes that theory is not demography's strong suit, and that progress in the area of household demography is particularly difficult because of the complexity of the subject. Burch offers several suggestions for future theoretical work, including the suggestion to start taking time seriously. This suggestion should create new hope in the hearts of the demographers of the world, since demography *par excellence* is the social science in which time plays a central role. His contribution ends with a plea for stating theories with sufficient rigour and formalization that at least some of the many competing views can be falsified and subsequently dropped.

Demographers can do more, however, than just taking (more) seriously the temporal dimension of family and household formation in their efforts of theory building. As mentioned by Burch, household decision-making (including decision-making with respect to family formation and dissolution) should be seen as a *process*, something taking place over time, with important temporal orderings among choices and observable events (including the traditional 'vital events' of demography). A person's current household and/or family status depends to a large extent on past demographic behaviour of both that person and the other household members. At the aggregate level, current household and family structures reflect the distribution of momentary stages in which all these individual-level processes are.

The implication for a specifically *demographic* contribution to theory building has to do with the two-way relationship that can be deduced from this insight. On the one hand, as was so aptly demonstrated by Wachter *et al.* (1978) for historical England and Wales, demographic conditions, first of all mortality and fertility regimes, act as constraints for households and family structures. For example, in a high-mortality situation, one simply cannot expect proportions of three-generation households, or average households sizes, to be as large as the norms and values in force might suggest. Similarly, to take an example from the present-day Western societies experiencing the so-called Second

Demographic Transition, in a long-lasting below-replacement fertility situation one should expect that proportions of one- and two-person households, of couple-only households headed by aged persons, and of households headed by widows or widowers, will rise anyway just because of population ageing, without any connection with the often emphasized processes of individualization and self-actualization assumed to be responsible for the observed declines in overall average household size.

On the other hand, and precisely because "the insight that distinguishes family demography from conventional individual-oriented demography is that the behaviour of an individual is dependent on familial relationships" (Ryder 1992, p. 166) (and, we insist to add, on household situations as well), family and household structures co-determine aggregate levels of the vital events from traditional individual-oriented demography. Perhaps national levels are too general to demonstrate this, national data being too much 'averaged-out', but we may assume that at least a part of current spatial differences in levels of demographic indicators can be looked upon as being the result of spatial sorting-out processes of living arrangements at lower regional levels, to a large extent driven by the selectivity of internal and international migration in terms of age, sex and life style preferences. Such a two-way relationship between the set of family-demographic indicators and that of indicators from the field of individual-oriented demography, as demonstrated for a selection of countries and sub-national regions in the URBINO project (Strohmeier & Matthiessen 1992) might lead to the emergence of self-perpetuating feed-back systems of family demographic developments, particularly at regional and local levels. The study of this would constitute a new research agenda in itself, the results of which would contribute to specifically demographic theory building in the field of household and family demography.

1.3.2. Data and Analysis

Few disciplines are as dependent on data as demography. In fact, we already entered the world of data analysis in Part I, when describing trends in household structures. This is essentially a form of time series analysis of historical household data. In Part II, the time period of observation lies not in the past but concerns the present. Now household demography is not only concerned with time series of household stock data, but especially with the study of transitions and events, which implies that, in order to study these processes, detailed data at the individual level over a certain period of time should be available. Not surprisingly, the issue of longitudinal data is of major importance in most chapters in this section.

Household data may be used for different purposes. Basically, a distinction can be made between descriptive and analytic purposes. While this distinction appears to be obvious for most researchers, its consequences for the observation plan are not always completely understood. If data are collected with the purpose of giving a reliable picture of the household structure at one point in time, conducting a large cross-sectional survey is often the optimal strategy. If attention is focussed on the (causal) analysis of household processes, a longitudinal observation plan is much more appropriate. The advantages of pure cross-sections for other than descriptive analyses are limited if no retrospective questions are included in the survey. On the other hand, longitudinal data are usually not reliable for providing cross-sectional descriptive information, unless much attention is given to attrition and refreshment strategies. In principle, longitudinal data are much more suited for the causal analysis of household processes. One problem, as Ott observes in Chapter 7, is the relatively low incidence of some household events, resulting in sample size problems. A stratified sampling scheme might offer a solution in these cases.

Longitudinal data may be collected in two ways: either by a panel or by a retrospective survey. Both have advantages and disadvantages, which are discussed by De Jong Gierveld in Chapter 6, and by Ott in Chapter 7. Here, the link between the observation plan and analysis is very important. The observation plan should be taken into account when analysing household processes. For instance, in a retrospective sample design, persons are observed conditional on remaining in the population until the time of observation. Consequently, inferences on processes such as mortality and outmigration cannot be made from these data, and rates pertaining to other events are conditional rates. In a panel, selective attrition and length bias should be taken into account when making inferences about the population.

Nowadays, in many countries, a detailed picture may be given of the distribution of persons over household categories and household positions, based on one or more of three sources of information: registers, censuses, and surveys. Definition and measurement issues are crucial in the proper study of household processes. In Chapter 5, Keilman focusses on household definitions, and observes that international comparisons of household structure are difficult due to the differences in definitions used. His second conclusion is more positive: trends in household structures are more comparable, since the bias resulting from differences in definitions is generally stable over time.

Keilman's conclusions are based on comparisons of macro trends. At the micro level of the individual, the picture is more complex. The combination of fuzziness and the required longitudinal nature of household data at the micro level makes life of a household demographer difficult. If definitions are not clear, a surveyed person is likely to give different answers at different times when in fact no change occurred. The opposite may happen as well: no change is reported when in reality an event *did* occur. Therefore, the measurement of individual change, in the form of transitions and events, may become heavily polluted with noise if these measurement problems are not taken seriously. A number of examples of inconsistencies in panel data are revealed by Ott, who deals with potentials and pitfalls of using longitudinal data in the study of household structures. Her chapter relies on experiences with the German Socio-Economic Panel Study, offering interesting views not only in the sphere of data issues, but also in terms of applied longitudinal analysis of household data. De Jong Gierveld differentiates between data obtained from censuses and surveys, and specifically focusses on the merits of household surveys in the collection of household data. Due to the complexity and fuzziness of the household concept, a well-designed survey, conducted by well-trained surveyors, is superior to other means of collecting household data.

The consequences of the complexity of the household concept and household processes are not only felt when *collecting* household data, but apply even more forcefully in the *analysis* of household processes. Both Courgeau in Chapter 8, and Galler in Chapter 9, show the implications of having to deal with multiple events and processes in a dynamic setting. Unfortunately, no simple methodological answers can be given here. Galler gives an overview of the main issues when studying multiple events simultaneously; his interest is in competing events, a notion that can fruitfully be applied to the analysis of the interdependence between different household processes. Courgeau deals with household histories, which at the simplest level include the processes of household formation and dissolution. Both authors have to deal with the dimension of time, although in different ways. For Courgeau, the question is basically a theoretical one: What is the role of time in household formation and dissolution processes? Is it the time elapsed since the start of the household that matters, or the length of the time spell since the last event? For the moment, the second assumption seems the more plausible, given the available data, but no definite answers can be given yet. For Galler, the issue of time is methodological: how to represent time in dynamic models. But of course in the choice between a continuous and a discrete representation of time in the analysis and modelling of household processes, theoretical arguments are important as well.

In summary, part II offers a number of highly relevant contributions for the collection and analysis of household data and processes. In response to Burch's suggestion in Chapter 4, to start to take time seriously, the chapters in part II show that, at least from a methodological point of view, time *is* taken seriously. However, it is also clear that the results of the types of complex models presented here have serious implications for simulation purposes. We will come back to this issue in the epilogue.

1.3.3. Models

The third part of this book is concerned with household models. Since the term 'model' is notoriously ill-defined, it may be helpful to indicate more clearly what these household models are.

For the purpose of the five chapters in part III, we define a model as a simplified (stylized), *quantitative* description of reality. This definition excludes several meanings in which the term 'model' is also commonly used. All models are simplifications, most models describe reality (with varying degrees of success), but only some models are quantitative. In particular, the household models to be presented in the following chapters do not contain any unknown parameters. This is in contrast to theoretical models, which describe how variables are linked without specifying the exact functional form of the relationship. This is also in contrast to estimation models, in which the functional form is specified but the parameters of the function are not.

All household models in part III of this volume are simplified, quantitative descriptions of the processes that determine household structures. They are simplified in the sense that not all variables affecting household structures are included in the model (they are also simplified in terms of functional form). They are quantitative in the sense that one set of numbers goes in and another set of numbers comes out. As a matter of fact, because of the obvious efficiency gains that are achieved if all necessary calculations are made by computer rather than by hand, the household models to be presented are also concrete computer programs. Strictly speaking, the computer program that actually produces the numbers coming out of the model should not be confused with the model itself. In practice, the term 'model' is frequently also applied to the computer program, and, admittedly, the dividing line is not always easy to discern.

Compared to other social sciences, demography is rather heavily future-oriented. A relatively large share of the total research efforts is devoted to forecasting and projection. Household demography is no exception. The household models that have recently been developed are primarily intended for projection purposes. Although the performance of a projection model can naturally be verified by checking whether the model accurately describes observed trends in the past, its main interest lies in the future. All five household models to be presented in the next chapters share this 'forward-looking' orientation.

Household models can be classified according to several criteria. One classification is between *static* and *dynamic* models. Static household models extrapolate proportions of household heads ('headship rates') for different population categories. These static models focus on changes in household structures at subsequent points in time, thus essentially treating the underlying processes of household formation and dissolution as a black box. In contrast, dynamic household models explicitly focus on the flows that underlie household changes. All the models in this book are dynamic models, with the possible exception of the hybrid dynamic-static model presented by Joop de Beer.

Another distinction is between *macro* and *micro* models. In macro models, numbers of demographic events (e.g. 'male entries into first marriage') are projected by applying rates (e.g. 'first marriage rate') to a *group* of individuals of a certain size (e.g. 'number of male bachelors'). Macro models in demography are almost universally of the multistate type, and the macro models in this book are no exception. In micro models, events are projected by applying probabilistic decision rules to individual persons, separately for a number of individual persons. Both types of models rely on the Law of Large Numbers, but in a different way: macro models assume that the groups are so large that projected number of events may be set equal to their expected values, whereas micro models assume that the number of repetitions of the probabilistic experiment is so large that the resulting projected number of events will approximately equal their expected values.

If the number of individual characteristics included in the model becomes larger, micro models have a competitive advantage: macro models tend to become unmanageable quite rapidly as the dimensions of the state space increase. On the other hand, it is often claimed that micro models are more powerful in taking account of interaction effects between variables. However, this claim is only valid for the model equations themselves, not for the parameters that go into the equations. Whenever a model-builder wants to include such interaction effects into the model, the parameters should be estimated in such a way that the interaction effects are properly taken into account. Thus, the estimation phase of model building is equally complex for micro and macro models alike. In this sense, micro models tend to give a false sense of security, since they make it very tempting to disregard interaction effects in the estimation phase, a practice which may lead to quite misleading results.

A third distinction is between *purely demographic* models and *mixed demographic / non-demographic* models. In purely demographic models, the future changes in household structure are explained from characteristics of the current population only, i.e. from the current population classified by age, sex, marital status, and/or household position. In mixed models, other non-demographic variables have a role to play as well. To this category belong models with a housing market module, where the availability of housing acts as a limiting or enabling factor for household formation and dissolution.

The first three models in this last part of the book are purely demographic, multistate macro models. Chapter 10 by Christopher Prinz, Åke Nilsson & Hakan Sellerfors is concerned with the choice of states in a multistate household model. They show how alternative levels of disaggregation in generalized marital status models affect the models' results. In a way, their models are not household models in the strict sense of the word, since they only describe the relationship between individuals and their spouse (if any). Thus, for instance, dependent children remain out of focus. However, the results of their sensitivity analysis are directly relevant for other, more elaborate household models.

Chapter 12 by Evert van Imhoff presents the multistate household projection model LIPRO. This model classifies the population by *household position*. This classification has been specified in such a way that a projection of the population by household position can be easily transformed into a projection of households by household type. LIPRO integrates the most recent ideas on dynamic household modelling in a single software package. However, the advantage of having a fully dynamic model is bought at the expense of huge data requirements.

Chapter 11 by Joop de Beer presents the model that is used by Statistics Netherlands (NCBS) for producing the official household forecasts for the Netherlands. It combines elements of the marital status type models investigated by Prinz *et al.*, and the dynamic household models such as Van Imhoff's. The NCBS model is a two-stage model, starting with a dynamic projection by marital status, followed by a static (headship rate) projection over household positions. As the distribution by household position is conditional on marital status, the NCBS model is dynamic to the extent that household changes are associated with nuptiality behaviour.

The last two models in part III of this book are mixed demographic / non-demographic models. Chapter 13 by Pieter Hooimeijer and Hans Heida presents a dynamic macro (multistate) household model with an endogenous housing market module. The inclusion of the housing market into the model makes it particularly suitable for household projection on a sub-national spatial scale. Finally, Chapter 14 by Jan Nelissen is about micro household models. His contribution gives an overview of existing micro models, showing their strong and weak points when compared to macro models. As already mentioned, the main strength of micro models is that they offer much more scope for introducing additional variables into the model. These potentials are illustrated by Nelissen for the case of the interaction between household variables and labour market variables.

Household demography has been, and always will be quite a complex business. This is due to the inherent variety in household patterns, as well as to the large number of determinants of household formation and dissolution behaviour. Except for the most simple, i.e. almost trivial, types of household models, therefore, the data requirements are so large that data availability and/or reliability is a seriously binding constraint. As the various model presentations in Part III of this volume illustrate, each model has to find a balance between comprehensiveness, on the one hand, and data requirements, on the other hand. However, within these limitations, there are still quite a number of degrees of freedom in model specification, of which some are clearly more preferable than others. This is nicely illustrated by Prinz, Nilsson, and Sellerfors, who show that some model specifications produce information that is not only irrelevant but even simply wrong. A main challenge for household demography thus seems to be to produce models that are relevant and correct. We trust that this volume will make a contribution to this challenge. In the closing chapter to this volume, we will look back and evaluate how much progress current household demography has made in this respect.

REFERENCES

BERQUÓ, E., & P. XENOS (eds.) (1992), *Family Systems and Cultural Change*, Clarendon Press: Oxford.

BOH, K., M. BAK, C. CLASON, M. PANKRATOVE, J. QVORTRUP, G.B. SGRITTA, & K. WAERNESS (eds.) (1989), *Changing Patterns of European Family Life: A Comparative Analysis of 14 European Countries*, Routledge: London/New York.

BONGAARTS, J. (1983), The Formal Demography of Families and Households: An Overview, *IUSSP Newsletter*, No. 17, pp. 27-42.

BONGAARTS, J., T.K. BURCH, & K.W. WACHTER (eds.) (1987), *Family Demography: Methods and their Application*, Clarendon Press: Oxford.

BRASS, W. (1983), The Formal Demography of the Family: An Overview of the Proximate Determinants. *In: The Family*, British Society for Population Studies Occasional Paper 31, Office of Population Censuses and Surveys: London, pp. 37-49.

BURCH, T.K. (1979), Household and Family Demography: A Bibliographic Essay, *Population Index*, Vol. 45(2), pp. 173-195.

BURCH, T.K. (1982), Household and Family Demography. *In:* J.A. ROSS (ed.), *International Encyclopedia of Population*, Vol. I, Free Press: New York/London, pp. 299-307.

CICRED (1980), *A New Approach to Cooperative Research in the Population Field*, CICRED: Paris.

CICRED (1984), *Demography of the Family*. Inter-Center Cooperative Research Programme, Project No. 2, Final Report, CICRED: Paris.

COUNCIL OF EUROPE (1990), *Household Structures in Europe*, Population Studies, Vol. 22, Council of Europe: Strasbourg.

DEVOS, S., & A. PALLONI (1989), Formal Models and Methods for the Analysis of Kinship and Household Organization, *Population Index*, Vol. 55(2), pp. 174-198.

ERMISCH, J., & N. OGAWA (eds.) (1994), *The Family, the Market and the State in Ageing Societies*, Clarendon Press: Oxford.

GREBENIK, E., C. HÖHN, & R. MACKENSEN (eds.) (1989), *Later Phases of the Family Cycle: Demographic Aspects*, Clarendon Press: Oxford.

HÖHN, C. (1992), The IUSSP Programme in Family Demography. *In:* BERQUÓ & XENOS, pp. 3-7.

KAUFMANN, F.-X., A.C. KUIJSTEN, H.-J. SCHULZE, & K.P. STROHMEIER (eds.) (forthcoming), *Family Life and Family Policies in Europe*, 2 Volumes, Oxford University Press: Oxford.

KEILMAN, N., A. KUIJSTEN, & A. VOSSEN (eds.) (1988), *Modelling Household Formation and Dissolution*, Clarendon Press: Oxford.

LE BRAS, H. (1979), *Child and Family: Demographic Developments in the OECD Countries*, OECD: Paris.

MASON, K., & A.-M. JENSEN (eds.) (forthcoming), *Gender and Family Change in Industrialized Countries*, Clarendon Press: Oxford.

PRIOUX, F. (ed.) (1990), *La Famille dans les Pays Développés: Permanences et Changements*, Congrès et Colloques No. 4, INED: Paris.

RYDER, N.B. (1985), Fertility and Family Structure, *Population Bulletin of the United Nations*, Vol. 15, pp. 207-220.

RYDER, N.B. (1992), The Centrality of Time in the Study of the Family. *In:* BERQUÓ & XENOS, pp. 161-175.

STROHMEIER, K.P., & C.W. MATTHIESSEN (eds.) (1992), *Innovation and Urban Population Dynamics: A Multi-Level Process*, Avebury: Aldershot.

UNITED NATIONS (1994), *Ageing and the Family*, United Nations: New York.

WACHTER, K.W., E.A. HAMMEL, & P. LASLETT (eds.) (1978), *Statistical Studies of Historical Social Structure*, Academic Press, New York etc.

XENOS, P., & S. KONO (eds.) (forthcoming), *Thinking about Family Change*, Ordina: Liège.

Part I

Trends and Theories

2. HISTORICAL DEVELOPMENT OF THE HOUSEHOLD IN EUROPE

Richard Wall

Cambridge Group for the History of Population
and Social Structure
27 Trumpington Street
Cambridge CB2 1QA
England

Abstract. In this chapter the attempts of John Hajnal and Peter Laslett to conceptualize the family and household systems of Europe as they existed in the past are reviewed and then extended on the basis of the following criteria: the welfare capability of the family, the household as a work unit, the status of women within the family, the patterns of marriage and household formation, the household as a kin group, and inequalities between households. National and local censuses are then analysed to reveal the extent of the variation in household and family forms that prevailed across much of Europe in the past, a diversity that, when measured at the national or regional level, appears to have lessened in recent years, without entirely vanishing. Finally, the process of change in the composition of households in England between the seventeenth century and the present day is examined in detail, and causes for those changes are suggested. It is argued that economic factors largely account for the decline between pre-industrial times in the frequency of co-residence with non-relatives (principally servants), and the decline between 1891 and 1981 in the frequency of living with relatives in the absence of a spouse and child. There is also evidence that membership of a family group, comprising a couple or a parent and child, was more common in 1981 than in either 1891 or before 1800. The differences, however, are not large, leading to the conclusion that in England there has been a considerable degree of continuity between past and present in the frequency with which individuals reside in family groups.

2.1. INTRODUCTION

In 1883, Samuel Langhorne Clemens indulged in some humorous speculations about the future and past of one of America's greatest rivers, the Mississippi:

Household Demography and Household Modeling
Edited by E. van Imhoff *et al.*, Plenum Press, New York, 1995

"In the space of one hundred and seventy-six years the Lower Mississippi has shortened itself two hundred and forty-two miles. That is an average of a trifle over one mile and a third per year. Therefore, any calm person, who is not blind or idiotic, can see that just a million years ago next November, the Lower Mississippi River was upward of one million three hundred thousand miles long,.... And by the same token any person can see that seven hundred and forty two years from now the Lower Mississippi will be only a mile and three quarters long.... There is something fascinating about science. One gets such wholesome returns of conjecture out of such a trifling investment of fact." (Mark Twain, *Life on the Mississippi*, 1883/1990, p. 93).

Unfortunately, social scientists have at times been inclined to make some almost equally cavalier assumptions about the form households and families might take in the future or what they might have looked like in the past. Concerning the process of recent change, two mutually contradictory positions have been adopted, each of which will be critically reviewed in this chapter. The first is that the increase, since the end of the Second World War, in the number of smaller and less complex households represents a continuation of a trend whose origins lie in the distant past. The second and contrasting hypothesis is that contemporary family patterns, and specifically the frequency with which people live alone, and in the incidence of lone parenthood and cohabitation in place of formal marriage, represent an abrupt break with the past. In opposition to these views, it will be argued that there has been a rather greater degree of continuity in family and household forms, broadly defined, than has been assumed. The opportunity will also be taken to reassess the role of the various processes which, it has been claimed, produced the family systems of the past and of today.

In section 2.2 of this chapter, the various attempts that have been made to define these family systems will be reviewed, with the emphasis on the need to identify a broader range of demographic, social, and economic factors as critical attributes of family systems. The diversity in family and household patterns in Europe in the past will then be assessed in section 2.3, followed by a discussion of the extent of temporal change in household and family patterns since pre-industrial times, based on evidence from England. The significance of the major finding that, at most ages, persons in the 1980s were more likely to reside in a family than was the case either in pre-industrial or late nineteenth century England, will be considered in a brief final section.

2.2. CONCEPTUALIZATIONS OF FAMILY AND HOUSEHOLD SYSTEMS

Demographic, economic, and cultural factors all have their part to play in determining the characteristics of a family and household system. To take one particular example, increases in fertility and nuptiality and decreases in mortality raise the possibility of the co-residence of parents and married children (Laslett 1979), while improvements in the standard of living may reduce the necessity for such co-residence and cultural changes may make it less desirable. Demographic change may also alter the age structure of the population, increasing or decreasing the numbers in particular age groups where, to give another example, to live alone is either more or less likely. Nevertheless, it appears that many 'explanations' of present-day family and household patterns give pride of place to cultural factors. Despite the presence of certain constraints (e.g. housing shortages), these are seen as very weak barriers indeed in the face of the drive for individual autonomy and the adult's right to choose what commitments to undertake and later break (Council of Europe 1990, pp. 7-11 and 32; Van de Kaa 1987; Lesthaeghe 1992). In other words, the argument is that individuals have seized on the opportunities presented by technological and economic change to take control of their own lives.

Economic, cultural, and demographic factors also underpin conceptualizations of the family and household systems of past societies. When Laslett first formulated what he perceived as the key characteristics of the West European family, he identified the predominance of nuclear or simple family households in conjunction with a relatively late age at childbearing, minimal age differences between husband and wife with a relatively high proportion of wives older than their husbands and, finally, the presence of a significant proportion of life cycle servants. The latter were young persons working for the households in which they lived, although unrelated to any of the members (Laslett 1977, pp. 13-20). These characteristics were considered by Laslett to exert a profound impact on the social fabric of society. The similarity in age of husband and wife he associated with companionate marriage, while the distance in age between parents and children gave parents, if not more authority in their handling of their children, then at least more maturity. The dominance of the nuclear family household also had important implications both for the child, from whose home kin other than their parents and unmarried siblings were largely absent, and for the elderly who would be living independently of their married children (Laslett 1977, pp. 37, 42, 49). The presence of servants introduced non-relatives into the family circle.

Yet, at first sight, the imperatives underpinning the West European family system appear to be economic rather than social or cultural. Parents from the lower social classes were unable to afford to keep their children at home once they reached working age (mid-teens), as their labour could not be profitably exploited while they remained in the parental home. Employers were able to direct where that labour should reside and accommodated it in their own households in the form of servants both for productive and domestic tasks. For their part, the children, as servants, would move between the households of various employers, gaining experience and accumulating savings which they would then use as the basis for launching their own marriage and household. Servants, it has been argued, may have delayed marriage as they profited from a higher standard of living while they remained in service (Schellekens 1991, p. 150). Yet clearly underpinning these patterns are sets of cultural preferences: a willingness, even on the part of the wealthy, to tolerate strangers (servants) in their own houses, and a preference for the residential independence of successive generations that was strong enough to endure over several centuries and survive a number of economic vicissitudes.

Similar considerations underlie other attempts to conceptualize family and household systems, including those of John Hajnal in 1983 and of Peter Laslett (again) in the same year (Hajnal 1983; Laslett 1983). The strengths and weaknesses of their two approaches will now be considered before a revised and expanded set of defining characteristics is elaborated. Hajnal's sketch of two family and household systems, one for Northwest Europe, and one a joint household system operating in India and China, involved sets of household formation rules (Hajnal 1983). For Northwest Europe, these involved men marrying on average later than age 26 and women marrying on average later than age 23. On marriage the younger couple would take over the headship of a household which might either be newly created or continue one from the parental generation (presumptively that of the groom's parents). This rule therefore clearly departs from Laslett's earlier (and later) conceptualization of the West European family, which envisaged one of its distinguishing characteristics as the creation of an independent household by the younger generation. In effect, Hajnal considered as one type a household newly formed at the moment of marriage and one when the son continued his father's household, providing always that his father ceded the headship on the occasion of the son's marriage. On the other hand, Hajnal's third rule, the circulation of labour in the form of life cycle servants, had already been incorporated into Laslett's conceptualization of a West European family and household system. The contrasting joint household system, as depicted by Hajnal, was principally the converse of the Northwest European household system. Young married couples were supposed to start their married lives dependent on the parental generation, men to marry on average before the age of 26 and women before the age of 21. There was also a 'splitting' rule which permitted the fission of some households which might otherwise become unmanageably large.

The greatest asset of Hajnal's conceptualization was that it focussed attention on a key moment in the life cycle of the individual: the formation of a new family. The weaknesses of his classification are, however, also apparent. In the first place, as he himself recognized, the rules of the two systems did not allow for all possibilities. There was, for example, no provision for the stem family household which was created by the marriage of the son and heir, but when the father did not immediately retire. As only one son married and remained in the parental household, the formation rules being followed were not those of the joint household system, yet as headship of the household did not pass to the younger generation, practice was not in conformity either with the household formation system of Northwest Europe. Nor did Hajnal make any provision for the possibility of the adoption of 'incompatible' rules for men who married after the age of 26 but formed joint households, or for women who married on average at age 22. Such populations certainly existed in the European past (cf. Benigno 1989). More serious, however, is the exclusive focus on marriage which represents just one of the transition points in the life cycle. Such a formulation also effectively excludes from consideration, for the purposes of defining a family and household system, that proportion of the population who would never marry. This proportion could at times be significant both in terms of numbers and for its social impact on a given community. For example, a quarter of the women in southeast Bruges aged between 45 and 49 in 1814 remained unmarried (Wall 1983a).

Hajnal's conceptualization of two family and household systems was accompanied in the same volume of essays by a schematic representation by Peter Laslett of four family and household systems (Laslett 1983). In a departure from his previous position (cf. Laslett 1977, p. 15), he now argued that it is possible to identify distinct family and household systems as dominant, if not entirely exclusive, to particular areas of Europe: West, West and Central, Mediterranean and East. The criteria selected by Laslett as defining characteristics of a family and household system included not only household formation rules, procreational and demographic characteristics, and types of kin present in the household, but also aspects of the role of the household in the area of work and welfare. The effect, possibly unintended, was to more firmly anchor each family system within a particular type of economy and within a broader social structure, the latter reflecting in particular the extent and nature of community or state support for any disadvantaged group in the population.

One of the weaknesses of Laslett's classification is the degree of intercorrelation that exists between features of the system that are presented as though they were independent. This applies particularly to the kinship criteria where, for example, there is an obvious association between the proportion of multi-generational households and the proportion of joint households, as well as inverse correlations with the proportions of the population living alone or in simple family households. Laslett also failed to specify whether certain characteristics of the family systems were more important than others. However, the main challenge to date has come from those who feel that the bundles of characteristics do not adequately describe the family and household systems of those areas of Europe with which they are supposed to be associated. In particular a great deal of evidence has been assembled on Mediterranean Europe which reveals a variety of family forms, some in flat contradiction to Laslett's model scheme (e.g. early marriage in combination with the formation of simple family households and late marriage with the formation of complex households). The notion of there being just one Mediterranean family and household system clearly has to be abandoned.

Nevertheless, Laslett's decision to incorporate the work and welfare functions of the household into the definitions of distinct family and household systems provides a useful basis for further work. One of two main concerns was to determine the use by the household of any spare space within the dwelling it occupied to recruit additional labour, whether in the form of resident relatives or non-relatives such as servants or inmates. Households hoarding labour were identified by the mean number of adults or workers who were present. A related issue was the frequency with which the households might primarily be suppliers of labour represented by the proportion of heads of household who were described as labourers. The welfare role of the household, by contrast, was much less effectively measured. The criteria suggested were simply the presence within the same house of two or more non-related households and the frequency with which heads of households were identified as paupers, or in other words as persons requiring the financial support of the community in order to maintain the viability of their households. Each of these welfare criteria draws attention to characteristics of a society which are likely to have a profound impact on the process of household formation and dissolution: the former signalling a probable housing shortage which may constrain, temporarily or otherwise, the creation of new households, while the latter may help to keep some households in existence whose members would otherwise be dispersed among other households or housed in a variety of institutions. Yet neither the presence of non-related households within the same house or houseful, to use the term coined by Peter Laslett (Laslett 1972, pp. 36-39), nor the identification of household heads who were paupers provides any direct comment on the role of the household as a welfare agency. This role ought to be measured more directly and considered in conjunction with other attributes of a family system concerning kinship, marriage and headship, the status of women, the role of the household as a work unit, and inequalities among households. The full set of criteria are set out in Table 2.1 and some comments are in order.

In the first place, it should be emphasized that historians have not yet assembled all this information, at least not systematically. Secondly, even as an ideal standard at which to aim, it is evident that the list of criteria, long as it is, is by no means exhaustive and can provide no substitute for a detailed examination of the household and family patterns of a given area. On the other hand, the criteria are intended to encompass all of the major functions of the household and the family. Inevitably this guarantees that a less clear-cut impression will emerge of family and household patterns across Europe than when fewer criteria are specified. In effect this marks a return to the position Peter Laslett took in 1977, that the literal identification of areas of Europe with discrete family systems is likely to prove puzzling and complex (Laslett 1977, p. 15). Different facets of the roles of families and households as work, welfare, or as kin groups may be more in evidence in some places than others, and may gain or lose their importance over the course of time. The third point that needs to be made is that no one criterion should be considered as more important than another. This in turn will increase the measured degree of geographical and historical variability. Detailed investigations of the households of specific communities are in fact required to measure the impact of certain features of the family system, for example, on the proportions of dependent, working, and service kin (cf. Wall 1986, pp. 284-285). Fourthly, given the focus on the processes that give rise to the co-residence of individuals within a household, it is clear that the principal source will be the high quality census, official or unofficial, which details the ages, relationships, marital statuses, and ages of the inhabitants. Occasionally, however, reliance has had to be placed on other sources, or combination of sources. Parish or civil registers, for example, provide the information on fertility; the linking of successive censuses makes it possible to determine those children who returned to the parental home; and Frederic Le Play recorded the nature of the work of married women, within and without the household (Le Play, 1855, 1877-9: *Ouvriers des deux mondes*, 1857-85, 1885-99).

For each attribute of the family system, the welfare capability of the family, the ability of the household to function as a work unit, the status of women within the family, and so on, a range of measures is proposed and a target population specified, representing those individuals most obviously benefitting when a family fulfilled a particular role or, alternatively, having to seek aid elsewhere when such support failed to materialize. One issue that is particularly difficult to resolve is how common a particular attribute would have to be in order to be deemed significant. For example, it could be argued that once one unmarried woman in a given community was able to set up a household on her own, and without a resident servant, perceptions of the range of living arrangements that were available to women who were as yet unmarried and might never do so, would be broadened. On the other hand, unless a substantial minority of

Table 2.1. Defining characteristics of family and household systems in the European past

Attributes of the family system	Proportions and means	Target population
I		
Welfare capability of the family	1. Children raised by non-kin	Aged 0-1, 1-4, 5-9
	a. Household care	
	b. Institutional care	
	2. Children raised by lone parent	Aged 0-4, 5-9, 10-14
	3. Adult children ever-returning to parental household	Aged 20+
	4. Lone parents living in own household	Lone parents with all children aged <5, <15
	5. Elderly residentially isolated	Aged 60+, 65-74, 75+
	a. Living alone	
	b. Institutional care	
II		
Household as a work unit	1. Mean number of adults and employed persons per household	Adults aged 20-69, employed persons (including servants, excluding inmates)
	2. Children not resident in the parental household	Sons, daughters, 10-14, 15-19, 20-4
	3. Households using non-familial labour	Servants, inmates sharing occupation of head
	4. Household heads in the labour force who were self-employed	Self-employed household heads
	5. Family members in the labour force sharing the same occupation or employed in the same economic sector	Male, female members sharing occupations or employment sector
	6. Family members working in the home	Male, female family members whose primary workplace was in the home

(continued on next page)

Attributes of the family system	Proportions and means	Target population
III		
Role and status of women within the family	1. Never-married women able to form own households	Alone, others co-resident (servants, other non-relatives, children, other relatives)
	2. Other never-married women not living with relatives	Servants, inmates
	3. Wives older than husbands on first marriage	Married couples. Wife older 5+, 1-4, 0-1 years
	4. Level of marital fertility	Married women 15-49
	5. Married women employed outside the family economy	Married women employed by persons who were not members of their own household.
	6. Widows living alone or with children	Alone, co-resident child (widow head, child head)
IV		
Marriage and household formation and dissolution	1. Never-married	Males, females aged 45-9
	2. Ever-married	Males, females aged 20-4, 25-9
	3. Currently married	Males, females aged 50-9, 60-9
	4. Headship rates by age, marital status, sex	Aged 20-4, 25-9, 35-44, 45-59, 60+
V		
The household as a kin group	1. Non-nuclear relatives within the household	Male, female relatives of the head (other than spouse and unmarried child)
	2. Range of kin types	Number of male, female relatives with given relationship to head
	3. Mean number per household of vertical and lateral kin	Male, female relatives of same generation as head (lateral kin), different generation (vertical kin)
	4. Dependent, working and service kin	Dependent relatives (parentless grandchildren, lone parents, persons aged 65+). Working relatives (apprentices, occupation shared with head). Service relatives (servants, non-married male head with female relative aged 15+)
	5. Individuals living with kin and non-kin	Males, females co-residing with any relatives or non-relatives
VI		
Inequalities between households	1. Distribution of households by size	Household sizes 1-13+
	2. Distribution of households by number of adults	Aged 20+
	3. Households headed by non-married persons	Males, females (never-married, widowed, divorced, separated) heading households

never-married women lived alone, it would not be appropriate to consider that this was a major feature of the female life course.

Another apt illustration concerns the interpretation of the incidence of child abandonment in much of southern Europe in the past. It is well known that large homes for foundlings were established much more frequently and at a much earlier date in southern than in northern Europe (Pullan 1989). The obvious inference to draw is that different child care practices were followed in southern Europe, particularly a greater willingness to resort to wet nurses, and that this in turn reflected the existence of either differences in attitude on the part of the parents in southern Europe towards the care of their own children, or in their ability to provide the necessary care. However, such generalizations are misleading. This is evident, for example, in the fact that although all major Italian cities had a foundling home which could receive considerable numbers of abandoned children, equivalent to 23 per cent of all births in Rome in the second decade of the eighteenth century, 22 per cent of births in Florence in the 1640s, and a third of all legitimate children born in Milan between the 1840s and 1860s, the incidence of abandonment at the level of the region or in Italy as a whole was very much lower. Overall, no more than three per cent of Italian children were being abandoned in the early nineteenth century, a rate of abandonment similar to that experienced by children in France (all data are from Kertzer, 1994, pp. 74-81). However important the presence of numerous abandoned children might be for a particular city and for the economy of the regions to which they were sent to be nursed, a three per cent abandonment rate is clearly too low to serve as a defining characteristic of a family system or as a significant stage in a 'standard' life course, either of the abandoned child or of the abandoning parent.

Another difficulty is occasioned by the fact that much of the information that is now being sought on family systems in the past has not yet been assembled on a systematic basis for Europe. Even the number and type of relatives resident in the household (section V of Table 2.1), which has received extensive, even excessive, attention from historians of the family, has not been fully researched. Debates have centred instead on the presence or absence of certain types of families, particularly the stem family, which involves the co-residence of one married son, and the patriarchal family, which is formed by the presence of several co-resident married sons. The focus on such complex household forms arises from an interest in the extent to which the transfer of property between successive generations might promote extended periods of co-residence between parents and a married son. The result, however, has been the relative neglect of other forms of co-residence, such as the presence of unmarried relatives of the same generation of the head and the role played by female relatives. In the English past, the majority of relatives living within the household were female (Wall 1979, pp. 112-113). Evidence as to whether kin co-residence in Europe in the past occurred primarily for the purpose of welfare, employment or inheritance is, therefore, at present suggestive rather than conclusive. However, one detailed study of households in a village in the southwest of England in the mid-nineteenth century has suggested that the welfare of the individual relative or of other individuals in the household was an important motive for kin co-residence. Nearly two thirds of the kin were to be found in households most likely to need the support of a relative, as the household heads were either elderly, non-married, or childless. Many of the relatives were themselves elderly, lone parents or children living apart from their parents (Wall 1986, pp. 282-287). This experience can be contrasted with that of the inhabitants of a series of villages in West Flanders early in the nineteenth century, where employment of a relative by a household head was more usual. However, in other parts of Europe, where there were fewer opportunities for employment off the land other than through migrating, temporarily or permanently some considerable distance, co-residence prompted by inheritance considerations is likely to have been more important (cf. on communities in the French Pyrenees, Fauve-Chamoux 1987, pp. 245-248).

For certain other attributes of a family system, as set out in Table 2.1, there is some Europe-wide evidence, but available in insufficient quantities to enable the variation across Europe to be mapped with any degree of accuracy. Information concerning the married women in the past who were employed by persons who were not members of their own households is a case in point. These women worked outside the family economy, as it has sometimes been defined (Tilly & Scott 1978, pp. 12 and 14), and would be seen as participating in the adaptive family economy: diversifying the family's sources of income (cf. Wall 1986, p. 265; Tilly & Scott 1987, pp. 44-60 and 123-136). Analysis of the 36 budgets of the lower class families collected by Le Play and his supporters prior to 1877 indicates that almost 80 per cent of married women had worked outside the family economy in the year prior to their being surveyed. The average of 65 days work outside of the family economy, although little more than half of the time they devoted to housework and child-rearing, was not significantly less (only 15 per cent) than they spent on income-generating activities within the family economy (Le Play, 1877-9; Wall 1993; cf. Tilly & Scott 1987, p. 54 for estimates of women's work time in the eighteenth century). This is a useful corrective to interpretations of the nature of the work of married women in the past which have stressed how rare it was for married women to be employed other than by members of their immediate families. Yet, on the basis of the budgets of only 36 families, almost half of them French, it would clearly be rash to embark on a search for evidence of systematic variation from one country to another in the nature of the work performed by married women.

In order to pursue the issue of variation in family forms across Europe and their development over the centuries, it is necessary to rely on such information as can be obtained from more widely available sources, such as parish registers and the censuses, using a standard classification scheme. One such standard classification scheme (more limited in scope than that set out in Table 2.1) will now be used to establish the degree of variation in the composition of households across Europe in the past, based on the distribution of members of the household according to their relationship to the head of the household. As this represents the only classification of household membership currently available for many larger populations of the past, groups of villages, large towns, and even on occasion as entire countries, there is no practical alternative to its use. The classification suffices to show, however, the considerable variation in the composition of the household, even within the area of Europe where, Hajnal contended, only the Northwestern European household formulation system was in operation (Hajnal 1983).

2.3. A GEOGRAPHICAL PERSPECTIVE ON MEMBERSHIP OF THE HOUSEHOLD IN THE EUROPEAN PAST

Table 2.2 sets out the membership of the household in a number of rural populations enumerated at a variety of dates between the late sixteenth and early nineteenth century, with membership of the household expressed in terms of the number of persons present of each type: heads of household (married and non-married), offspring, relatives, servants and, finally, any other persons not known to be related to the head of the household. The number of persons of all types found within the household varies considerably, and the variation would no doubt be greater if it had been possible to include more populations from southern and central, let alone from eastern Europe. In the case of servants, for example, in the selected rural populations the range is from 118 per hundred households in Iceland in 1703 to fewer than 20 per hundred households in the countryside around Gouda in the Netherlands in 1622, in certain Swiss communities between the mid-seventeenth and early eighteenth centuries, and in the Spanish province of Cuenca in the eighteenth century. For offspring, the range is from 279 per hundred households in Egislau in Switzerland, in the area around Gouda and in West Flanders in 1814, to 157 per hundred households in Cuenca in 1724.

Gauging the significance of the variation is more difficult, but can be considered from the following points of view. In the first place, there is the question of the smoothness of the distributions when the populations are placed in rank order from those with most offspring or kin or servants to those with least. Three populations, for instance, stand out as having an above average

Table 2.2. Persons per 100 households by relationship to household head, European rural populations

Country	Area	Dates	Household Heads Married couples	Household Heads Others	Offspring	Relatives	Servants	Other unrelated	All	Total households
Belgium	West Flanders	1814	167	16	272	20	84	26	585	656
Denmark	Rural sample	1787	177	12	194	24	99	16	521	1283
	Rural sample	1801	175	12	191	26	81	12	497	1483
England	5 settlements	1599-1749	134	33	189	19	51	9	435	481
	18 settlements	1750-1821	150	25	209	22	51	24	481	1900
Iceland	-	1703	139	31	191	38	118	99	616	8177
Netherlands	Rijnland	1622	177		252	6	26	7	468	2578
	Krimpenerwaard	1622	178		272	9	19	6	484	2073
	Noorderkwartier	1622-1795	155		176	5	24	13	373	3269
	West Noord Brabant	1775	142	28	230	10	70	10	490	2199
Norway	Rural districts	1801	175	12	217	34	68	40	546	144914
Spain	Cuenca	1699-1794	142	31	157	9	18	0	357	778
Switzerland	Egislau	1647-1671	166	17	279	13	11	16	502	441
	Flaach	1671-1736	157	19	234	12	29	14	464	557
	Volcken	1671-1736	165	17	237	20	18	28	484	191

Sources:
Belgium: Calculations from transcriptions of censuses of St Pieters-op-de-dijk, Uitkerke, Nieumunster, Vlissegem, Mariakerke, Stene, Wilskerke, Ettelgem and Snaaskerke, by Vlaamse Vereniging voor Familiekunde, Afdeling Brugge, (1976-7). For further information on these communities see Wall (1983b).
Denmark: Johansen (1975).
England: Calculated from lists of inhabitants in the Library of the Cambridge Group. For 1650-1749, Puddletown (Dorset) 1724; Goodnestone (Kent) 1676; Harefield (Middlesex) 1699; Clayworth (Notts) 1676; and Lower Heyford (Oxford) 1724. For 1750-1821 Binfield (Berks) 1801; West Wycombe (Bucks) 1760, Littleover (Derby) 1811; Mickleover (Derby) 1811; Corfe Castle (Dorset) 1790. Ardleigh (Essex) 1796; Forthampton (Gloucs) 1752; Barkway and Reed (Herts) 1801; Lower Heyford (Oxford) 1771; and Barton, Bampton, Kings Meaburn, Lowther, Hackthorpe, Great Strickland, Morland and Newby (Westmorland) 1787. See also Wall (1983d).
Iceland: Statistical Bureau of Iceland (1960).
Netherlands: Van der Woude (1980), supplemented by Klep (1973).
Norway: Calculated from Norges Offisielle Statistikk (1980).
Spain: Reher (1988).
Switzerland: Calculated from unpublished tabulations produced by Dr J. Ehmer from lists of inhabitants held in the Vienna Database.

number of servants (Iceland, Denmark, and West Flanders). Then follow a number of populations with more moderate numbers of servants (Norway, West Noord Brabant and England) and finally three populations with very few servants (the Swiss communities, Cuenca, and three of the rural areas in the Netherlands). In a similar vein, populations can be identified where very few households were headed by non-married persons (Denmark, West Flanders, and the Swiss communities) or which contained a large number of unrelated persons (Iceland and to a lesser extent Norway). On the other hand, there are very few occasions when any one population is sufficiently distinctive as regards a particular component of the household, in for example the number of its offspring, relatives, or servants to stand apart from all other populations. Of the various populations examined so far, the most distinctive in view of the large number of servants and of other unrelated persons is Iceland, but even Iceland may come to look less distinctive as investigations of other European populations are completed.

Indeed, already the differences between Iceland and the rest of Europe look quite modest when set alongside the structure of the household in some non-European populations, that of India for example, where there were 122 relatives per hundred households in 1951, and the Russian serfs of the nineteenth century with their 520 relatives per hundred households (Hajnal 1983). Nevertheless, it is obvious that there has been considerable variation in household structure even within the confines of northern and central Europe. For example, the Danish, Norwegian, and West Flemish populations had many married household heads and many offspring and servants, the number of offspring being boosted in West Flanders by the frequency with which both widowers and widows remarried (Wall 1983b). The populations of rural Holland stand out on an account of the relative rarity of kin in the household and of servants. The distinctiveness of Iceland, on the other hand, as has already been mentioned, was due to the large number of servants and other unrelated persons attached to its households. Finally, a fourth household pattern may exist, exemplified by the relative frequency with which non-married persons headed households. This pattern is found in England, West Noord Brabant, and Cuenca. Iceland also had many non-married heads of household, while in addition having special features of its own.

A place may also have to be reserved for a European urban household. Table 2.3, using the same classification scheme as in Table 2.2, shows that urban households were generally less likely than rural households to be headed by a married couple and that they contained fewer offspring but more relatives and many more unrelated persons, many of whom of course would be the lodgers and boarders traditionally associated with town life. Overall, urban households, even including lodgers, were generally smaller than rural households. As with the rural households, however, there is also evidence of

Table 2.3. Persons per 100 households by relationship to household head, European urban populations

Country	Town	Dates	Household Heads Married couples	Household Heads Others	Offspring	Relatives	Servants	Other unrelated	All	Total households
Belgium	Bruges south east	1814	134	33	190	42	2	99	524	1516
England	London St Mary Woolchurch	1695	142	29	146	9	213	161	700	69
	Southampton	1695-1696	116	42	162	12	42	20	394	317
Germany	Konstanz	1774	109	46	167	14	72	20	427	948
Italy	Rome	1653-1659	95	52	120	47	95	117	526	273
	Rome	1700-1701	104	48	128	20	27	140	467	762
	Rome	1800-1827	129	37	153	27	20	78	443	2333
Netherlands	Leiden	1581	124	38	159	18	25	23	386	2985
	Gouda	1622	171		207	9	9	28	425	3503
	Gouda	1674	161		171	7	10	6	355	2632
	Delft	1749	159		127	11	23	28	347	3433
	Leiden	1749	167		142	2	23	28	362	9778
Norway	Main towns	1801	152	24	142	18	78	55	469	8462
	Other towns	1801	146	27	160	21	53	45	452	10188
Spain	Cuenca	1724	140	30	159	20	45	3	395	1046
	Cuenca	1800	138	32	140	19	42	10	380	815
Switzerland	Zurich	1637	138	31	193	12	65	29	468	752
	Geneva	1720	134	33	153	19	60	37	436	3031
	Fribourg	1818	114	43	176	36	62	91	522	1194

Sources:
Belgium: Calculated from Vlaamse Vereniging voor Familiekunde, Afdeling Brugge, vol. vi (1977). For further information see Wall (1983d).
England: Calculated from lists of inhabitants in Library of Cambridge Group. Calculations for Southampton are based on lists of the inhabitants of the parishes of Holy Rhood, St John and St Lawrence.
Germany: Calculated from unpublished tabulations produced by Dr J. Ehmer from lists of inhabitants held in Vienna Database.
Italy: Calculated from unpublished tabulations produced by Dr J. Ehmer from lists of inhabitants held in Vienna Database. Parishes included: Sant' Agostino in urbe (1653, 1700 and 1800), Sant' Angelo in Pescheria (1659, 1700 and 1800), Santa Dorothea (1813), San Giovanni dei Fiorentini (1700 and 1800), San Marcello (1701 and 1827) and Trastevere (1827).
Netherlands: Van der Woude (1980), supplemented by Daelemans (1975).
Norway: Calculated from Norges Offisielle Statistikk (1980).
Spain: Calculated from Reher (1990).
Switzerland: Calculated from unpublished tabulations by Dr J. Ehmer of censuses of Zurich and Fribourg. For Geneva see Perrenoud (1979).

considerable variation from place to place. The households of the inhabitants of Norwegian towns were, for example, most likely to have married couples as heads. Households in Bruges, Gouda, and Zurich were more likely than those of other towns to contain offspring. Relatives, other than members of the heads' own nuclear family, were most often to be found in the households of the inhabitants of Bruges and Fribourg. Servants turn up most frequently in Norwegian and Swiss towns and in Konstanz, and unrelated persons in Rome, Bruges, and Fribourg.

Yet, despite this variation in the composition of the urban household, the association of specific types of household with particular towns is not an easy task. In part, this reflects the very fragmentary nature of the evidence currently available. London, for example, is only represented by one of its central and wealthier parishes, Rome by a handful and variable number of parishes at different points in time, and even Southampton by only half of its parishes. This should in fact increase the measured degree of variation from area to area, yet the reality is that households from different urban populations seem to differ somewhat less in certain key respects (number of households with married and non-married heads and in the number of offspring they contain) than households from different rural populations. Of the towns and sections of towns covered in Table 2.3, only those in Holland really stand out on account of their low numbers of relatives, servants, and other unrelated persons in their exceptionally small households.

The emphasis on variability is reinforced when we come to consider the sex ratio of particular categories of people within the household, non-married heads, offspring, relatives, servants, and other unrelated persons in the rural populations of Europe (Table 2.4). As might be expected, due to women generally being younger than spouses on marriage, and generally outliving them, most of the non-married persons heading households were women. Even here, however, Iceland is an exception, while two of the three Swiss communities lie at the other end of the distribution with more than three times as many non-married women than non-married men heading households.

Sons and daughters residing with their parents were usually present in almost equal numbers. Any marked departure from a sex ratio close to a hundred in a population of any size would indicate either a mortality differential by sex or, most probably in these populations, an earlier exit from the parental home either by sons or by daughters. In rural populations, other factors being equal, any desire to retain male family labour in farming and keep the heir in residence, assuming the heir was by preference a son, would tend to raise the sex ratio. However, the effect on the sex ratio of the entire offspring group (as opposed to offspring over the age of ten) is likely to be muted, since the vast majority of offspring would be of an age when both sons and daughters would normally still be in the parental home. Nevertheless, in the case of England, where local censuses giving ages have been analysed, it emerges that prior to the late eighteenth century, it was sons and not daughters who were first to leave the parental home (Wall 1987, pp. 92-95).

Whether this is the same elsewhere would merit investigation. What is already evident is that the majority of the young rural labour force recruited as servants were male. Many of these servants, of course, were the offspring who were 'missing' from the homes of their parents. The surplus of male servants shows up strongly in West Flanders in 1814 and in England before 1750. However, the surplus is less marked in Denmark at the end of the eighteenth century, and is reversed in two of the Swiss communities and in Iceland, indicating considerable differences in how these societies used service as a source of labour. Comparable differences occurred in the sex ratio of the groups of related and unrelated persons in the household. In a number of the populations, for example, such as rural Denmark in 1787, female relatives outnumbered male relatives by more than two to one, whereas in West Flanders there were almost 1.5 as many male relatives as female.

A glance at the sex ratio of the urban populations (Table 2.5) suffices to show that females predominated in the majority of towns that it was possible to examine (unfortunately, few data sets were available). Females were generally in the majority amongst the non-married heads of households, amongst relatives and servants, and other unrelated persons to a much greater extent than in the rural populations (see Table 2.4). The effect of this excess of females (in Bruges, for example, there were only six adult males to every ten females) on the economic and social life of certain towns was considerable (Wall 1983a; cf. Fauve-Chamoux 1983). Through their preponderance in the population, these women made a major contribution to the economic vitality of these towns, and their networks of contacts with other women, both relatives and non-relatives, were an important feature within the social structure.

Of course, not all urban populations were like this. The city of London parish of St Mary Woolchurch, for example, had a marked surplus of men among the non-married household heads. However, of all the urban populations examined, it is those from Rome which are most distinctive on account of the relative preponderance of males in all constituent parts of the household, at least in the first two of the periods studied (1650s and 1700s). In the case of Rome, there is no reason to doubt the representative nature of these results since, year by year, through the course of the seventeenth century, a marked surplus of males was recorded in the total population of the city (Schiavoni & Sonnino 1982).

Since the early nineteenth century, the diversity in household forms across Europe has lessened considerably. Admittedly, certain sociologists have detected a plethora of family types in developed populations of the present-day. One estimate is of 2000 distinct 'family types' just in the countries represented in

Table 2.4. Number of males per 100 females of specified relationship to household head, European rural populations

Country	Area	Dates	Non-married household heads	Offspring	Relatives	Servants	Other unrelated	Total population
Belgium	West Flanders	1814	64	96	141	158	94	105
Denmark	Rural sample	1787	71	103	48	116	57	98
	Rural sample	1801	50	105	64	134	50	101
England	5 settlements	1599-1749	46	101	79	167	24	97
	18 settlements	1750-1821	77	102	61	116	112	100
Iceland	-	1703	122	98	40	78	57	83
Switzerland	Egislau	1647-1671	30	100	54	67	64	93
	Flaach	1671-1736	26	90	31	125	108	90
	Volcken	1671-1736	68	110	73	94	130	104

Sources: See Table 2.2.

Table 2.5. Number of males per 100 females of specified relationship to household head, European urban populations

Country	Area	Dates	Non-married household heads	Offspring	Relatives	Servants	Other unrelated	Total population
Belgium	Bruges south east	1814	44	93	48	25	52	73
England	London St Mary Woolchurch	1695	186	84	50	116	73	96
	Southampton	1695-1696	40	72	46	81	82	76
Germany	Konstanz	1774	67	93	33	66	433	90
Italy	Rome	1653-1659	155	129	163	291	229	165
	Rome	1700-1701	154	101	119	263	299	151
	Rome	1800-1827	79	108	63	126	116	102
Switzerland	Zurich	1637	19	98	20	79	98	85
	Fribourg	1818	28	102	58	54	122	89

Sources: See Table 2.3.

the OECD, while a cross-classification of household type by family type and number of resident children, it is claimed, could yield as many as 100,000 different 'situations', if all elements of the classification scheme were counted (Bernandes 1986; Le Bras 1979). This variation, however, appears to be largely contained within specific populations, as differences between countries, and even between regions of the same country, appear to be much more muted today than was the case in the past (cf. Schürer 1992, p. 266; Wall 1982, pp. 92-96 on regional variations in the family and household patterns in England in pre-industrial times and in the 1960s and 1970s). Nevertheless, in southern and eastern Europe, vestiges of the differences of earlier times are still visible in the lower proportions of one-person households and higher proportions of households containing relatives from beyond the immediate family of the household head (Wall 1983c, p. 48; Wall 1989, pp. 375-377).

2.4. THE EVOLUTION OF HOUSEHOLD FORMS IN ENGLAND SINCE THE SEVENTEENTH CENTURY

The process of change in household forms over the centuries has been most clearly delineated in the case of England. Progressively, the household has become less diverse as first servants and then other non-relatives (principally lodgers and boarders) decreased in numbers. After 1891 the household shrank

Table 2.6. Mean number of persons per 100 households: England seventeenth to twentieth century

Date	Head and spouse	Unmarried children	Relatives	Servants	Total	Other unrelated	Total	Households
1650-1749	167	189	19	51	419	9	435	481
1750-1821	175	209	22	51	457	24	481	1900
1851	168	201	31	25	425	35	460	4428
1891	173	220	34	21	448	22	470	17372
1901	173	214	33	17	439	19	458	20148
1911	174	205	32	16	428	19	447	22947
1921	172	180	39	10	401	16	417	24471
1947	180	134	42	2	358	9	367	5997
1971	170	96	12	0	279	5	284	174562
1981	167	88	10	0	265	5	270	193574

Sources: 1650-1749 and 1750-1821, Table 2.1 above; 1851 and 1947 national samples of the population of Britain listed in Wall (1983), 497; 1891-1921 anonymized data on 13 local populations from censuses of England and Wales 1891-1921; 1971-1981 calculated from the national samples of the population of England and Wales taken by the Office of Population Censuses and Surveys for the purpose of the Longitudinal Study.

in size in response to the fall in fertility, but relatives who were not members of the immediate nuclear family of the head were present in some numbers as late as 1947 (see Table 2.6). Some earlier changes in the composition of the household, the increase in the number of resident children during the eighteenth century, were also promoted by demographic change, in this case the fall in marriage age and the consequent rise in fertility (Wall 1983d). In this particular case, the influence of economic factors on household forms was indirect, as marriage rates and changes in the age at first marriage were themselves responding to improvements in the level of real wages after a delay of some 30 years (Smith 1981, p. 601). But direct economic influences were also possible as, for example, during the later eighteenth century, when farmers came to favour the employment of day labourers in preference to live-in farm servants. The result was a marked fall in the proportion of servants, particularly male servants, and a later departure by sons from the parental home (cf. Wall 1987, p. 94, and note the decline in the latter half of the eighteenth century in the relative number of male servants documented in Table 2.4 above). Changes in the number of relatives in English households can also reflect economic pressures of various sorts, the number increasing following the First and Second World Wars, for instance, due to a shortage of housing, although the rise after 1750 in the number of relatives present in the household appears to owe more to the direct and indirect consequences of a more youthful age structure (see the arguments outlined in Wall 1983d, pp. 507-510).

A more detailed perspective can be provided by considering how living arrangements varied across the life cycle at different points in time. Three such points have been selected, rural England before 1800, England and Wales in 1891 when the fertility transition was getting underway, and England and Wales in 1981, the latest available census when the analysis was undertaken. Only the latter provides a random sample of the population, although it is known that the 13 local populations of 1891 for which information is available are representative of the national population in terms of age structure and marital status (Wall *et al.*, forthcoming). For 1981, the sample taken for the purposes of the Longitudinal Study has been analysed in cross-sectional form, in preference to the summary tables published by the Office of Population Censuses and Surveys (see OPCS 1984 and 1988; Social Science Research Unit 1990). The representativeness of the pre-1800 populations is more open to question, as only six such enumerations included information on age and relationship, a smaller set than could be used earlier for Table 2.1, and there is no national population count available to establish the extent of their typicality. What is clear is that, like many other rural populations in England prior to 1800, many of the inhabitants were not directly employed in agriculture. In this particular group of villages, for example, fewer than a fifth of the heads of households were farmers, while more than a fifth were involved in various service occupations and more than a quarter in manufacturing (Wall, forthcoming).

For the purpose of analysis, a new classification of living arrangements has been developed, derived largely from that implemented by Statistics Canada and the Central Statistics Office of Ireland as part of their census programme of the 1970s and 1980s (Central Statistics Office, Dublin, 1983, Part II, Table 1; Statistics Canada 1987, Table 7). The essence of the classification is the identification of a hierarchy of relationships within the household, beginning with the couple (including the cohabiting couples where these can be identified), and then proceeding in turn to identify persons with at least one co-resident never-married child, persons living with relatives other than spouse or never-married child, persons living with non-related people only and, finally, persons living alone. Persons resident in institutions have been excluded. In this scheme, couples and parents with their never-married child constituted families and all others are considered as living outside families. Those living outside the families comprise persons living alone, or with non-relatives only, whether servants, lodgers or friends, or with relatives in the absence of a spouse or unmarried child. These relatives could be married, divorced, or widowed children, as well as siblings (married or unmarried) of the householder or his wife or more distant relatives. The strength of the classification lies in its identification of core relationships, couples and parents and their children which, for the majority of individuals in most populations, will be more important than additional ties they may have to more distant relatives or non-relatives.

2.4.1. Membership of a family group

Figure 2.1 sets out the percentage of the male population who belonged to a family group as thus defined. Surprisingly, even in rural England before 1800, membership of a family group was the experience of the vast majority at most ages, although it might have been expected that with the level of mortality higher than in the present many would have been deprived of the opportunity to live as a family. The one significant exception occurred in early adulthood, due, it will be shown later, to a combination of life cycle service and lodging. Before 1800 membership of a family by adult males peaked fairly late, within the age group 55-59. Thereafter, the rate fell only slightly, so that the majority of men aged 75-79 were still members of a family. The other feature is that not all children under 5 were actually members of a family. Most of these were infants in their first or second year of life placed with wet nurses in Ealing, situated close to London, and the first local population to be enumerated in 1599.

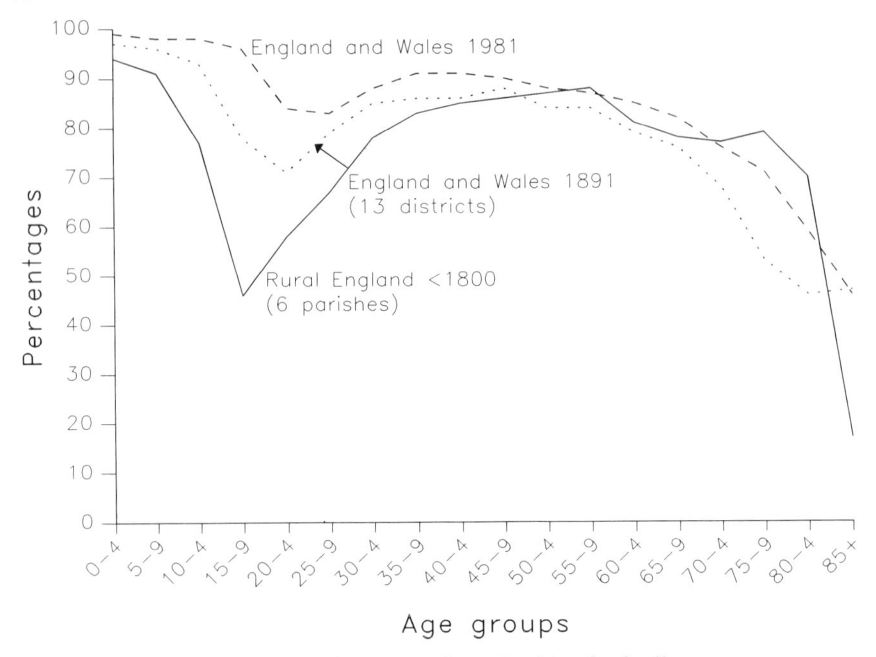

Figure 2.1. Relative frequency of membership of a family group: males aged 0-85+ in England and Wales from pre-1800 to 1981

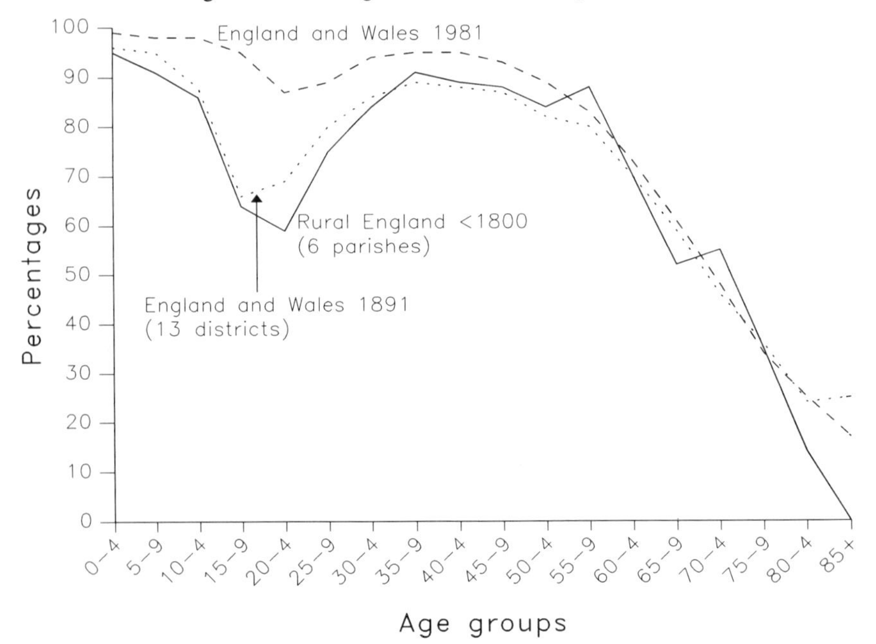

Figure 2.2. Relative frequency of membership of a family group: females aged 0-85+ in England and Wales from pre-1800 to 1981

In 1891, membership of a family by males under 50 was more likely, and considerably more likely during earlier adulthood, than it had been before 1800. The opposite, however, was the case during late middle age and in old age. By 1981, this decline in membership of a family by men over 50 had been reversed, although it was still the case that in 1981 fewer old men in early old age (aged 70-84) were members of a family than in the rural populations enumerated before 1800. Another difference is that, by 1981, the maximum level of family membership was achieved by adults earlier than in 1891, within the age group 30-39. Finally, and most significantly, given the frequent references to a recent decline in the family as an institution, at most ages a higher proportion of the 1981 male population lived with at least one other closely related member of their family than was the case either in 1891 or in the pre-1800 populations. The reasons for these changes are complex and will require fuller consideration, but it is worth emphasizing at this point that membership of the family, as it has been defined here, is determined by a wide range of factors, the chances of remaining married, or remarrying in the case of widowhood or, in later times, divorce; or if widowed or never-married, living as a lone-parent with a never-married child. On occasion, such factors may have reinforced each other; on another, been in part self-cancelling.

Comparison with Figure 2.2 indicates that, in all periods, family membership by females declined much more markedly with age than was the case for males. This reflects the fact that higher male mortality ensured that men were more likely to die when still married and that women would die as widows. Nevertheless, in all three time periods, more than half of all women who reached their late sixties were still members of a family group formed by the presence of a spouse and/or an unmarried child. Indeed, in the rural populations enumerated before 1800, a majority of women in their early seventies belonged to such a family group. For women, as for men, an increased incidence of family membership is indicated by 1891, although limited on this occasion to childhood and early adulthood, and an even higher frequency of family membership in 1981 spread across most age groups. In 1981, men and women also experienced the maximum level of family membership when approximately the same age. Earlier, however, there had been a considerable difference in this respect. In 1891 the maximum had been reached when women were aged 40-44 and men 45-49 and before 1800 when women were aged 35-39 and men as much as twenty years older.

A further perspective on these patterns can be provided by considering one of the constituents of the family group that has been under observation, male and female lone parents (i.e. any adult without a partner co-residing with a child regardless of age). In the pre-1800 populations, the incidence of male lone parenthood rose steadily with age to a peak in the age group 65-69 (Figure 2.3). Thereafter the level is quite variable due to the small number of older men in

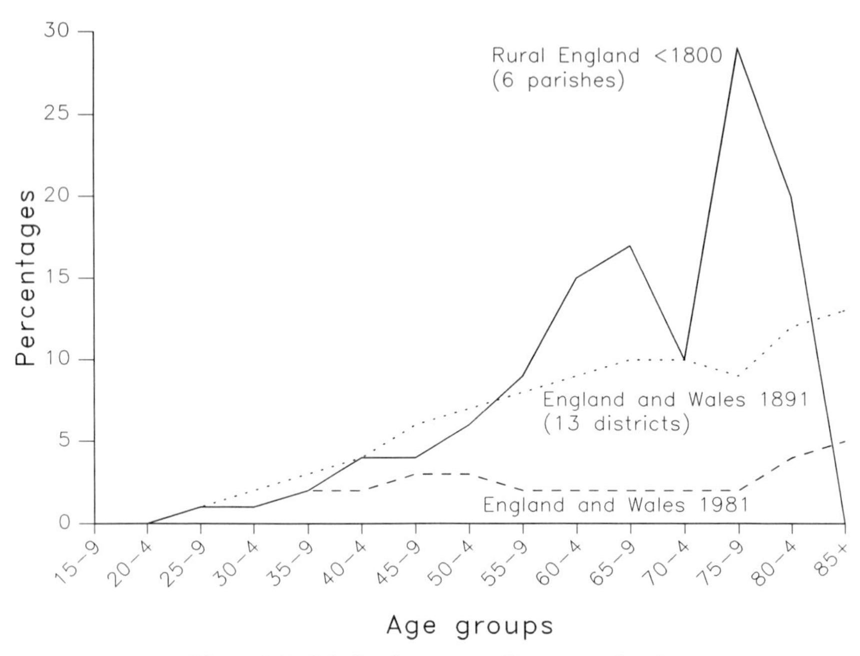

Figure 2.3. Relative frequency of lone parenthood:
males aged 15-85+ in England and Wales from pre-1800 to 1981

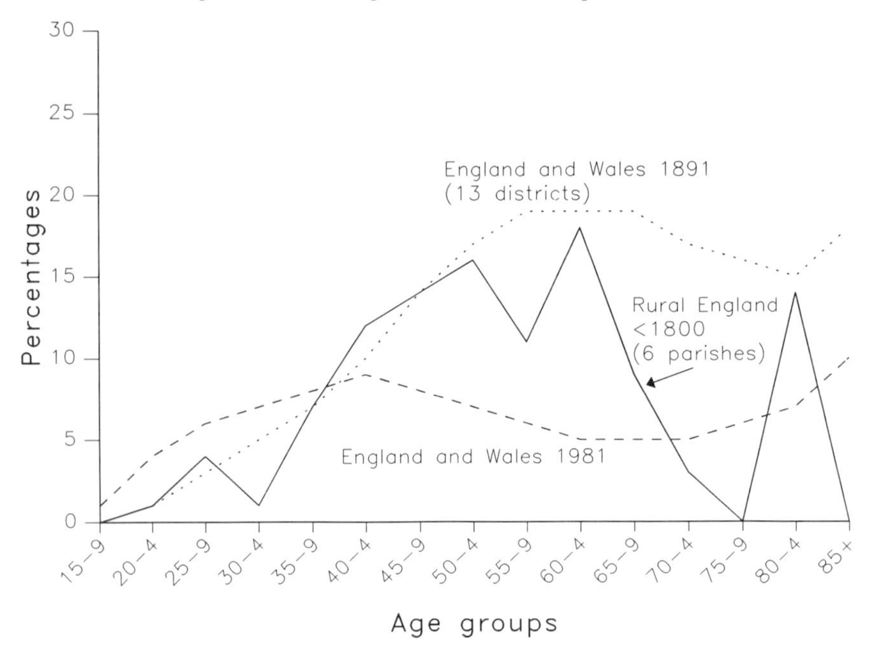

Figure 2.4. Relative frequency of lone parenthood:
females aged 15-85+ in England and Wales from pre-1800 to 1981

the populations. At all ages between 25 and 85, there were some male lone parents. The English and Welsh populations of 1891 also evidence this rise in the frequency of lone parenthood through to age 65-69 although the rates of lone parenthood are much lower for men in their sixties and more stable after age 65 than they had been before 1800. The real difference is with male parenthood in 1981 which was minimal at all ages, and rarer for men over 40 than it had been in seventeenth and eighteenth century England. The incidence of male parenthood, considered in relation to the trends in membership of a family as delineated in Figure 2.1, makes it clear that the decline in male lone parenthood must be more than counterbalanced by an increase in the proportions of older men who were married. This in turn is primarily the result of the greater improvement in female than in male life expectancy, which made it more probable that the wives of those men who did survive into old age would still be alive.

A parallel profile of female lone parenthood is set out in Figure 2.4. In all periods, the possibility that a woman would be a lone parent increased much faster during early and middle adulthood than was the case for a man. The incidence of lone parenthood also peaked earlier for women, in the age group 50-54 or 60-64 in the pre-1800 populations, in the age group 60-64 in 1891, and in the age group 40-44 in 1981. The abrupt decline in the number of women over the age of 60 in the pre-1800 populations who were lone parents implies that older women before 1800 had other residence patterns, either living alone or moving in or taking into their own households more distant relatives or non-relatives. Later analysis (cf. below, Figures 2.6, 2.8, and 2.10) shows that it was the latter two roles that were of particular importance. In 1891, the rate of female lone parenthood did not decline during old age to anything like the same extent, indicating that the presence of the unmarried child was a major source of support for the older women. The situation in 1981 differs in that the frequency of lone parenthood varies within much narrower limits across the age spectrum than in either of the earlier periods. Whereas in 1891 and before 1800 the tendency was for fewer women under than over 40 to be lone parents, in 1891 the rates of female lone parenthood were considerably higher during early adulthood (under the age of 40), although they increased again in extreme old age.

These findings, assessed in conjunction with the information on membership of a family as set out in Figure 2.2, reveal that but for the increased number of women over the age of 50 who were lone parents in 1891, the proportion of older women who were members of a family in 1891 would have been considerably below the proportion of older women in the pre-1800 populations who were family members. With the lone parents included, depending on the particular age group, family membership by older women in 1891 was sometimes a little below, sometimes a little above the proportions of family members in the pre-1800 populations. Conversely, the lower proportions of women over the age of 45 who were lone parents in 1981 indicates that family membership could only have achieved such a high level in 1981 through an increase in the proportion of older women who were still married.

2.4.2. Non-family members

Changes between pre-industrial times and 1981 in the residence patterns of persons without any member of their immediate family co-resident will now be considered. In all three time periods (before 1800, 1891, and 1981) for men or women to live with a relative in the absence of a spouse or unmarried child was unusual except in old age. There are differences, however, between one period and another in both the timing and pace of this increase. The frequency with which men lived with such relatives (Figure 2.5) increased by at least five per cent between the age groups 60-64 and 65-69 in the pre-1800 populations, in 1891 between the age groups 55-59 and 60-64, and in 1981 between the age groups 75-79 and 80-84. It is also clear that the increased frequency of living with relatives in old age is much less marked in 1981, and that it was in 1891 that living with relatives was most probable. Turning to the women (Figure 2.6), it is clear that differences over time in the frequency with which women lived with relatives in the absence of a spouse or unmarried child follow a broadly similar pattern, while in all periods increasing a little earlier in old age for women than for men.

Prior to 1800, however, a substantial minority of both males and females lived with non-relatives only. For males this was most common when in their late teens, although more than one in ten males were in this situation between the ages of 10 and 50 (Figure 2.7). In 1891, despite the decline in male farm servants, the frequency of co-residence only with non-relatives still peaked in early adulthood and maintained a level usually just under one in ten of the male population over the rest of the life cycle. Even in 1981, when to live just with non-relatives had become much rarer at all ages, the highest frequency still occurred during early adulthood. However, the peak on this occasion is much less marked and arises, at least in part, from the presence of a certain number of undetected *de facto* spouses, as well as from instances of non-related persons sharing a household.

Females too might live with non-relatives in the English past, mainly as servants but also as lodgers (Laslett 1977, p. 34; Wall 1987, pp. 84-85). However, the maximum level of living with non-relatives in the absence of any relatives was reached a little later by females than by males, in the pre-1800 population, within the age group 20-24 (Figure 2.8). A second distinguishing characteristic of these residence patterns was the increase which occurred in the frequency with which women lived with non-relatives during old age. In

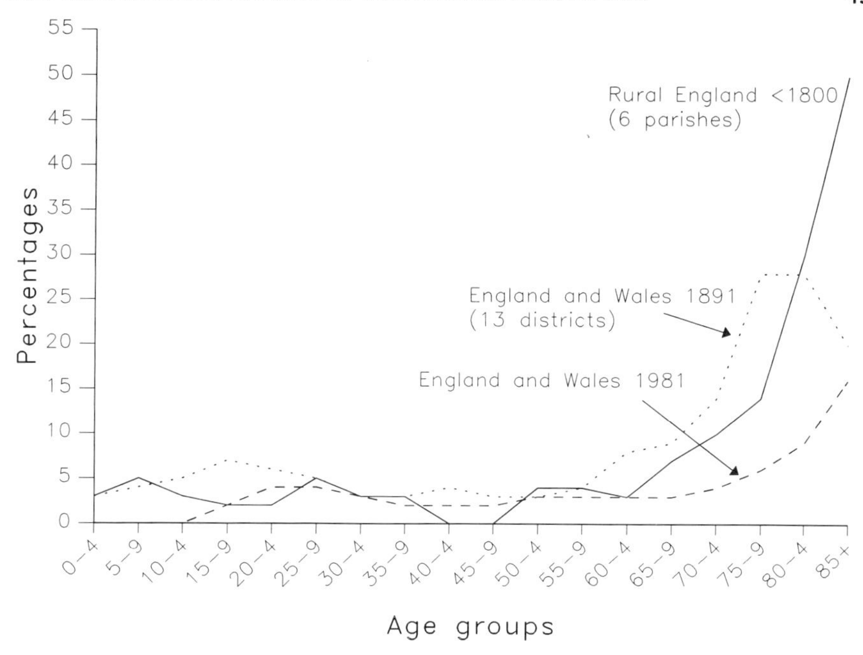

Figure 2.5. Relative frequency of co-residence with relatives in the absence of spouse and unmarried children: males aged 0-85+ in England and Wales from pre-1800 to 1981

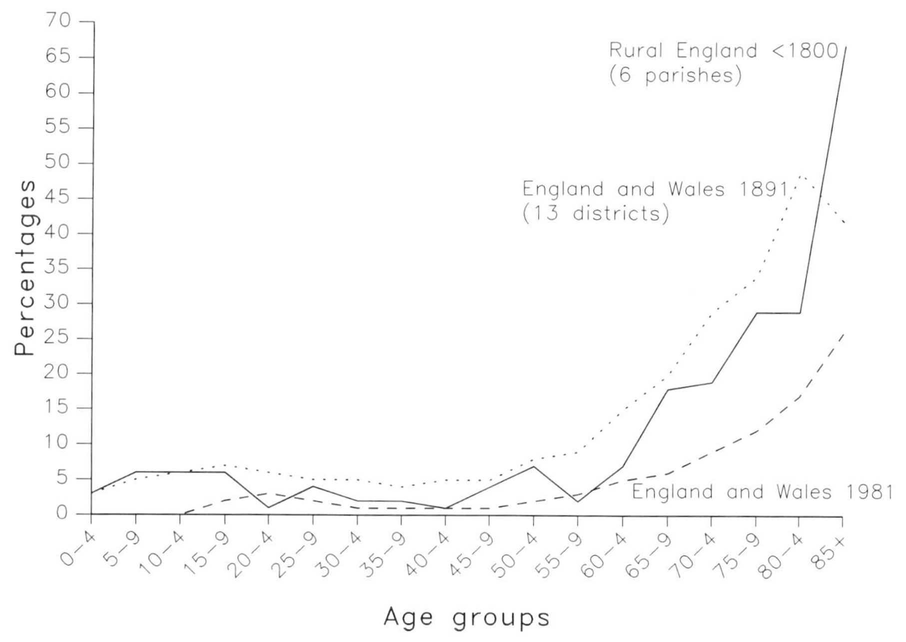

Figure 2.6. Relative frequency of co-residence with relatives in the absence of spouse and unmarried children: females aged 0-85+ in England and Wales from pre-1800 to 1981

Figure 2.7. Relative frequency of co-residence with non-relatives only: males aged 0-85+ in England and Wales from pre-1800 to 1981

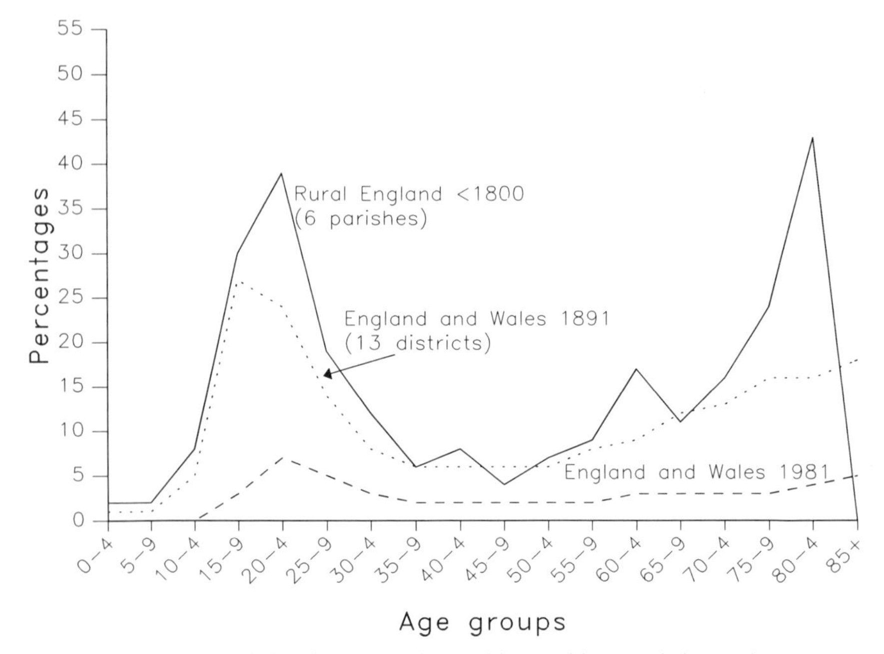

Figure 2.8. Relative frequency of co-residence with non-relatives only: females aged 0-85+ in England and Wales from pre-1800 to 1981

the pre-1800 populations, the 50-54 age group was the first to show evidence of such a rise, which is likely to have been occasioned by women needing to seek out alternative and cheaper accommodation as they aged, and particularly after widowhood. Residential arrangements on these lines could also on occasion be engineered by Poor Law authorities anxious to economize on their expenditure by arranging for a younger woman to care for an unrelated older one (Erith 1978). There were still signs of a higher frequency of residence by older women with non-relatives only in 1891, but by 1981 this pattern has disappeared. Moreover, the frequency with which women lived with non-relatives in 1981 was not only much lower, but also much less variable across the life cycle.

Finally, some males, and rather more females, lived on their own. For males to live alone in the pre-1800 populations was extremely unusual (Figure 2.9). In 1891 it was also unusual, although there are signs of an increased frequency of living alone beginning in the age group 50-54. A comparable trend, although at a higher level, is also visible in 1981 but beginning in the age group 45-49, and with a suggestion of an earlier peak when men were in their late twenties and early thirties.

For females, there is evidence for the pre-1800 populations of an increased incidence of living alone in old age (from the age group 60-64, see Figure 2.10). This broadly complements the increased incidence of living with non-relatives, but was not characteristic of the living arrangements of older men (cf. Figures 2.8 and 2.9). It is also evident that a greater proportion of older than of younger women lived alone in 1891 and in 1981. Clearly, what distinguishes the residence pattern of older women in 1981 is therefore not that more older than younger women lived alone, but the magnitude of the rise in the incidence of living alone during old age.

2.5. CONCLUSION

By way of conclusion, two general comments will first be advanced. The first is that, at most ages, a higher proportion of the population of England and Wales in 1981 co-resided with at least one other closely related member of their family than was the case in pre-industrial times or in 1891. As a residential unit, it is clear that the family today is not in the process of long-term terminal decline, popular accounts to the contrary notwithstanding. Secondly, apart from the period of early adulthood, when residence outside of a family group was considerably more common in the past because of the extent of live-in service and lodging, it is remarkable how little variation there has been between pre-industrial times, the end of the nineteenth century, and the latter part of the

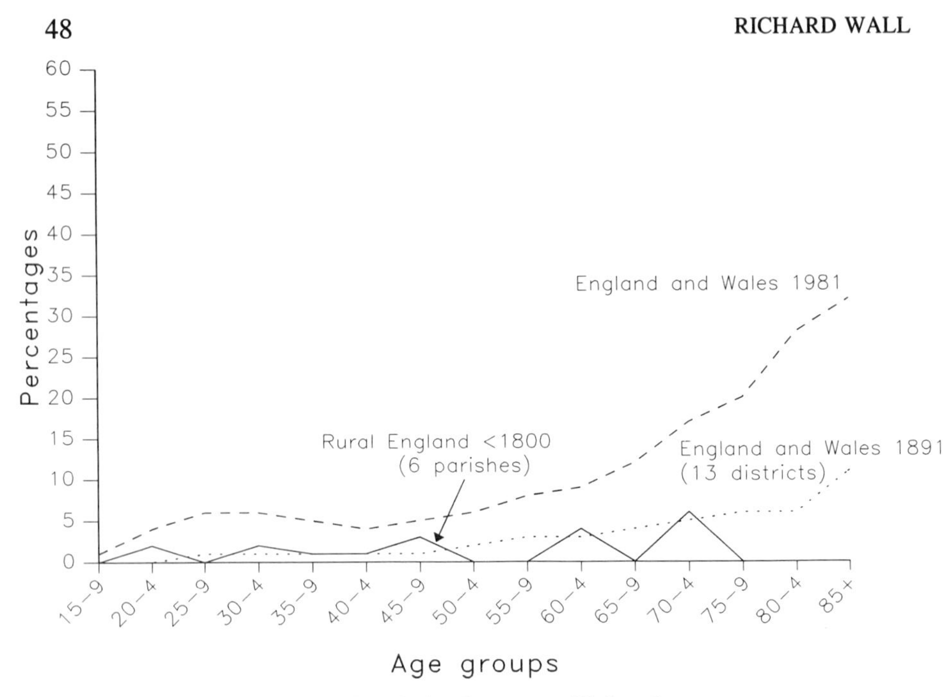

Figure 2.9. Relative frequency of living alone: males aged 15-85+ in England and Wales from pre-1800 to 1981

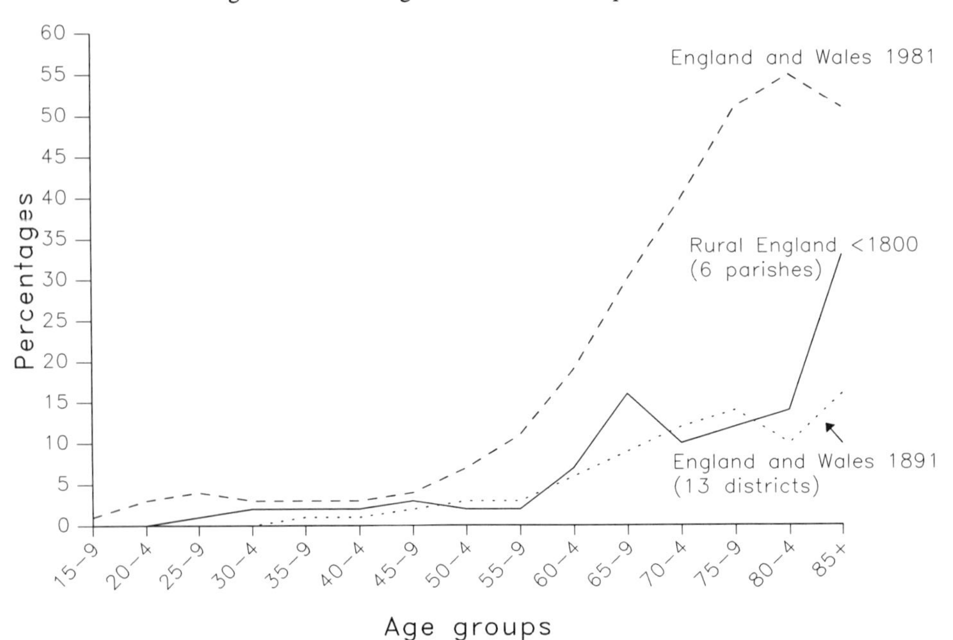

Figure 2.10. Relative frequency of living alone: females aged 15-85+ in England and Wales from pre-1800 to 1981

twentieth century in the frequency with which individuals reside in family groups.

All this is not to deny that there has not been some change. The above account has attempted to chart this in considerable detail. Particularly significant is the decline between pre-industrial times and 1891 in the frequency with which both males and females lived only with non-relatives, and between 1891 and 1981 in the incidence of living with relatives other than a spouse and unmarried children. The primary influence in each case is undoubtedly economic in nature, with changes in the allocation of labour requiring fewer workers to be attached to the particular households in the form of servants, while improvements in living standards during the course of the twentieth century, and particularly since the 1950s, have enabled more people to live alone who otherwise would have lived with relatives or shared a household with non-relatives. Nevertheless, the overriding impression that remains is how resilient the family has been to the great demographic and economic changes of the past two centuries, with membership of the core family group, couple or parent-child, more widespread at most ages in 1981 than had been the case either in 1891 or before 1800.

ACKNOWLEDGEMENTS

In the course of the preparation of this chapter, I have benefitted greatly from discussions with my colleagues at the ESRC Cambridge Group and at the Netherlands Interdisciplinary Demographic Institute. Section 2.3 on the European family is a revised version of a paper that appeared originally in the volume of essays dedicated to Etienne Helin (Wall 1991). Particular thanks are due to Kevin Schürer who provided the tabulations from the censuses of 1891-1921 and to Erik Beekink who redrew all the figures when all the originals disappeared along with my briefcase in Amsterdam Central Station. Bruce Penhale, then of the Social Statistics Research Unit of City University, London, collaborated with me on the analysis of the 1971 and 1981 Longitudinal Study. I was also fortunate in being able to read, in advance of publication, David Kertzer's *Sacrificed for honor. Italian infant abandonment and the politics of reproductive control.*

REFERENCES

BENIGNO, F. (1989), The Southern Italian Family in the Early Modern Period: A Discussion of Co-Residence Patterns, *Continuity and Change*, Vol. 4(1), pp. 165-194.

BERNANDES, J. (1986), In Search of the Family: Analysis of the 1981 UK Census, a Record Note, *The Sociological Review*, Vol. 34(4), pp. 828-836.

CENTRAL STATISTICS OFFICE (1983), *Census of Population of Ireland 1979, Volume 3*, Stationery Office: Dublin.

COUNCIL OF EUROPE (1990), *Household Structures in Europe*, Population Studies Vol. 22, Strasbourg.

DAELEMANS, F. (1975), Leiden 1581: Een Socio-Demografisch Onderzoek, *A.A.G. Bijdragen*, Vol. 19, pp. 137-215.

ERITH, F.H. (1978), *Ardleigh in 1796. Its Farms, Families and Local Government*, Hugh Tempest Radford: East Bergholt.

FAUVE-CHAMOUX, A. (1983), The Importance of Women in an Urban Environment: the Example of the Rheims Household at the Beginning of the Industrial Revolution. *In*: R. WALL, J. ROBIN & P. LASLETT (eds.).

FAUVE-CHAMOUX, A. (1987), Le Fontionnement de la Famille-Souche dans les Baronnies des Pyrénées avant 1914, *Annales de Démographie Historique*, pp. 241-261.

HAJNAL, J. (1983), Two Kinds of Pre-Industrial Formation Systems. *In*: R. WALL, J. ROBIN & P. LASLETT (eds.).

JOHANSEN, H.C. (1975), *Befolkningsüdvikling og Familiestruktur i det 18. Arhundrede*, Odense University Press: Odense.

KERTZER, D.I. (1994), *Sacrificed for Honor: Italian Infant Abandonment and the Politics of Reproductive Control*, Beacon Press: Boston.

KLEP, P.M. (1973), Het Huishouden in Westelijk Noord-Brabant. Struktuur en Ontwikkeling 1750-1849, *A.A.G. Bijdragen*, Vol. 18, pp. 23-94.

LASLETT, P., & R. WALL (eds.) (1972), *Household and Family in Past Time*, Cambridge University Press: Cambridge.

LASLETT, P. (1977), *Family Life and Illicit Love in Earlier Generations*, Cambridge University Press: Cambridge.

LASLETT, P. (1979), The Stem Family Hypothesis and its Privileged Position. *In*: K.W. WACHTER, E.A. HAMMEL & P. LASLETT (eds.) *Statistical Studies of Historical Social Structure*, Academic Press: New York and London.

LASLETT, P. (1983), Family and Household as Work Group and Kin Group: Areas of Traditional Europe Compared. *In*: R. WALL, J. ROBIN & P. LASLETT (eds.).

LE BRAS, H. (1979), *L'Enfant et la Famille dans les Pays de l'OCDE*, Organization for Economic Cooperation and Development: Paris.

LE PLAY, P.G.F. (1855, 1877-9), *Les Ouvriers Européens: l'Organisation des Familles*, first edition, Imprimerie Impériale: Paris; second edition, 6 volumes, Alfred Mame et fils: Tours.

LESTHAEGHE, R. (1992), The Second Demographic Transition in Western Countries: an Interpretation, IPD-Working Paper, Free University of Brussels.

NORGES OFFISIELLE STATISTIKK (1980), *Foketeljinga 1801. Ny Beardeiding*, Statistisk Sentralbyrå: Oslo.

OFFICE OF POPULATION CENSUSES AND SURVEYS [OPCS] (1984), *Census 1981. Household and Family Composition England and Wales*, HMSO: London.

OFFICE OF POPULATION CENSUSES AND SURVEYS [OPCS] (1988), *Census 1971-1981. The Longitudinal Study, England and Wales*, HMSO: London.

OUVRIERS DES DEUX MONDES (1855-85, 1885-99), first series vols. 1-5; second series vols. 1-5: Société Internationale des Etudes Pratiques d'Economique Sociale.

PERRENOUD, A. (1979), *La Population de Genève du Seizième au Début du Dix-Neuvième Siècle*, Vol. 1, Librairie A. Jullien: Genève.

PULLAN, B. (1989), *Orphans and Foundlings in Early Modern Europe*, The Stenton Lecture 1988, University of Reading: Reading.

REHER, D.S. (1988), *Familia, Población y Sociedad en la Provincia de Cuenca 1700-1970*, Centro de Investigaciones Sociológicas - Siglo XXI: Madrid.

REHER, D.S. (1990), *Town and Country in Pre-Industrial Spain. Cuenca 1550-1870*, Cambridge University Press: Cambridge.

SCHELLEKENS, J. (1991), Determinants of Marriage Patterns among Farmers and Agricultural Labourers in two Eighteenth Century Dutch Villages, *Journal of Family History*, Vol. 16(2), pp. 139-155.

SCHIAVONI, C., & A. SONNINO (1982), Aspects Généraux de l'Evolution Démographique à Rome: 1598-1826, *Annales de Démographie Historique*, pp. 91-109.

SCHÜRER, K. (1992), Variations in Household Structure in the Late Seventeenth Century: Toward a Regional Analysis. *In*: K. SCHÜRER & T. ARKELL (eds.) *Surveying the People. The Interpretation and Use of Document Sources for the Study of Population in the Later Seventeenth Century*, Leopard's Head Press: Oxford.

SMITH, R.M. (1981), Fertility, Economy and Household Formation in England over Three Centuries, *Population and Development Review*, Vol. 7(4), pp. 595-622.

SOCIAL SCIENCE RESEARCH UNIT (1990), *OPCS Longitudinal Study. User Manual*, City University: London.

STATISTICAL BUREAU OF ICELAND (1960), *Population Census 1703*, Gefid út af Hagstofu Íslands: Reykjavík.

STATISTICS CANADA (1987), *The Nation: Families, Part I*, Minister of Supply and Services: Ottawa.

TILLY, L., & J. SCOTT (1978, 1987), *Women, Work and Family*, first edition, Holt, Rinehart and Winston: New York; second edition, Methuen: London.

TWAIN, M. [pseud = S.L. Clemens] (1883, 1990), *Life on the Mississippi*, first edition 1883; Bantam Classic edition 1990: New York and London.

VAN DE KAA, D.J. (1987), Europe's Second Demographic Transition, *Population Bulletin*, Vol. 41(1), Population Reference Bureau: Washington D.C.

VAN DER WOUDE, A.M. (1980), Demografische Ontwikkeling van de Noordelijke Nederlanden 1500-1800, *Algemene Geschiedenis der Nederlanden*, Vol. 5.

VLAAMSE VERENIGING VOOR FAMILIEKUNDE, AFDELING BRUGGE (1976-7), *Volkstelling 1814*, Vols. i, iii, v, vi: Brugge.

WALL, R. (1979), Regional and Temporal Variations in English Household Structure from 1650. *In*: J. HOBCRAFT & P. REES (eds.) *Regional Demographic Development*, Croom Helm: London.

WALL, R. (1982), Regional and Temporal Variations in the Structure of the British Household since 1851. *In*: T. BARKER & M. DRAKE (eds.) *Population and Society in Britain 1850-1980*, Batsford: London.

WALL, R. (1983a), The Composition of Households in a Population of 6 Men to 10 Women: South-east Bruges in 1814. *In*: R. WALL, J. ROBIN & P. LASLETT (eds.).

WALL, R. (1983b), Does Owning Real Property Influence the Form of the Household? An Example from Rural West Flanders. *In*: R. WALL, J. ROBIN & P. LASLETT (eds.).

WALL, R. (1983c), Introduction. *In*: R. WALL, J. ROBIN & P. LASLETT (eds.).

WALL, R. (1983d), The Household: Demographic and Economic Change in England, 1650-1970. *In*: R. WALL, J. ROBIN & P. LASLETT (eds.).

WALL, R. (1986), Work, Welfare and the Family: an Illustration of the Adaptive Family Economy. *In*: L. BONFIELD, R.M. SMITH & K. WRIGHTON (eds.) *The World we have Gained*, Basil Blackwell: Oxford.

WALL, R. (1987), Leaving Home and the Process of Household Formation in Pre-Industrial England, *Continuity and Change*, Vol. 2(1), pp. 77-101.

WALL, R. (1989), Leaving Home and Living Alone: an Historical Perspective, *Population Studies*, Vol. 43, pp. 369-89.

WALL, R. (1991), European Family and Household Systems. *In*: SOCIÉTÉ BELGE DE DÉMOGRAPHIE, *Historiens et Populations: Liber Amicorum Étienne Hélin*, Academia: Louvain-la-Neuve.

WALL, R. (1993), The Contribution of Married Women to the Family Economy under different Family Systems: some Examples from the Mid-Nineteenth Century from the Work of Frédéric Le Play, Paper presented to Session 9 of XXIInd IUSSP General Conference, Montreal.

WALL, R. (forthcoming), Elderly Persons and Members of their Households in England and Wales from Pre-Industrial Times to the Present. *In*: D. KERTZER & P. LASLETT (eds.) *Ageing in the Past. Demography, Society and Old Age*, University of California Press.

WALL, R., J. ROBIN & P. LASLETT (eds.) (1983), *Family Forms in Historic Europe*, Cambridge University Press: Cambridge.

WALL, R., K. SCHÜRER, E. GARRETT, & A. REID (forthcoming), *Demographic Transition, Life Cycle and the Household in England and Wales 1891-1921*.

3. RECENT TRENDS IN HOUSEHOLD AND FAMILY STRUCTURES IN EUROPE: AN OVERVIEW

Anton Kuijsten

Department of Planning and Demography
University of Amsterdam
Nieuwe Prinsengracht 130
1018 VZ Amsterdam
The Netherlands

Abstract. Recent trends in household structures in European countries can be viewed as reflecting the basic demographic changes that have been summarized in the concept of the Second Demographic Transition. On the basis of a typology of leading and lagging countries in this transition process, first some basic demographic indicators relevant for family and household structures are discussed. Then, this typology is used as a structuring framework for describing recent trends in average household size, in household size distributions, and in family type distributions. Taking the example of the Netherlands, changes in household structures are then summarized by presenting changes over the 1980s in 'top-6' lists of life style patterns for women both at the beginning and at the end of their family formation career. The chapter concludes with expectations regarding the future course of household and family trends in Europe.

3.1. INTRODUCTION: LEADERS AND LAGGARDS IN DEMOGRAPHIC TRENDS

In the past decade, a number of useful overviews of household and family trends in Europe have been published (Council of Europe 1990; Hall 1986; Hoffmann-Nowotny 1987; Höpflinger 1991a; Keilman 1987; Kuijsten & Oskamp 1991; Schwarz 1988). Unfortunately, none of these overviews deals with all European countries, including those in Central and Eastern Europe that recently experienced dramatic political, social, and economic changes. The new political map of Europe justifies inclusion of these countries in comparative demographic research, if only because several experts expect increasing convergence of trends in countries all over the region (e.g. Vichnevsky *et al.* 1993; for a more

Household Demography and Household Modeling
Edited by E. van Imhoff *et al.*, Plenum Press, New York, 1995

cautious assessment, see Coleman 1991). Moreover, these overviews often lack a clear and rigid framework for giving structure to the data. My contribution to this volume is one of the results of the preparatory phase of a research project aimed at overcoming these drawbacks. At this stage, at best a general and incomplete picture can emerge, subtlety and nuances having to await further work on each and every aspect.

In his article on 'The Family in Western Europe: Differences and Similarities', Roussel (1992) proposed a typology that can very well serve the purpose of a data structuring framework. He uses the metaphor of the clown's robe for describing the European family map: it looks like a piece of patchwork with its different colours. But, Roussel says, during the past decades these colours have first become more diversified, and later have grown more similar again, and he expects this process of convergence to continue in the next two decades.

The determinants of this more or less cyclical process of divergence and convergence must be found in spatial and temporal differences in trends in basic demographic phenomena and, behind these, in socio-economic indicators that predominantly have to do, in Roussel's view, with the changing position of women, both in society at large and in the micro realm of the family and household. To demonstrate his thesis, Roussel develops a typology of 'European families' according to their demographic characteristics. I reproduce his typology of four groups in Table 3.1, in a slightly adapted way through tentatively extending it, particularly with a fifth group which mainly contains the East European countries.

Group A (NORTH) contains Scandinavian countries only. Here, people have the longest tradition of considering their behaviour in the relevant aspects as purely 'private'. Marriage or cohabitation, births within or outside marriage: all are private decisions that are hardly influenced anymore by legal considerations. Group B (WEST) contains countries in which there still is somewhat more reluctance: divorce is relatively high, but cohabitation is still predominantly *pre*marital. Nevertheless, childbearing outside marriage becomes more and more accepted. Countries in Group C (CENTRE) seem to be even more reluctant still: cohabitation and extramarital births are less frequent. Group D (SOUTH) contains the countries that have waited longest with joining the trends in the other groups, but, since joining them, have proceeded very quickly in the modern trends of what may be called a 'de-institutionalization' of the family. Their current fertility levels are extremely low, but they are still lagging in the other indicators concerning marriage, divorce, and births out-of-wedlock. Group E (EAST & FRINGE) is still lagging most. Fertility and nuptiality are still at high levels, cohabitation is still low, and with respect to levels of divorce and extramarital births, the group seems rather heterogeneous.

Table 3.1. A typology of European families

Group	Level of Indicator *		Country	Demographic indicators (around 1988)			
				Total Fertility Rate (TFR)	Total First Marriage Rate	Divorce per 100 Marriages	Percentage Births outside Marriage
Group A	F	= relatively high	Denmark	1.62	572	46	45
NORTH	D	= high	Sweden	2.02	601	41	50
	C	= high	Finland	1.78	592	38	25
	EMF	= moderate/high					
Group B	F	= low	France	1.81	540	31	28
WEST	D	= high	Norway	1.89	558	37	28
	C	= low	Netherlands	1.55	601	28	10
	EMF	= moderate	United Kingdom	1.81	665	42	25
Group C	F	= very low	Austria	1.45	599	30	23
CENTRE	D	= high	Belgium	1.58	718	31	8
	C	= moderate	FRG	1.39	598	32	10
	EMF	= moderate/high	Luxembourg	1.52	579	37	12
			Switzerland	1.51	663	33	6
Group D	F	= very low	Italy	1.29	695	8	6
SOUTH	D	= low	Greece	1.50	870	12	2
	C	= low	Portugal	1.53	787	11	14
	EMF	= low	Spain	1.30	640	-	8
Group E	F	= high	Ireland	2.11	710	forbidden	12
EAST	D	= heterogeneous?	Iceland	2.21	470	36	52
&	C	= low	Poland	2.15	840	17	6
FRINGE	EMF	= heterogeneous?	Czechoslovakia	2.02	900	32	7
			Hungary	1.81	750	28	12
			Bulgaria	1.96	852	18	11
			Romania	2.31	870	21	4

* Explanation:
F = Level of Fertility
D = Level of Divorce
C = Level of Cohabitation
EMF = Level of Extramarital Fertility

Sources: Roussel 1992, pp. 135 and 137; Council of Europe 1993; Total Fertility Rates for Poland, Czechoslovakia, Hungary and Bulgaria: Monnier 1990, p. 236; Total First Marriage Rates for Poland, Czechoslovakia, Hungary and Bulgaria: Sardon 1992, Table 1; other data for Finland: Yearbook 1992, p. 111; additional divorce indicators not given by Roussel: De Guibert-Lantione & Monnier 1992.

In this chapter, I will use this typology as a structuring framework for the data to be presented. My theoretical framework, that of the Second Demographic Transition, is presented in section 3.2. Section 3.3 discusses recent trends in some demographic indicators that are basic determinants for the changes in household structure presented in section 3.4. In section 3.5, conclusions are drawn and outlooks for the future are discussed.

3.2. THE SECOND DEMOGRAPHIC TRANSITION

In order to explain the changes alluded to in these group characterizations, the demographic transition model is a good starting point.

Those who originally formulated this model of demographic transition envisaged that this transition process would come to an end in a situation in which, again, mortality and fertility would be more or less in balance. This would mean a situation of long-lasting very slow or even quasi-zero population growth. But they also clearly envisaged the possibility of negative growth, at least temporarily, for *structural reasons*, that is, reasons of age structure imbalances, caused by developments as they were expected to occur during the last part of the middle transitional phase (e.g. Thompson 1929; see also Schubnell 1966, pp. 64-65 when speaking of a coming 'second demographic revolution').

Despite these warnings, nobody has expected the far-reaching reversals of tendencies that have happened all over the modern, industrialized western countries, the vast majority of European countries included, from the middle of the 1960s. It would last two decades, till 1987, before in the Low Countries a new concept for describing these post-1965 developments was cradled: the *Second Demographic Transition* (Lesthaeghe & Van de Kaa 1986; Van de Kaa 1987).

The developments in the Netherlands, as depicted in Figure 3.1, can be taken as an example. In fact, the third phase of the demographic transition occurred in the Netherlands from the middle of the 1960s. But it did so in such a high tempo and, as was the case in most other modern, industrialized countries, it was connected with so many other changes in demographic behaviour that have not been foreseen in the original model of demographic transition, that it seemed justified to distinguish these developments by this new and separate concept of the Second Demographic Transition. These changes were viewed by Van de Kaa (1988) as the consequences of the mutual influences of a number of factors that can be categorized in three dimensions:

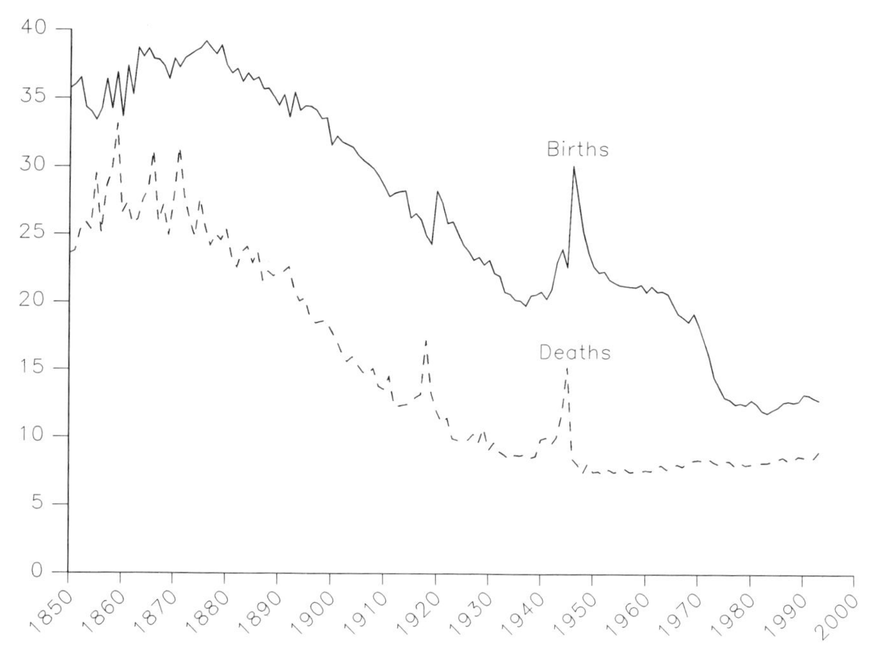

Figure 3.1. Birth and death rates (per 1000 population), The Netherlands, 1850-1993
Source: Statistics Netherlands

- **Structural factors** that relate to the completion of post-industrial society and of the so-called welfare state: increased standard of living, social security, functional differentiation, structural complexity, social and occupational mobility, education, and female labour force participation;
- **Cultural factors** as summarized by Inglehart (1977) under the concept of the *Silent Revolution* and embracing changes in values and political styles in western societies: factors such as a decrease of social inequalities, an increase in personal freedom and in democracy, value pluralism, individualization and secularization: changes that started along the lines of birth generations (Becker 1985, 1990), but after a short time were adopted by the vast majority of population (De Feijter 1991);
- **Technological factors**, including the introduction of perfect contraceptive means and techniques, and improvements of means of transport, communication, and health care.

Van de Kaa positions this Second Demographic Transition as opposite to the other one, from now on called the *First* Demographic Transition, from the perspective of the conceptual dualism between *altruism* and *individualism*. Thus, the first transition was an altruistic one, since the changes in fertility were meant in the first place to increase the standard of living of the couple and the family, mainly its children. These could be better fed, better housed, better educated, etc. when there were fewer of them in the family. The second transition is individualistic or egoistic, since it first of all starts from the right to personal well-being of adult individuals, the right to self-actualization of individuals, women in particular.

From a demographic point of view, the most important changes that characterize the era of the Second Demographic Transition have been summarized in four slogans (Van de Kaa 1987, p. 11):

- shift from the 'Golden Age of Marriage' to the dawn of cohabitation,
- shift from the era of the king-child with parents to that of the king-pair with a child,
- shift from preventive contraception to self-fulfilling conception,
- shift from uniform to pluralistic families and households.

The demographic manifestations of these changes were:

- postponement of marriage and increasing unmarried cohabitation,
- a decline of fertility to well below replacement level,
- rising proportions of illegitimate births, and
- declining average size of households and a growing variety of household types.

Before taking a closer look at these demographic trends, we must first pay attention to their most important driving forces: value changes. Again, the Netherlands is taken as an example, but the general trend holds for most European countries, starting in the 1960s in the West and North European ones, and followed from the late 1970s by those of Southern Europe. Evidence of such value changes has been presented by Van de Kaa (1987, p. 8). He gives an overview of changes in proportions in agreement to attitude questions in a series of surveys of adults, usually 16-74 years old, surveys held in the Netherlands from 1965 onwards. Over the years, one observes continuously growing proportions, with the big upward jumps always from 1965/1966 to 1970, responding affirmatively to statements such as 'There is no objection to sexual relations for people intending to marry'; 'Voluntary childlessness of a couple is acceptable'; 'In a bad marriage it is better to divorce even if there are still children living at home'; or 'Labour force participation of a married woman with school-age children is acceptable'. What these data show, basically, is a growing permissiveness with respect to one's own behaviour and that of other people. Special features of post-industrial society are mentioned in the literature as obviously weakening the individual's enthusiasm for long-lasting relationships, parenthood, and childbearing. Decisions about these matters that involve long-range planning necessarily are made within 'an atmosphere of apprehensive uncertainty'. In summary, there is a growing literature that suggests a basic incompatibility between the family on the one hand and capitalist industrial economies on the other, a basic incompatibility that was already predicted long ago by the economist Schumpeter (1988).

Demographically, these and other societal reflections of changing partnership and child rearing find their expression, at least partially, in decreasing expected and actual family size, in a decreased emphasis on the rewards of children in the overall life plans of young people, *and especially through commitments of decreasing proportions of the life course to partnerships, to childbearing and to child rearing*. For instance, according to the so-called utility model in the 'New Home Economics', the decision to have a child, or an additional child, is based on an evaluation of the costs and benefits of children and the potential reward value of childbearing and child rearing, relative to alternatives within a much more widely defined preference field.

3.3. RECENT TRENDS IN BASIC DEMOGRAPHIC INDICATORS

3.3.1. Marriage

We will now take a look at the most important demographic changes, and start with marital behaviour. Comprehensive reviews of these developments have been presented by Höpflinger (1991a), Roussel (1989), and Hoffmann-Nowotny (1987). Haskey (1993) has summarized recent trends in formation and dissolution of unions in several countries of Europe. He states that these changes have perhaps resulted from a change in attitudes to relationships, both to marriage and cohabitation (Haskey 1993, p. 211). According to Haskey (1993, p. 215), around 1970, a turning point occurred in both timing and prevalence of marriage in most European countries. The major causes are:

- the emergence of cohabiting unions;
- the reaching of marriageable age by the first generations born after the War;
- changing aspirations concerning family size and lifestyle; and
- changing economic circumstances of the countries of Europe.

Whatever the difficulties may be in deciding upon accelerations or upon trend breaks as the major characteristic of these demographic changes, a series of graphs provided by Cliquet (1991, pp. 92-94) nicely shows that first marriage evolution is one of the clearest examples of obvious trend breaks during the past decades, at least in the Member Countries of the Council of Europe. Between 1970 and 1975, the total period first marriage rates started to fall in most countries, although they continued to rise in Eastern Europe, and in three countries in Southern Europe - Portugal, Spain, and Greece. In the next five years, however, the rate was declining in virtually every country, and this trend continued, albeit at a slower pace, during the 1980s, apart from a few policy-induced and temporary upsurges, such as in Austria in 1983 and 1987 (see Prioux 1992), and in Sweden in 1989. By the end of the decade, period first marriage rates were generally highest in Eastern Europe, and lowest in Northern Europe, with those in Western and Southern Europe at intermediate levels. In some countries, marriage has merely been postponed, whilst in others, for a proportion at least, it has been replaced either by cohabitation or by living alone.

At the same time, a reversal took place in the trend of the mean age at marriage. The trend of decline in the mean age at first marriage reversed in the middle of the 1970s, showing a steady increase since then. As a result, at the end of the 1980s, high total period first marriage rates are almost invariably associated with low mean ages at first marriage, and *vice versa* (Figure 3.2).

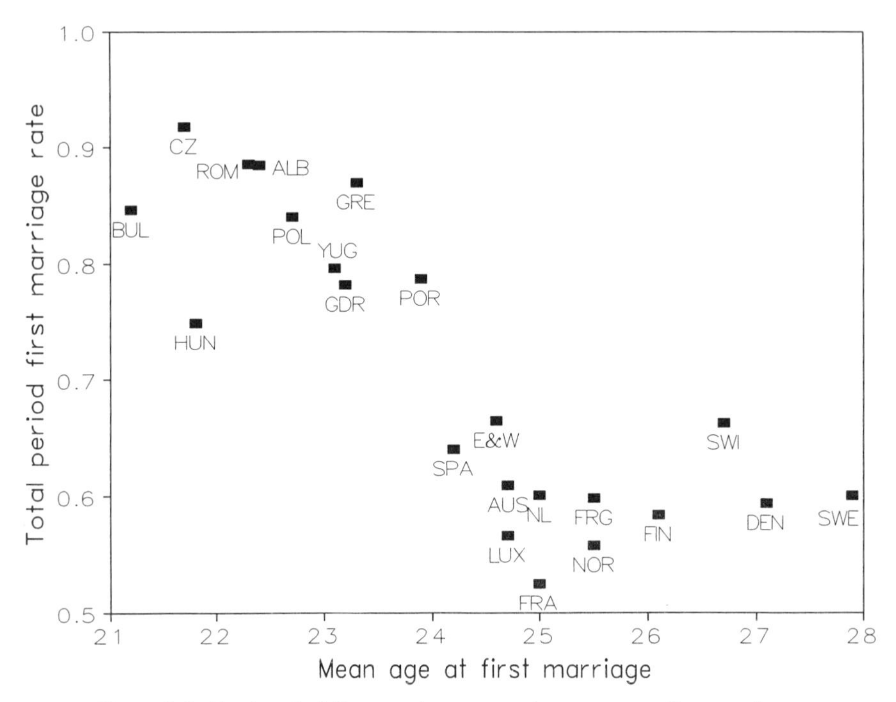

Figure 3.2. Total period first marriage rate and mean age at first marriage, females, 1988, selected countries in Europe
Source: Sardon 1992

When looked upon over time, a general tendency for large decreases in nuptiality seems to be associated with large increases in mean ages. This evidence suggests that, for most countries, postponement of marriage has at least played a part in the decline of the total period first marriage rate.

Sardon (1992) recently proposed an interesting graphical presentation method for showing the relationship between trends in level and trends in timing of first marriage. His method implies a plotting of the time evolution of the total first marriage rate (on the horizontal axis) against the time evolution of mean age at marriage (on the vertical axis). I have made two such plots in one graph, comparing developments in Hungary and the Netherlands for the period 1950-1990 (Figure 3.3). The line for the Netherlands shows the circular path that is typical for countries which have fully participated in the Second Demographic Transition. First, mean age at first marriage declines without a drop in total first marriage rates, which remain at a relatively high level. This phase lasts till 1970. Then total first marriage rates drop rapidly without any further decline in the mean age at first marriage. Finally, since the end of the 1970s, mean age at first marriage rises again, whereas total first marriage rates stop

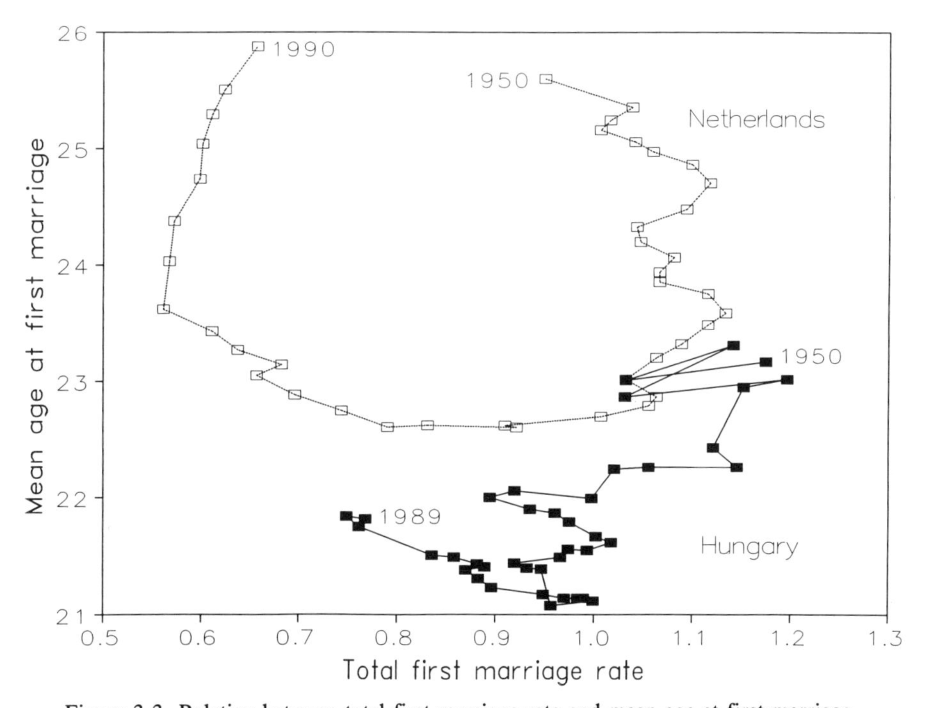

Figure 3.3. Relation between total first marriage rate and mean age at first marriage, females, Hungary and The Netherlands, 1950-1991
Source: Sardon 1992

declining, and even rise a little bit. In contrast, Hungary is still far from having completed this circular movement.

These changes in marital behaviour, including those in divorce and remarriage, have led to profound changes in marital status structure which show significant differences between groups of countries that have more or less far proceeded on the road of the Second Demographic Transition. On the basis of a cluster analysis of proportions by age and marital status categories, Rychtarikova (1993) has distinguished five groups of countries with a very different marital status structure at the end of the 1980s. These five groups of countries, which nicely move from Eastern through Southern and Western to Northern Europe in a way that highly resembles the typology of Roussel discussed above, showed significant differences in 1985 in their group averages of patterns of proportions married and proportions divorced by age. As a general rule, proceeding clockwise from East through South and West to North, for almost all age groups and both sexes, proportions married shade over from higher into lower and proportions divorced from lower into higher.

Of course, when looking at such overall proportions in a marital status (for the population aged 15 and over), age structure is a disturbing factor. However, a map for 1980 of age-standardized proportions married (Decroly & Vanlaer 1991, p. 144), a map with much more regional detail, clearly visualizes this general pattern too, with standardized proportions married above 75 in many regions in Eastern Europe and the former Soviet Union, and in almost all of Portugal, and proportions married around 50 only in most Scandinavian regions, Austria, Switzerland, and in Ireland where the traditional pattern of late marriage and low marriage frequency still exercises its influence. And this picture is almost reversed when one looks at a map of age-standardized proportions single in the population of 15 years and older (Decroly & Vanlaer 1991, p. 143). To some extent, proportions married and proportions single are complementary if one only looks at the legal marital status criterion. But here one must not forget that, especially in Western and Northern Europe where we have the highest proportions of singles, many of these singles do not live alone, but cohabit with somebody of the opposite sex in an informal relationship that nowadays in all aspects but the legal one resembles the situation of married people. So 'Hajnal's line' (Hajnal 1965), the old historical division in traditional nuptiality regimes, which in the 19th century cut Europe into two parts along either side of an imaginary line from Trieste to Saint Petersburg, is back indeed, albeit for different reasons (see also Coleman 1991).

3.3.2. Cohabitation

In all West and North European countries, the proportion of women and men cohabiting before their first marriage has increased steadily over the past two decades. As observed by Haskey (1993), on a cohort basis, one can see that a majority of men and women today cohabit before marriage. On a period basis, however, proportions of *young* people in consensual unions are still relatively modest, except in the Scandinavian countries, and of course proportions cohabiting among *all* unions are still smaller. Although there was an increase between 1975 and 1980 in the proportion of young women living in cohabiting unions, the *overall* proportion of women living in cohabiting, rather than married, unions in each country was quite low in 1980, and it was still less than 10 per cent of all unions in most countries in 1985: 1.3% in Italy, 2.5% in Austria, 4.4% in Belgium, 7.4% in the Netherlands and in France; 9.1% in Germany, but between 15 and 20% in Denmark and Sweden. This suggests a wide variation in the extent of cohabitation amongst the different countries of Europe in the latter half of the 1980s. The difficult question to judge is whether West European countries are in the process of following the lead set by Sweden or Denmark, or whether they pursue a pattern of their own, in which, for example, although the prevalence of cohabitation has been rising, it is almost always a prelude to marriage rather than a replacement of it (Haskey 1993, p. 225). For example, at present, the union formation intentions of young adults who have a steady dating relationship in the Netherlands show a great diversity (Liefbroer & De Jong Gierveld 1991, pp. 10-11). Some opt for the more 'traditional' pattern of marrying straightaway. Others opt for rather 'new' arrangements, like the Living Apart Together arrangement, in which partners plan to keep their separate households for a considerable period of time. Notwithstanding this diversity, it is clear that unmarried cohabitation is viewed by most young adults as a natural element in their union formation plans. To most of them, cohabitation is viewed as a prelude to marriage. A much smaller category intends to cohabit, without subsequent marriage, as seems to be the more general pattern in the Scandinavian countries.

3.3.3. Divorce

Rising divorce rates make another contribution to the growing proportions of cohabitees in Western and Northern European countries. According to Haskey (1993, p. 222), in general terms, based on the total period remarriage rate in 1965, about 60 to 70 per cent of divorced women in most Western European countries, and about 55 per cent in the Nordic countries, might have been expected to remarry, whereas all these proportions had fallen by about 20 per cent by the mid-1980s. Haskey considers it reasonable to conclude that the drop in the rates of remarriage after divorce is due to the increased proportions cohabiting, and also to the increased proportions cohabiting for longer periods. Whether or not divorced men and women remarry, often depends on their income situation and on the presence of small children.

3.3.4. Fertility

Basically because of these growing proportions of cohabitees before marriage and in between marriages, the rates of childbearing outside marriage have increased in most European countries. Although proportions of live births born out-of-wedlock are not the most sensitive indicators of trends in extra-marital fertility (see Trost 1992), a map of regional proportions of births out-of-wedlock around 1980 (Decroly & Vanlaer 1991, p. 50) shows that such proportions were highest in the Scandinavian countries, the former German Democratic Republic, Austria, France, and the United Kingdom. As shown in Table 3.2, generally speaking, between-country differences in percentages of births outside marriage have persisted over time since the 1960s. The same

Table 3.2. Proportion of children born out-of-wedlock, selected European countries, 1960-1989

Country	1960	1965	1970	1975	1980	1985	1986	1987	1988	1989
Iceland			29.8	32.9	39.7	48.0			52.0	52.9
Sweden	12.8	16.0	27.7	32.4	39.7	46.4	48.4	49.9	50.9	51.8
Denmark	7.8	9.4		21.7	33.2	43.0	43.9	44.5	44.7	46.1
Norway		4.6	6.9	10.3	14.4	25.8	27.9	30.9	33.7	36.4
GDR		9.8	13.3	16.1	22.8	33.8			33.4	33.6
France	6.1	5.9	6.8	8.5	11.4	19.6	21.9	24.1	26.3	28.2
United Kingdom	5.2	7.3	8.0	9.0	11.5	19.0	21.0	22.9	25.1	27.0
Finland		4.6	5.8	10.1	13.1	16.4	18.0	19.2	20.6	22.9
Austria		11.2	12.8	13.5	17.8	22.4	23.3	23.4	21.0	22.6
Portugal	9.6	7.8	7.3	7.2	9.2	12.3	12.8	13.3	13.7	14.6
Ireland	1.6	2.2	2.7	3.7	5.0	8.5	9.7	10.9	11.7	12.6
Hungary		5.2	5.4	5.6	7.1	9.2			11.9	12.4
Luxembourg	3.2	3.7	4.0	4.2	6.0	8.7	10.2	11.0	12.1	11.8
Bulgaria		9.4	9.4	9.4	10.9	11.7			11.0	11.5
The Netherlands	1.4	1.8	2.1	2.2	4.1	8.3	8.8	9.3	10.2	10.7
FRG	6.3	4.7	5.5	6.1	7.6	9.4	9.6	9.7	10.0	10.2
Spain	2.3	1.7	1.4	2.0	3.9	8.0	8.0		9.1	9.4
Belgium	2.1	2.4	2.8	3.1	4.1	7.1	7.8	9.0		
Czechoslovakia		5.1	5.7	4.8	5.7	7.0			7.3	7.7
Italy	2.4	2.0	2.2	2.6	4.3	5.4	5.7	5.8	5.8	6.1
Switzerland		3.9	3.8	3.7	4.7	5.6	5.7	5.9	6.1	5.9
Poland		4.5	5.0	4.7	4.7	5.0			5.8	5.8
Romania			3.5	3.5	2.8	3.7			4.2	4.3
Greece	1.2	1.1	1.1	1.3	1.5	1.8	1.8	2.1	2.1	2.1

Sources: Rabin 1992, p. 4; Yearbook 1992, p. 113; Council of Europe 1993, pp. 56 and 62; United Nations 1992, p. 121-128.

holds for within-country regional differences. In fact, at this moment, the proportion of extra-marital births has become the demographic phenomenon with the greatest regional variability in Europe (Decroly 1992, p. 262). This is the more remarkable since out-of-wedlock births have currently obtained quite another meaning than they had before. Nowadays, growing numbers of cohabitees see no reason at all to marry if a child is expected, and postpone this 'ritual' of a civil marriage until later when perhaps better reasons for marrying present themselves.

Besides, although 'illegitimate' fertility may have risen, overall fertility has dramatically dropped, to levels generally below replacement level, thus providing one of the basic characteristics of the Second Demographic Transition. Time series of fertility reveal a basic East-West division, by showing different evolutionary models for socialist Eastern Europe and capitalist Western Europe, in a way that reminds one again of 'Hajnal's line' (Grasland 1990). The 1988 European fertility map (Decroly & Vanlaer 1991, p. 39) shows above-replacement fertility in Iceland, Ireland, Poland, Romania, and parts of the former Soviet Union only. Sweden joined this category in 1989 and 1990, but we still have to see whether this will be permanent. Everywhere else, current period fertility is below replacement level, most of all in some South European countries (Italy and Spain in particular) which joined the countries entering the Second Demographic Transition rather late but, after having entered it, did so with a tempo of fertility decline that is astonishing. In the late 1980s, Italy had the lowest period total fertility rates in Europe, rates which had declined from 2.3 in 1974, over 1.8 in 1979 and 1.34 in 1988 (Sorvillo & Terra Abrami 1991, p. 5) to as low as 1.23 today! One should bear in mind that these are averages: there are still large differences between North and South Italy. The period total fertility rate in 1988 was 1.113 in North and 1.187 in Centre, as against 1.659 in South (which equals the North and Centre mid-1970s values.). And there are regions in the North of Italy where period total fertility rate has dropped to below one child per woman.

Simultaneously, the age schedule of fertility has shifted towards higher ages. This process of 'aging of fertility' can be connected with the time evolution of period TFR in such a way that clusters can be detected (Bosveld *et al.* 1991) that come close to Roussel's classification mentioned in section 3.1, albeit with significant exceptions.

Finally, till now childlessness does not seem to have emerged as a factor varying significantly between the countries, at least not in Central and Western Europe, and certainly not to such an extent that it may be regarded as the expression of a new, 'post-modern' life style strongly related to the other between-country differences discussed above (Höpflinger 1991b, p. 97).

3.4. RECENT TRENDS IN HOUSEHOLD STRUCTURE

All these changes in fertility and marital behaviour, together with the ageing of population, have caused far-reaching changes in the structure and composition of households, the fourth and last demographic characteristic of the Second Demographic Transition in Van de Kaa's classification. When attempting to document and illustrate these household and family changes, for the sake of simplicity, I will ignore the effects of the important problem of data comparability. When making international comparisons, one needs to be very cautious due to differences in definitions and classifications, as explained by Keilman in Chapter 5 of this volume. However, the present chapter focusses on trends, rather than on cross-sectional comparison, and Keilman aptly demonstrates that different definitions have little impact on overall household trends.

Changes in household and family structures can be illustrated in two different ways:

- by pointing to changing cross-sectional structural data, at the level of households or at the level of individuals by household position; and
- by pointing to trends in the individual life course data with respect to household position.

Below, I will restrict myself to the first approach. Although the second approach would be much better, my focus here is European and comparative, and truly comparable life course data for a large number of countries are still very hard to collect. Further, in reviewing these cross-sectional data, a second limitation is to present some basic trends only.

3.4.1. Average Household Size

Since the 1950s, average household size of private households has dropped in almost all countries (see Table 3.3), except in Turkey between 1950 and 1970, and in Iceland between 1950 and 1960. In this table, I have classified the countries according to Roussel's typology. And indeed, the classification shows that, in terms of trends in average household size, the typology makes sense, although in some cases between-group differences are rather gradual and within-group variance is substantial (e.g., Federal Republic of Germany and German Democratic Republic in groups C and E, and of course Turkey in group D). Generally speaking, group A countries have the lowest average household sizes during the entire period, followed by group B and group C countries which do not differ very much from each other. In group D and group E countries, current average household size more or less equals that in the countries of the

Table 3.3. Average household size, selected European countries, 1950-1990

		1950	1960	1970	1980	1990
A	Sweden	2.9	2.8	2.6	2.3	2.3
	Denmark	3.1	3.0	2.8	2.5	2.3
	Finland	3.6	3.3	3.0	2.6	2.4
B	France	3.1	3.1	2.9	2.7	2.7
	Norway	3.3	3.3	3.0	2.7	2.4
	The Netherlands	3.5	3.2	2.8	2.5	
	United Kingdom	3.2	3.1	2.9	2.7	2.5
C	Austria	3.1	3.0	2.9	2.7	2.6
	Belgium	3.0	3.0	2.9	2.7	
	FRG	2.9	2.8	2.7	2.4	2.3
	Luxembourg		3.3	3.1	2.8	
	Switzerland		3.3	2.9	2.5	
D	Italy	4.0	3.6	3.3	3.0	2.8
	Greece	4.1	3.8	3.4	3.1	
	Portugal	4.1	3.8	3.7	3.4	
	Spain		3.8	3.6	3.4	
	Turkey	5.3	5.7	5.7	5.2	5.2
E	Iceland	3.8	4.0			
	Ireland		4.1	4.1	3.8	3.5
	Bulgaria		3.9	3.2	3.2	
	GDR		2.6	2.7	2.6	
	Yugoslavia	4.4	4.0	3.8	3.6	
	Hungary		3.2	3.1	2.9	2.7
	Poland		3.6	3.5	3.2	3.1

Source: Council of Europe 1990, and updated with figures for around 1990.

other groups around 1950. The same picture emerges from more detailed maps, displaying regional average household sizes around 1980 (Decroly & Vanlaer 1991, p. 153; Council of Europe 1987) . Average household size was still relatively high in most regions within Ireland, Spain and Portugal, Poland, the Balcan countries, and of course Turkey, and relatively low in those within the countries of Western and Northern Europe. The lowest values (below 2.5, sometimes below 2 even) can be found in highly urbanized regions in Germany and in Scandinavia. At these regional levels, there is a high positive correlation between average household size and proportions of households with five or more persons, and an equally high negative correlation between these and proportions of persons living alone in the population of 20 years and older. Proportions of 30% or more of the population aged 20 and over living alone (which can mean that about *half* of the households are one-person households!) can again be found in the highly urbanized regions of Northern and Western Europe.

This decline in average household size over ten-year periods almost everywhere resulted simply from the fact that the number of households grew faster than the population size. There have only been few exceptions to this rule: Belgium, France, Iceland, Norway, and Turkey in the 1950s, and Turkey and the German Democratic Republic in the 1960s. But, the number of households has grown as well, and average household size also declines, because of basic shifts in household size distribution. Kuijsten and Oskamp (1991) have tried to quantify the relative contributions of population growth and of size distribution shifts to household growth and to the decline in average household size in a large number of European countries, calling these contributions the *demographic effect* and the *structure effects*, respectively.

The demographic effect (*DE*) is calculated on the basis of the simple assumption that if the growth rate of the total population is r^p, the number of households would have grown with the same rate in a situation where population growth would have been evenly distributed over members of all household size classes, so that average household size would not have been affected by population growth or decline:

$$DE = (G'_{t,t+1} / G_{t,t+1}) \times 100 \qquad (1)$$

where

- $G_{t,t+1} = H_{t+1} - H_t$ is the real growth of the number of households,
- $G'_{t,t+1} = H_t \times r^p_{t,t+1}$ is the hypothetical growth of the number of households,
- H_t is the number of households at time t, and
- $r^p_{t,t+1}$ is the rate of population growth between t and $t+1$.

Under this assumption, *DE* gives the proportion of total household growth that is caused by mere population growth, a proportion that is always below 100 percent if the number of households grows or declines more than the number of population grows or declines.

Table 3.4 shows the time evolution of DE on the increase in the number of households for 23 European countries, arranged according to Roussel's typology, during the four decades between 1950 and 1990. Note that the time evolutions depicted should not be regarded as straightforward and direct reflections of time evolutions of population growth as such. What is shown is

Table 3.4. Time evolution of DE on household growth, selected European countries, 1950-1990

		1950	1960	1970	1980
A	Sweden	74	43	19	57
	Denmark	60	48	28	2
	Finland	57	21	21	32
B	France	115	60	33	84
	Norway	105	45	34	33
	The Netherlands		50	38	21
	United Kingdom	59	44	8	37
C	Austria	37	53	13	49
	Belgium	116	73	19	
	FRG	78	77	22	-5
	Luxembourg		39	53	
	Switzerland		55	13	
D	Italy	39	44	29	
	Greece	53	29	49	
	Portugal	34	466	54	
	Spain			57	
	Turkey	141	100	70	94
E	Ireland		69	64	-0
	Bulgaria		28	77	
	GDR		-13	-125	
	Yugoslavia	59	67	59	
	Hungary		36	35	
	Poland		74	47	

Source: Kuijsten & Oskamp 1991

the time evolution of changes in the relative importance for household growth of the factor population growth (expressed as *DE*). On the other hand, a relationship between population growth and magnitude of this demographic effect is not completely absent. There is indeed some tendency that low population growth is associated with a relatively low demographic effect. It is for that reason, probably, that a general trend of declining demographic effects from the 1950s to the 1970s emerges in many countries. In six cases where data for the 1980s were available (Austria, Sweden, United Kingdom, France, Turkey, and Finland), a rise in DE for the 1980s as compared with the 1970s is revealed. From the figures in Table 3.4, it becomes clear that there is indeed a tendency for countries in groups A to C, where the Second Demographic Transition started earlier, to have lower values of this demographic effect, especially for the 1970s.

I will now proceed to shifts in household size distributions, the complementary determinant of changes in average household size. In the analysis mentioned earlier, Kuijsten and Oskamp calculated these structure effects (SE_i) as:

$$SE_i = (G'_{i,t,t+1} / G_{t,t+1}) \times 100 \quad \text{for } i = 1,2,3,4,5+ \tag{2}$$

where $G'_{i,t,t+1} = H_{i,t+1} - [H_{i,t} \times (1+r^p_{t,t+1})]$ is the difference between hypothetical and real number of households of size class i at $t+1$.

The basic assumption is the same as when calculating *DE*: it is based on the difference between the actual number of households of a given size i and the hypothetical number that could be expected if population growth would have been evenly distributed over all household size classes, thus without the disturbing factor of the structure effects. Please note that the sum of *DE* and the SE_i-values for the five size classes 1 to 5+ always equals 100, and so explains in statistical terms the entire growth of the number of households in a specific period t to $t+1$. Defined in this way, this kind of analysis can easily be transposed to other types of structural effects as well: e.g. shifts in the distribution by household type, by sex and/or age and/or marital status of the household head, and so on.

For details of the results of this analysis of structural effects of the shifting household size distributions, I refer to Kuijsten and Oskamp (1991). On the basis of their results, the authors propose a phase model of household size distribution trends:

- **phase 1**: The number of households in size class 5+ increases faster than population grows. Both *DE* and SE_{5+} display the most pronounced effects. This phase can be considered as pre-transitional, a phase in which the process of individualization has not really had its take off and in which population growth, while relatively evenly distributed over the household size classes, causes the proportion of large households to remain high;
- **phase 2**: The transition from relatively high to relatively low average household size starts with households of size 4 and below increasing faster than does the population, while numbers of 5+ households increase less or decline;
- **phase 3**: The driving force behind the transition is taken over by an overproportional growth in the numbers of one- and two-person house-

holds: SE_1 and SE_2 become very pronounced, DE and SE_{5+} are very low or negative, whereas SE_3 and SE_4 have low to moderate values.

The above-mentioned current tendency for *DE* to rise in some of the most advanced countries, in combination with a tendency for the decline in average household size to level off in many of these cases, makes the authors wonder whether these phenomena are the precursors to an end in the process of shifts in the household size distribution, and therefore an indication that the basic processes behind the household trends during the period of the Second Demographic Transition are approaching a point of saturation (Kuijsten & Oskamp 1991, p. 131-132). I will return to this in the conclusions.

As far as the changing household size distributions are concerned, it turns out that their evolutions during the period 1950-1990 show some tendency of rather systematic differences between Roussel's classificatory types. As an illustration, one representative of each of these types is included in Figure 3.4.

Sweden exemplifies the Group A countries. Proportions of one-person households, already relatively high in the 1950s, have increased by almost half of their 1950s value, in order to become the largest class today. Proportions of households with two persons have increased continuously but slightly, and proportions of households with three or with four persons have decreased continuously but slightly, those with three more than those with four persons. Finally, proportions of households with five or more persons have relatively speaking decreased most, and are currently among the lowest in Europe.

The Netherlands represents the Group B countries. Again, one sees proportions of one-person households growing, but at levels lower than those in Group A. Developments in proportions of two- and three-person households do not differ significantly from those in Group A. The major differences with Group A are for the four- and five+-person classes. Four-person households have higher proportions and show more stability over time than is the case in Group A. Just like in Group A, proportions of 5+ households are declining, but still have higher levels than in the group A countries.

As can be seen from the graph for the Federal Republic of Germany, the basic tendencies in the Group C countries are the same as those in Groups A and B, be it that developments seem to be more similar to those in the Group B countries than to those in the Group A countries.

Quite different is the situation in groups D and E. First, what is striking when looking at the graph for Italy as a representative of Group D countries are the relatively high proportions in the 5+ size class. Whereas in groups A to C the distributions have been skewed to the left since 1950, with the highest proportions currently for one- and two-person households, in Southern Europe they are in a process of transformation from being skewed to the right in the 1950s and 1960s to being much more balanced afterwards, with the one-person

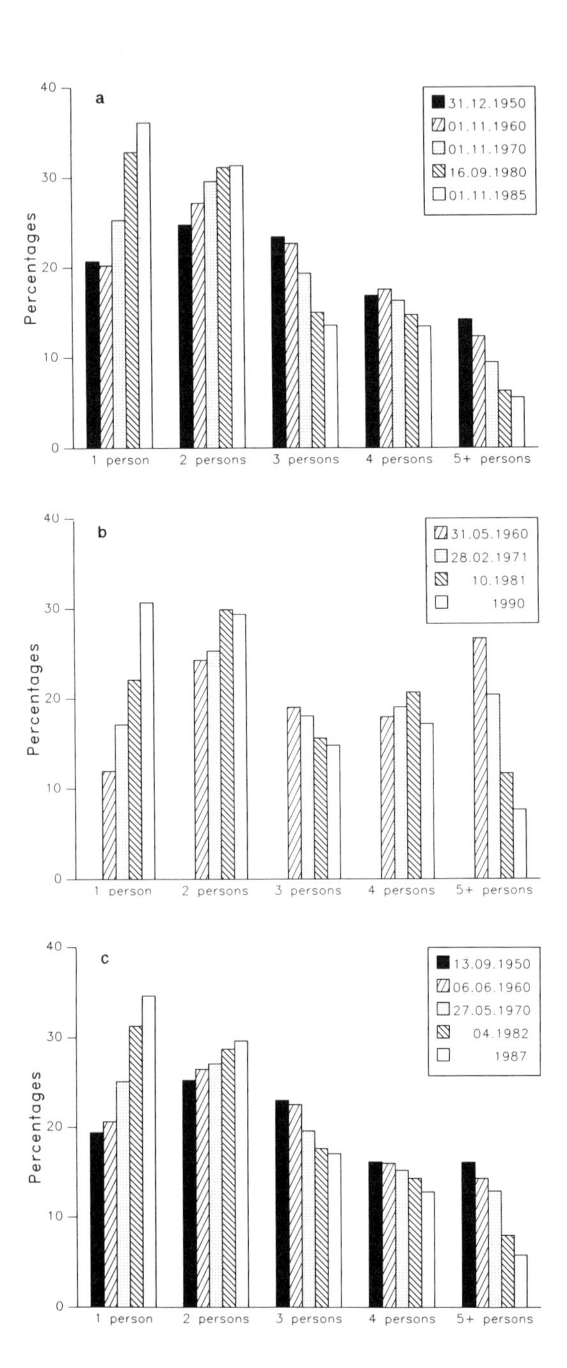
a
40
30
20
10
0
Percentages
31.12.1950
01.11.1960
01.11.1970
16.09.1980
01.11.1985
1 person
2 persons
3 persons
4 persons
5+ persons
b
40
30
20
10
0
Percentages
31.05.1960
28.02.1971
10.1981
1990
1 person
2 persons
3 persons
4 persons
5+ persons
c
40
30
20
10
0
Percentages
13.09.1950
06.06.1960
27.05.1970
04.1982
1987
1 person
2 persons
3 persons
4 persons
5+ persons

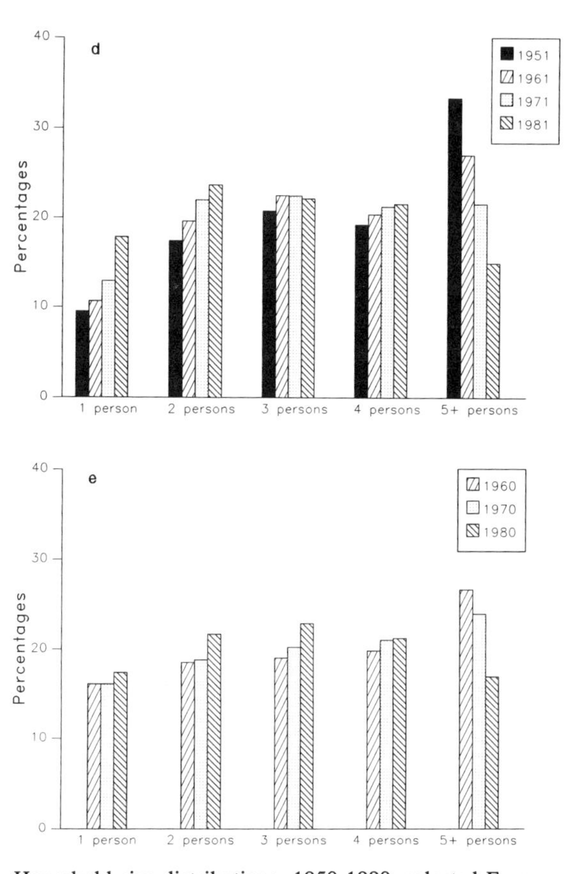

Figure 3.4. Household size distributions, 1950-1990, selected European countries
a. Group A countries
b. Group B countries
c. Group C countries
d. Group D countries
e. Group E countries
Source: Kuijsten and Oskamp 1991

households still being the smallest category, despite the tendency of rising proportions here as well.

Finally, as mentioned before, the Group E countries have a much greater within-group heterogeneity. The graphs for Ireland and Yugoslavia, for instance, are similar to those of Group D countries, except for having significantly higher proportions of one-person households, particularly Ireland. Of the other East European countries, Hungary shows a pattern that is quite similar to that of Group C countries, whereas both Bulgaria and Poland (the graph for this country is included in Figure 3.4e) show household size distributions that have been both remarkably even and constant over time during the 1960s and 1970s.

What are the basic causes of these shifts in household size distributions? Höpflinger (1991a) distinguishes three factors affecting several changes in family and household structures, including the above-mentioned changes in household size distributions:

- the basic demographic changes observed in almost all European countries in the field of declining fertility levels combined with higher life expectancies, resulting in substantial population ageing;
- the rapid and sometimes spectacular shifts in family formation;
- the increasing number of family dissolutions and reconstitutions.

The first factor of demographic change seems the prominent one in the changing household and family structures in the top half of the age pyramid: more and more households of elderly people that are almost by definition one- and two-person households nowadays, at least in Western and Northern Europe. To this cause, I would like to add the factor of increasing de-habitation of three-generation households as a behavioural factor. At the bottom of the age pyramid, other behavioural factors (cohabitation, leaving home, low fertility, and union dissolution) are the more prominent agents of change, with substantial effects of fertility decline on household size and plummeting proportions of households of sizes 4 and higher.

The Council of Europe's Select Committee of Experts on Household Structures has summarized the most important socio-demographic changes within the period 1950-1980, with regard to size and structure of households and families, as follows (Council of Europe 1990, p. 10):

- decrease in nuptiality;
- increase in the divorce rate;
- decline in the number of children within marriages;
- increase in the number of consensual unions;

- increase in the number of one-person households, in particular, owing to the increase in the share of the elderly people living alone and of young people leaving their parents' home relatively early.

The graphs with shifting household size distributions in Figure 3.4 lend some further support to Roussel's typology, although in several cases one would prefer to see more clear distinctions. Such a clear distinction emerges much better if we take one indicator that seems to be basic: the proportions of one-person households presented in Table 3.5. The general outlines support Roussel's typology, in the sense that Groups A and D are clearly the two

Table 3.5. Proportions one-person households, selected European countries, 1950-1990

		1950	1960	1970	1980	1990
A	Sweden	21	20	25	33	36
	Denmark	14	20	21	29	34
	Finland	19	22	24	27	32
B	France	19	20	22	25	24
	Norway	15	14	21	28	34
	The Netherlands		12	17	22	31
	United Kingdom	11	13	18	22	26
C	Austria	18	20	26	28	27
	Belgium	16	17	19	23	
	FRG	19	21	25	31	35
	Luxembourg		12	16	21	
	Switzerland		14	20	29	
D	Italy	10	11	13	18	17*
	Greece	9	10	11	15	
	Portugal	8	11	10	13	
	Spain			7	10	11*
	Turkey	7	3	3	6	5
E	Ireland		13	14	17	18
	Bulgaria		6	17	17	
	GDR		27	26	27	
	Yugoslavia	13	14	13	13	
	Hungary		15	18	20	
	Poland		16	16	17	18*

Sources: Council of Europe 1990; Kuijsten & Oskamp 1991.

extremes. But for the other groups there are at least some problems with some countries. Apart from the fact that Groups B and C again do not discriminate as much as hypothesized by Roussel, West Germany in Group C and the Netherlands and Norway in Group B in the current situation could easily be added to group A, and the same perhaps is true for the former German Democratic Republic in Group E. East European countries seem to hold an intermediary position between Groups A/B/C and Group D. What should be remembered, moreover, is that Roussel's typology is based on a cross-sectional comparison around 1988, whereas Table 3.5 contains time series from around 1950.

3.4.2. Distributions by Family Type

Shifting household size distributions are just one example of changing structural characteristics of households. Other examples are shifts in the distribution of households by type. This trend has been documented by Council of Europe (1990), Keilman (1987), Höpflinger (1991a), and Schwarz (1988).

The growing importance of one-person households has already been discussed in section 3.4.1. It goes hand in hand with a gradual decrease in the proportions of family households, however defined, both with or without children. In most countries of Europe, the archetypical 'Western' family of a married couple with dependent children has lost its traditional overwhelming dominance, and has diminished in numerical importance. Furthermore, there is a tendency for proportions of 'couples with children' to drop more rapidly than proportions of 'families' do: not only has the traditional family household lost ground, but fewer of them have children, and those that have children have fewer of them on average. At the same time, there have been sharp rises everywhere in the numbers and proportions of one-parent families, a trend very largely ignored in Roussel's typology.

The data in Table 3.6 on proportions of women by age, living in a consensual union, are not time series, but when looking at the latest available data, Roussel's typology fits the data much better for this aspect of household structure. What emerges very clearly is not only the vanguard position of the Scandinavian Group A countries, but also a differentiation between Groups B and C that is more convincing than in other tables presented before. When looking at time series, one can see that the basic distinction with respect to the prevalence of cohabitation between Groups A and B already existed in the 1970s. In the few countries for which time series on cohabitation are available, one can see that proportions of women living in consensual unions have risen from the 1970s to the 1980s in all young age groups, except that of 30-34.

Table 3.6. Proportions of females living in a consensual union, by age, selected European countries

		18-19	20-24	25-29	30-34	35-44
A	Sweden 1981		44	31	14	10
	Denmark 1981		37/45	23/25	11	
	Finland 1989	-------71-------		36	19	9
B	France 1985		19	11	9	
	Norway 1986	12	28	16	9	
	Netherlands 1988		19	16	7	
	United Kingdom 1987		11			
C	Belgium		7			
	FRG		14	8		
	Switzerland		12	9	3	3
D	Italy		1		1	
E	Ireland		2			

Sources: Keilman 1987; Hoffmann-Nowotny & Fux 1991; Yearbook 1992, p. 115. In cases where two sources give two different figures, both figures are presented.

3.4.3. 'Top-6' Life Style Patterns

To conclude this section, I would like to show some preliminary results of a recent comparative research project, coordinated by the Institut für Bevölkerungsforschung und Sozialpolitik (IBS) of the University of Bielefeld. Participants in this project, from ten European countries, have tried to follow a more integrative approach towards the study of changing household and family structures. Besides demographic data on type of household in which people are living, they added the socio-economic dimension of labour force participation of men and women, and then, restricting themselves to two age groups (25-29 and 45-49), asked what were the top-6 lists of living arrangements for women in these age groups at the beginning and at the end of the 1980s. The results are quite interesting, as can be seen from those for the Netherlands presented in Table 3.7 (source: Kuijsten & Schulze 1992), and constructed from linking Housing Demand Survey data to data from other sources.

Table 3.7. Top-6 'Life style' ranking, females 25-29 and 45-49 years, the Netherlands

a. females 25-29

1981/1982			1989/1990		
rank	%		%	rank	diff.
1	23.5	married, 2+ children, one-income	13.1	4	-10.4
2	19.3	married, no children, two-income	18.5	1	- 0.8
3	16.6	married, one child, one-income	9.2	5	- 7.4
4	10.0	single	16.4	2	+ 6.4
5	5.5	cohabiting, no children, two-income	14.6	3	+ 9.1
6	4.9	married, 2+ children, two-income	(4.0)	-	- 0.9
-	(4.8)	living with parents	7.3	6	+ 2.5
	15.4	other	16.9		

b. females 45-49

1981/1982			1989/1990		
rank	%		%	rank	diff.
1	36.8	married, 2+ children, one-income	22.7	1	-14.1
2	15.2	married, 2+ children, two-income	16.	2	+ 0.9
3	12.5	married, one child, one-income	12.3	3	- 0.2
4	7.2	married, no children, one-income	(8.0)	-	+ 0.8
5	5.4	married, one child, two-income	9.9	4	+ 4.5
6	5.1	single	8.2	6	+ 3.1
-	(4.2)	married, no children, two-income	8.7	5	+ 4.5
	13.6	other	14.1		

Source: Kuijsten & Schulze 1992.

The results for women aged 25-29 are given in panel a. of Table 3.7. The table shows the six life style types with the highest proportional shares in 1981/1982 and in 1989/1990, plus their gains or losses in percentage points over the decade. In 1981/1982, the traditional family life style (married, two or more children, one income) is still the largest category even in this young age group, but at the end of the decade it has already tumbled down to rank 4, with a relative loss of ten percentage points. At the end of the 1980s, the top position has been taken over by the more modern life style type of the childless two-income couple (although with a slightly smaller proportional share than it already had in 1981/1982), with the equally 'modern' life style types 'single' and 'childless, two-income cohabitees' in ranks 2 and 3. These are the two types which show the highest gains in proportional shares, with 6.4 and 9.1 percentage points, respectively. A further point of interest is the 2.5 percentage point rise in the proportion of 25-29 year old women still (or again) residing in their parents' home, which reflects the rise in the mean age at leaving the parental home as witnessed in the 1980s. Finally, one might interpret the slight rise in the proportional share of the category 'other', from 15.4 to 16.9 per cent, as a sign of still ongoing pluralization in this age group in the middle of its family formation period.

Age group 45-49, in panel b. of Table 3.7, shows a much smaller amount of dynamics over the 1980s in its top-6 list. Positions 1 to 3 have remained the same, although in a relative sense the traditional family type in position 1 has lost most of all precisely in this relatively old age group (a percentage point loss of 14.1!). When combined with the smaller gains for the two-income, married couple without children, with one child or with two or more children in ranks 5, 4, and 2, respectively, I suppose that this significant drop in proportional contribution of the top-1 life style type has to do with the post-1965 overall fertility drop, as much as it has to do with more women successfully looking for (part-time) jobs when the children are old enough to take care of themselves. Pluralization of life styles seems less pronounced in this age group than in the 25-29 age group.

3.5. CONCLUSIONS

Of course, the picture I have presented is still far from complete and far from conclusive. Roussel's typology seems to make some sense as far as a number of recent trends and current structures are concerned. He clearly presents his typology as fitting the transitional nature of the present situation in many European countries. The model behind that reasoning, saying that in former days countries were much more similar, that now there is a temporary period of greater diversity because all countries are following the path of the Second Demographic Transition, but at different speeds and with time lags in the moment when they start the journey, and that, in the future, when they all have reached the same destination, there will again be a period of much more similarity, may be too simple. As Trost (1992) recently indicated, a lot of countries have a long tradition with respect to *some* of the indicators that together and in combination characterize the Second Demographic Transition. And there is no reason to believe that this will be different in the future. On the other hand, the driving forces behind the Second Demographic Transition and the way it manifests itself in family and household structures are much the same all over Western Europe, and may be extended to Eastern Europe quite easily in the near future.

Vichnevsky *et al.* (1993) have recently expressed the latter view with respect to fertility, but their thesis has a much broader significance. They do not exclude the possibility of a future rise in fertility because of religious and/or nationalist/ethnic revivals in the East European countries and in the now independent countries that were part of the former Soviet Union (Vichnevsky *et al.* 1993, p. 68). Nevertheless, as a general evaluation, they say:

> "All in all, it seems that the principal direction in the demographic development of the Eastern European countries will be different. Their modernization, the reinforcement of the role of the free market and other similar changes will contribute to a rise in the role of values such as the liberty of individual choice, self-actualization, hedonism etc. The most probable outcome will be an acceleration here too of the 'second demographic transition' which until now has proceeded relatively slowly, a greater matrimonial and sexual liberty, more perfect contraceptive practices and methods, and a rise of the births out of wedlock. Fertility will therefore fall till the low levels already reached in Western Europe." (quotation from the original English-language conference paper text).

The future will show whether or not Vichnevsky and colleagues are right in their expectations for Eastern Europe. But, supposing they are right, the answer to the question whether or not there will be further convergence at the all-European level will depend then on the answer to the question whether or not the countries in the other parts of Europe will continue on that road followed some decades now. In its report, the Council of Europe's Select Committee of Experts on Household Structures concludes that the most important tendencies, as emerging from the relatively few current household and family projections for European countries, are basically expectations of continuity in the current trends (Council of Europe 1990, p. 72):

- an expected further decline in the average household size;
- an increase in the proportion of smaller households, particularly one-person households, and a decrease in the proportion of larger households;
- an overproportional growth of households with a reference person of middle age; and
- an increase in household types other than the traditional family household.

As an example, I refer to the main results of the latest household projections for the Netherlands (see De Beer, chapter 11 in this volume):

- between 1990 and 2010, according to the medium projection variant, the total number of households will grow by about 20 per cent;
- but the number of unmarried couple households will grow by more than 50 per cent, and their proportional share will therefore increase from 6.5 to 8.4 per cent of all private households;
- the number of one-person households will grow by almost 50 per cent, and their proportional share will increase from 30 to 36 per cent of all private households;
- the 20 per cent increase in the total number of households can be explained for 50 per cent by population growth, for the other 50 per cent by ageing, individualization, and women's emancipation.

All this justifies the expectation that *diversity* or *pluralization* will remain the key words for the decades to come. Hoffmann-Nowotny and Fux (1991) expect that "an increasingly wide range of experiments and short-term developments, movements and socio-cultural 'fashions' will come up and disappear again". Höpflinger (1991a) does not think that the cultural diversity within Europe will disappear to be replaced by a boring uniformity of household and family structures within a politically and economically integrated Europe. But, on the other hand, he does not think either that 'new' family patterns will fully replace 'traditional' marriage and the family. Instead, he expects an increasingly complex combination of traditional and modern patterns. These views need not be contradictory to the expectations of Roussel, since the latter expects ongoing fading of the clown's robe of the *national* European family map, in the sense of declining between-country variation in national averages, without mentioning within-country variations between lower-level aggregates. Since the growing within-country diversity and pluralization will without any doubt be correlated with spatial determinants, it might therefore be possible that, in the future, the *regional* European family map will be more polychrome than ever before.

Or will other factors counterbalance these trends towards growing diversity? One possible counter-evidence might be the deep-set anthropological strata of national and regional family and household structures as revealed by Todd (1983; 1990), to the extent that at a given point these might constrain further diversification. If it is true indeed that this factor has operated for almost a millennium now, there is no reason at all to assume that it will cease to do so in the first decades to come. The other counter-proof will depend on the answer to the intriguing question, how far the tendencies of individualization can continue without running the risk of becoming counter-productive to people's ultimate aims and values in life. It is this question that will be the crucial one for future hypothesis-making in the field of household and family projections.

REFERENCES

BECKER, H.A. (1985), Generaties, *Hollands Maandblad*, Vol. 4, pp. 14-25.

BECKER, H.A. (ed.) (1990), *Life Histories and Generations*, ISOR: Utrecht.

BOSVELD, W., C. WIJSEN, & A. KUIJSTEN (1991), The Growing Importance of Fertility at Higher Ages in the Netherlands, PDOD-Paper No. 3, Postdoctorale Onderzoekersopleiding Demografie: Amsterdam.

CLIQUET, R.L. (1991), *The Second Demographic Transition: Fact or Fiction?*, Population Studies, Vol. 23, Council of Europe: Strasbourg.

COLEMAN, D.A. (1991), European Demographic Systems of the Future: Convergence or Diversity?, EUROSTAT International Conference on 'Human Resources in Europe at the Dawn of the 21st Century', Luxembourg, 27-29 November 1991.

COUNCIL OF EUROPE (1987), *Proceedings of the Seminar on Demographic Problem Areas in Europe*, (Strasbourg, 2-4 September 1986), Council of Europe: Strasbourg.

COUNCIL OF EUROPE (1990), *Household Structures in Europe*, Population Studies, Vol. 22, Council of Europe: Strasbourg.

COUNCIL OF EUROPE (1993), *Recent Demographic Developments in Europe and North America*, Council of Europe: Strasbourg.

DECROLY, J.-M. (1992), Les Naissances Hors Mariage en Europe, *Espace, Populations, Sociétés*, Vol. 1992(2), pp. 259-264.

DECROLY, J.-M., & J. VANLAER (1991), *Atlas de la Population Européenne*, Editions de l'Université de Bruxelles: Bruxelles.

DE FEIJTER, H. (1991), *Voorlopers bij Demografische Veranderingen*, NIDI rapport nr. 22, NIDI: Den Haag.

DE GUIBERT-LANTOINE, C., & A. MONNIER (1992), La Conjoncture Démographique: l'Europe et les Pays Développés d'Outre-Mer, *Population*, Vol. 47(4), pp. 1017-1036.

GRASLAND, C. (1990), Systèmes Démographiques et Systèmes Supranationaux: La Fécondité Européenne de 1952 à 1982, *European Journal of Population*, Vol. 6, pp. 163-191.

HAJNAL, J. (1965), European Marriage Patterns in Perspective. *In*: D.V. GLASS & D.E.C. EVERSLEY (eds.) *Population in History*, Edward Arnold: London, pp. 101-143.

HALL, R. (1986), Household Trends within Western Europe, 1970-1980. *In*: A. FINDLAY & P. WHITE (eds.), *West European Population Change*, Croom Helm: London, pp. 19-34.

HASKEY, J.C. (1993), Formation and Dissolution of Unions in the Different Countries of Europe. *In*: A. BLUM & J.-L. RALLU (eds.) *European Population. II. Demographic Dynamics*, John Libbey Eurotext: Paris, pp. 211-229.

HOFFMANN-NOWOTNY, H.-J. (1987), The Future of the Family. *In*: *European Population Conference 1987. Plenaries*, IUSSP and Central Statistical Office of Finland: Helsinki, pp. 113-200.

HOFFMANN-NOWOTNY, H.-J., & B. FUX (1991), Present Demographic Trends in Europe. *In*: COUNCIL OF EUROPE, *Seminar on Present Demographic Trends and Lifestyles in Europe*, Council of Europe: Strasbourg, pp. 31-97.

HÖPFLINGER, F. (1991a), The Future of Household and Family Structures in Europe. *In*: COUNCIL OF EUROPE, *Seminar on Present Demographic Trends and Lifestyles in Europe*, Council of Europe: Strasbourg, pp. 291-338.

HÖPFLINGER, F. (1991b), Neue Kinderlosigkeit - Demographische Trends und gesellschaftliche Spekulationen. *In*: G. BUTTLER, H.-J. HOFFMANN-NOWOTNY & G. SCHMITT-RINK (eds.), *Acta Demographica 1991*, Physica-Verlag: Heidelberg, pp. 81-100.

INGLEHART, R. (1977), *The Silent Revolution: Changing Values and Political Styles Among Western Publics*, Princeton University Press: Princeton.

KEILMAN, N. (1987), Recent Trends in Family and Household Composition in Europe, *European Journal of Population*, Vol. 3(3/4), pp. 297-325.

KUIJSTEN, A., & A. OSKAMP (1991), Huishoudensontwikkeling in Europa, 1950-1990, *Bevolking en Gezin*, Vol. 20(2), pp. 107-141.

KUIJSTEN, A., & H.-J. SCHULZE (1992), Changing Family Structures in the 1980s - The Case of the Netherlands. Country report prepared for the Second Meeting of the International Research Project 'Familiale Lebensformen, Lebenslagen und Familienalltag im europäischen Vergleich, Bonn - Bad Godesberg, April 8-12.

LESTHAEGHE, R., & D.J. VAN DE KAA (1986), Twee Demografische Transities? *In*: D.J. VAN DE KAA & R. LESTHAEGHE (eds.), *Bevolking: Groei en Krimp*, Van Loghum Slaterus: Deventer, pp. 9-24.

LIEFBROER, A.C., & J. DE JONG GIERVELD (1991), The Impact of Rational Considerations and Norms on Young Adults' Intentions Concerning Marriage and Unmarried Cohabitation. Paper European Population Conference, Paris, 21-25 October 1991.

MONNIER, A. (1990), The Demographic Situation of Europe and the Developed Countries Overseas: an Annual Report, *Population, English Selection*, Vol. 2, pp. 231-242.

PRIOUX, F. (1992), Les Accidents de la Nuptialité Autrichienne, *Population*, Vol. 47(2), pp. 353-388.

RABIN, B. (1992), De Plus en Plus de Naissances Hors Mariage, *Economie et Statistique*, Vol. 251, pp. 3-13.

ROUSSEL, L. (1989), *La Famille Incertaine*, Editions Odile Jacob: Paris.

ROUSSEL, L. (1992), La Famille en Europe Occidentale: Divergences et Convergences, *Population*, Vol. 47(1), pp. 133-152.

RYCHTARIKOVA, J. (1993), Nuptialité Comparée en Europe de l'Est et en Europe de l'Ouest. *In*: A. BLUM & J.-L. RALLU (eds.) *European Population. II. Demographic Dynamics*, John Libbey Eurotext: Paris, pp. 191-210.

SARDON, J.-P. (1992), La Primo-Nuptialité Féminine en Europe: Eléments pour une Typologie, *Population*, Vol. 47(4), pp. 855-892.

SCHWARZ, K. (1988), Household Trends in Europe after World War II. *In*: N. KEILMAN, A. KUIJSTEN & A. VOSSEN (eds.) *Modelling Household Formation and Dissolution*, Clarendon Press: Oxford, pp. 67-83.

SCHUBNELL, H. (1966), Die Entwicklung unserer Bevölkerung. *In*: H.-J. NETZER (ed.), *Die Gesellschaft der nächsten Generation*, Verlag C.H. Beck: München, pp. 11-75.

SCHUMPETER, J. (1988), Schumpeter on the Disintegration of the Bourgeois Family, *Population and Development Review*, Vol. 14(3), Archives, pp. 499-506.

SORVILLO, M.P., & V. TERRA ABRAMI (1991), Cohort Fertility in Italy: Some Preliminary Results of the Regional Reconstruction. Paper European Population Conference, Paris, 21-25 October 1991.

THOMPSON, W.S. (1929), Population, *American Journal of Sociology*, Vol. 34, pp. 959-975.

TODD, E. (1983), *La Troisième Planète, Structures Familiales et Systèmes Idéologiques*, Editions de Seuil: Paris.

TODD, E. (1990), *L'Invention de l'Europe*, Editions du Seuil: Paris.

TROST, J. (1992), The Last Decades and Matrimonial Changes. *In*: H. BIRG & F.-X. KAUFMANN (eds.), *Bevölkerungswissenschaft heute - Kolloquium anlässlich des 10jährigen Jubiläums des Instituts für Bevölkerungsforschung und Sozialpolitik*. IBS-Materialien Nr. 33, IBS: Bielefeld, pp. 91-106.

UNITED NATIONS (1992), *Patterns of Fertility in Low-fertility Settings*, United Nations: New York.

VAN DE KAA, D. (1987), Europe's Second Demographic Transition, *Population Bulletin*, Vol. 41(1), Population Reference Bureau: Washington D.C.

VAN DE KAA, D. (1988), The Second Demographic Transition Revisited: Theories and Expectations. Werkstukken PDI, no. 109, Planologisch en Demografisch Instituut Universiteit van Amsterdam: Amsterdam.

VICHNEVSKY, A., I. OUSSOVA, & T. VICHNEVSKAIA (1993), Les Conséquences des Changements Intervenus à l'Est sur les Comportements Démographiques. *In*: A. BLUM & J.-L. RALLU (eds.), *European Population. II. Demographic Dynamics*, John Libbey Eurotext: Paris, pp. 49-75.

YEARBOOK (1992), *Yearbook of Population Research in Finland XXX 1992*, The Population Research Institute: Helsinki.

4. THEORIES OF HOUSEHOLD FORMATION: PROGRESS AND CHALLENGES

Thomas K. Burch

Population Studies Centre
University of Western Ontario
London, Ontario N6A 5C2
Canada

Abstract. The demography of family, kinship, and household has made great strides in recent years, but as with demography generally, the theory of the sub-field has lagged behind. This may reflect the difficulty of the subject matter, but it also may reflect a general neglect of systematic theory by mainstream demography. Four frameworks for the study of household status decision-making are reviewed, and four challenges to future theoretical work identified: 1) giving adequate attention to supply factors; 2) exploring the subjective realm (perceptions, time horizons, preferences); 3) studying households and families in the broader context of networks and community; 4) taking seriously the temporal dimension of family and household formation.

4.1. INTRODUCTION

A balanced assessment of behavioural theory of household formation needs to be put into two broader contexts: 1) the state of theory in population studies generally; 2) recent progress in the field of household and family demography[1] as a whole.

[1] By *household and family demography* I mean the broad field of demography dealing with: union formation and dissolution; family and household structure and change; and kinship. For convenience, I shall often use the shorter phrase *family demography*.

Household Demography and Household Modeling
Edited by E. van Imhoff *et al.*, Plenum Press, New York, 1995

Theory, by any account, is not demography's strong suit.[2] And this is as true of family demography as of any other part of the discipline. Landmark writings attest to the problem, and hint at an underlying notion that demography may properly be in some sense a non-theoretical discipline. Vance's query in his Population Association of America presidential address four decades ago (Vance 1952) — "Is Theory for Demographers?" — continues to echo through the field and to be answered by many demographers in practice if not in principle with a workaday 'no.'

In the decades since, there has been progress in demographic theory, but not as much as one might have expected from a field so rich in firm empirical description over a wide swath of time and space, at both the macro and micro levels. 'Demographic transition theory,' often described as the centrepiece of demographic theory, has resisted the synthesis, formalization, and standardization that usually comes with such stature. It exists rather in several different versions, fairly informal and highly personal in expression, that is, associated with one author. The National Academy of Sciences synthesis (Bulatao & Lee 1983) represents a recent and welcome, but only partially successful, exception, so that McNicoll can still write of "transition theory" meaning many different things to different demographers (1992, pp. 9-10).

If demographic theory has not flourished, the same cannot be said for family demography, broadly defined. A couple of decades ago, only a handful of demographers specialized in one or more of its parts — Glick on the family life-cycle, Jacobson and Hajnal on marriage, Siegel and Kono on households, to mention some obvious examples. Texts and compendia typically lacked chapters specifically on marriage, family, or household (the chapter due to Kono in the 1973 edition of *Determinants and Consequences of Population Trends* is a landmark), and sessions devoted to the topic at national and international meetings were few and far between.

All this has changed, and by 1981 Lee (in his plenary address to IUSSP in Manila, 1981) could identify family demography as a "growth industry," a forecast that has been borne out by subsequent developments. Section 4.2 below gives a brief overview of some areas of progress in family demography.

This overall progress has included theory, but as with demography generally this has not been the area of greatest progress. Section 4.3 sketches

---

[2] *Theory* in this context is taken to mean general statements about social, economic, cultural, and demographic interrelations, or about the behavioural underpinnings of demographic events — in the phrase of the two well-known U.N. compendia, 'determinants and consequences.' It may be contrasted with the completely general but substantively more limited propositions of *formal demography*, recently defined by Ryder (1992) as "... the deductive study of the necessary relationships between those quantities serving to describe the state of a population and those serving to describe changes in that state, over time, *in abstraction from their association with other phenomena*" (p. 162, emphasis added).

four representative efforts to model household-formation decisions, efforts that may not qualify as full-blown *theories* in the textbook sense, but are at least serious moves in that direction.

Section 4,4 discusses areas where present-day family demography (including the theory noted above) falls short, sketching challenges for future work.

The chapter concludes with some thoughts on the prospects for theory in family demography and demography generally, and on procedural requirements for the realization of these prospects.

4.2. FAMILY DEMOGRAPHY: AREAS OF PROGRESS

Since the focus of this chapter is on theory, the present section is brief and impressionistic, with documentation and citations by way of illustration only. Detail can be found in one or more of the growing number of review articles or summary chapters on household and family demography.[3] The examples are largely from the English-language demographic literature, with which the author is most familiar.

4.2.1. Conceptualization

The scope of household and family demography is now fairly well defined, and many of its basic concepts clarified. A key step has been the full realization of the difference between *household* and *family*, a distinction somewhat blurred in early demographic work, due to a Western tendency to identify the two, and to resulting census conventions (for example, definition of *family* as two or more relatives sharing the same household).[4]

The distinction has long been standard in anthropology, where field experience made clear that *kinship* and *co-residence* (in the *domestic group*) could and often did represent two separate dimensions. Ryder (1987, 1992) has refined and clarified these dimensions for demographic use, with the terms *conjugal, consanguineal,* and *residential* dimensions:

---

3 Now old but still useful are: Sweet (1977); Burch (1979); Bongaarts (1983). See also the relevant chapters in Keyfitz (1985).
Recent overviews are found in the introductory or summary chapters of: Bongaarts *et al.* (1987); Keilman *et al.* (1988); Grebenik *et al.* (1989); Berquò & Xenos (1992); and in DeVos and Palloni (1989). For an anthropological perspective, see Yanagisako (1979).

4 One of the author's first published writings in the field (Burch 1967) made the necessary distinctions, but used the title 'The Size and Structure of Families' in an analysis that dealt solely with household data.

> "Residential distribution of personnel is implicit in the dual character of the family. Because it is a descent unit, it is exposed to the vagaries of demographic processes; because it is a residential unit, its cross-sectional composition must make economic sense day to day" (1992, p. 169).

It is this multidimensionality that gives rise to the relatively large number of units for study in family demography — the individual, the couple, the nuclear family, broader kinship units, the household. The number of different units of analysis, the fact that each has a somewhat different formal demography (in terms of demographic events affecting structure and change), and the fact that behavioural determinants of individual events of entry into or exit from the various units differ by age, sex and other basic characteristics — all combine to give family demography its vast scope, perhaps exceeding that of all the rest of demography as conventionally defined.

4.2.2. Measurement

Some progress has been made in the utilization of aggregate demographic data for the conventional measurement of household and family dimensions. Three examples can be cited. Burch has explored various measures of household composition, including age-standardized measures of the propensity to form households, demonstrating among other things that the measure *average household size* is largely a reflection of fertility, not so much of household complexity. The latter, however, is fairly well captured by such a simple measure as the average number of adults per household (Burch 1980; Burch *et al.* 1987). Preston (1987) has used generalized stable population analysis for the measurement of marriage and other family formation patterns.

Work by Laslett and Wall and others (Laslett & Wall 1972; Wall *et al.* 1983) has led to refined, empirically derived typologies of household/family structure (based on co-residential data), building on the older trichotomy of *nuclear, stem,* and *extended.*[5] This work faces the standard problems of typological analysis, namely, elements of subjectivity in constructing the typology, and a proliferation of types. Somewhat oddly, this line of work has seen very little exploration by means of modern data reduction techniques — for example, factor, cluster, or principal components analysis — even though it seems likely that the complexities of household structure might be reduced to variation on a few well-defined and interpretable dimensions (for a rare example of such analysis, see Benoit *et al.* 1983).

---

[5] The empirical work falls short of the geographically broader and even more refined typology due to Todd (1985), based largely on ethnographic data. Part of the problem relates to the lack of detail in census data on families. For some suggestions on implementing the Todd typology in censuses or surveys, see Burch (1989).

Conventional work of this sort has not developed as far as it might have, partly due to limitations of census and vital registration data pertaining to the family, and partly due to the increasing availability of richer data sets from surveys, including family event histories and data on kin networks. For the past, however, much remains to be done, especially with census manuscripts or machine-readable versions of these, from which the maximum detail as to household composition can be recovered.

4.2.3. Formal Modelling of Household and Family

Perhaps the greatest progress in this field has been in the area of modelling the interrelations between basic demographic processes and the formation, change, and dissolution of various household/family units. An early statement of the relevant events is by Bongaarts (1983; see also Burch & Matthews 1987, p. 496).

This work has benefitted greatly from earlier work on modelling the separate demographic processes involved, for example, model life tables as standardized and compact representations of mortality, or the parametric representations of first marriage by Coale and McNeil (1972) and by Hernes (1972). Less progress has been made with respect to cohabitation, divorce, remarriage, or home-leaving, not to mention the many other forms of entry and exit from households. Whether these events, less closely tied to biology than birth, death, or first marriage, will lend themselves as successfully to parametrization or model schedules remains to be seen, although work by Krishnan on model divorce tables (Krishnan & Kayani 1976) and by Rogers and his associates on migration schedules (Rogers & Castro 1981) suggests hope.

A major breakthrough (one of contemporary demography's real discoveries — of something not at all obvious before the fact) was the estimation of kin numbers in a stable population, given assumed levels of fertility and mortality (Goodman *et al.* 1974). A central finding of their work is the strong dependence of numbers of kin on fertility, for all kin categories except ascendants in the direct line (parents, grandparents, etc.). The implications of this fact can be hinted at by imagining what would have been the consequences of a perfect realization of China's one-child policy — in a generation, collateral kinship (a central institution in China as in all human societies) would virtually have disappeared: no siblings, no cousins, no aunts and uncles, no nieces and nephews. In the present context, it is useful to note also that kin numbers constitute important constraints on household status decisions.

Similar analytic approaches to household size and composition in relation to demographic determinants were developed by several authors (Coale 1965; Burch 1970; Brass 1983), reinforcing the views of Laslett (Laslett & Wall 1972) and of Levy (1965), based respectively on empirical and theoretical grounds, that extended family households had not been commonplace in the past, whatever the ideal structures.

Bongaarts family status life-table (1987) has modelled the development of the nuclear family (seen from the female point of view). The model has been used to good effect to trace broad outlines of family change in the United States since the late 1800s (Watkins *et al.* 1987). It has been further developed by Zeng Yi (1991) to take account of three-generation families, and has been published in a user-friendly PC version.

LIPRO (Van Imhoff, chapter 12 in this volume), developed at NIDI, focusses on modelling the distribution of population by detailed household types, given transition probabilities relating to marriage, death, fertility, home-leaving, and other relevant events. The data demands are heavy, requiring information on transitions that is only now becoming available for most populations. But the program is mature and well-documented and is gaining increasing use, especially in Europe.

Another well-developed macro-model of households, requiring more conventional data than LIPRO, is HOMES (Mason 1987).

In general, family demography has been a leading beneficiary of the increasing availability of event history data and of the development of multistate demographic techniques for their analysis. Beginning with analyses of marriage processes, this work has now been extended to most of the 'family life cycle,' covering the same ground as traditional family life-cycle analysis but without reliance on sometimes misleading synthetic cohort results. A recent example of such detailed description traces adult family life from respondent's leaving the parental home to the empty nest, the departure of his or her last child from the home (Ravanera *et al.* 1993).

Another approach to the formal demography of families and households and of kinship numbers has abandoned the analytic or macro-modelling approaches in favour of microsimulation. A number of microsimulation models have been constructed, of which the best-known and most widely used is SOCSIM (see Hammel 1990, for a complete description and bibliography of applications).

SOCSIM has evolved over many years, used by Hammel and Wachter to study a series of specific substantive issues (for example, the implications for household structure of a rule of primogeniture versus ultimogeniture). The authors have stressed its use as a tool for investigating implications of theories, rather than for trying to reproduce the concrete real world, (Wachter 1987; Hammel 1990). SOCSIM can model two or more populations simultaneously, with 'migration' between them, and provides output lists on marriages, households, and kinship relations, although the extraction of a particular type of information from the lists is often a non-trivial programming task.

Other microsimulation packages include FASIM (from the Department of Demographic Science, University of Rome; see Pinnelli & Vichi 1988, Bertino *et al.* 1988), and FADEP (from Statistics Canada, Ottawa; see Wright 1992), but these have not been widely distributed or used in the international demographic community.

Microsimulation appears to remain a 'hard sell' in many demographic quarters. Computer time costs have been reduced by several orders of magnitude since the first microsimulations, but they can still be substantial, as can the data storage requirements. Also, it has proven difficult to develop general-purpose packages that can be used off the shelf. And it is not easy to modify a program or customize output. Finally, both scientists and policy analysts often seem suspicious of what they see as the non-empirical character of simulations, in what is perhaps yet another manifestation of an atheoretical bias in demography.

4.2.4. Empirical Generalizations

There is a large and diverse literature on the determinants of the formation, structure, and change of households, most of it focussing on the individual in terms of his or her household status. A pervasive theme is that of an underlying tendency of adult human beings to live separately or together with others, what Kuznets (1978) once termed the "jointness or apartness of adults." Favourite dependent variables in these empirical studies have included living alone, headship, and various forms of co-residence.

Much of the literature focusses on specific sub-groups, such as never-married young adults, or older, formerly-married women, sub-groups in which variation in living arrangements is seen to be particularly interesting or problematic. Taken as a whole, the literature is somewhat schizoid, with housing economists focussing (sometimes almost exclusively) on income and the housing market (supply, price, etc.), and social demographers typically ignoring the latter set of factors.

The definitive review of extant literature seems not to have been done, but a few broad generalizations are possible.

Income appears to be positively associated with separate living, at least in contexts where separate living is valued. Separate living clearly is expensive (some would say a luxury good, at least to the extent realized in much of the developed West), both to individual and society. As real incomes have stagnated in the last decade, questions have been raised as to whether recent rises in separate living may not be over, or even whether current levels can be sustained.

Housing prices generally appear to discourage marriage and the formation of family households. In societies where departure from the parental home before marriage has become commonplace, this may have the somewhat paradoxical result of raising rates of non-family household formation. In societies where young adults tend to remain at home until marriage (e.g. Italy) high housing costs simply reduce the overall number of households.

Physical disability and *poor health* appear to discourage living alone, in general but especially among the elderly, among whom these conditions are frequent.

The *number of kin* an individual has limits various kinds of co-residence, insofar as the pool of eligibles or preferred 'partners' for co-residence consists largely of relatives, as it still does in many societies. The Kobrin hypothesis on availability of adult children as a determinant of household status of older women has been verified many times over. And there is evidence that step-children tend to leave a parental home earlier than others, perhaps because of the existence of another household in which they have some claims to residence (Mitchell *et al.* 1989).

Changing gender roles have had a major impact on all aspects of family formation and structure, including household formation and dissolution. A concrete generalization is that rising rates of female labour force participation and resulting economic independence have lowered marriage rates and raised divorce rates, with implications especially for the spread of one-person and lone-parent households. More abstractly, Burch (1985) has theorized that a breakdown of traditional age-sex roles, with the disappearance of age-sex-specific niches in the household, has tended to reduce family solidarity and to increase crowding and competition.

As always, the role of *culture* remains something of a mystery, with frequent assertions that it is important, but a lack of rigorous empirical demonstrations to this effect. There is a widespread conviction, for example, that many of the patterns of family and household formation observed in the West since 1970 (part of the so-called 'second demographic transition') may be bound up with a unique Western emphasis on the individual, and may not be repeated in the developing societies of Asia or Africa.

4.3. SOME THEORIES OF HOUSEHOLD FORMATION[6]

This section reviews four attempts to elaborate on and partially model the process of household formation: 1) Ermisch (1988); 2) Haurin, Hendershott, and Kim (1990, 1992a, 1992b); 3) Burch and Matthews (1987); 4) Duncan and Morgan, (1976b) and Hill and Hill (1976).

---

[6] The phrase *household formation* is shorthand for household formation, change, and dissolution, for several forms of movement into and out of households, and for the household status of individuals.

Notable by its absence from this short list is the relevant body of work by Gary Becker, a point which calls for comment. Although basic micro-economic ideas have become central to behavioural demography, the details of Becker's work have not. As McNicoll (1992) has pointed out, their use has been mainly heuristic. This has been due in part to a widespread view that Becker's rigorous economic approach was too narrow, making little room for 'softer' social-psychological determinants of behaviour. This view is epitomized by the classic Blake query "Are Babies Consumer Durables?"[7] Another reason is simply that to acquire a working knowledge of the details of Becker's work calls for a grasp of economic theory and mathematical economics well beyond that of the average demographer.

4.3.1. Ermisch's Two-Level Theory of Household Formation

Perhaps the best developed general scheme for studying the determinants of household formation/status is due to Ermisch and his colleagues. A recent synthesis is contained in Ermisch (1988), and is the basis of the description below.

In this scheme, the demography and economics of household formation are separated:

> "Analysis is easier if the units are such that demographic influences on household formation and composition can be separated from economic influences. In particular, it would be helpful to separate instances of *family* formation and dissolution from *household* formation and dissolution" (p. 24, italics in original).

This separation is based on the concept of *minimal household units (MHUs)*, defined as "...the smallest group of persons within a household that can be considered to constitute a *demographically* definable entity" (p. 24). The four MHU types are : 1) childless, unmarried adults; 2) lone parents with dependent children; 3) childless married couples; 4) married couples with dependent children.

The number of MHUs in the population is interpreted as the maximum number of households that *could* be formed (assuming that married people live in the same household). Movement of an MHU from one category to another

[7] This view has tended to persist despite the fact that sociological exchange theory, as represented by Homans, Blau, Emerson, and others (one of the leading schools of late-twentieth century *sociological* theory) is essentially microeconomic at its core, and despite the fact that Becker's theoretical structure has developed over time to relax earlier 'anti-sociological' assumptions, for example, unvarying tastes or impersonal exchange in the family.

is defined solely in terms of demographic events, namely, marriage, birth, death, and home-leaving of an adult child.

The second level of analysis considers how MHUs arrange themselves into 'household contexts' or 'household groupings,' of which Ermisch defines six as especially relevant to Britain: 1) living alone; 2) living with parents; 3) living with minor children; 4) living with adult children; 5) living with other unmarried adults; 6) other combinations.

The behavioural model of household formation relates the choice by an MHU of one or another household context to a vector of characteristics of the MHUs members.

Ermisch refers to the separation as involving "levels of analysis," and explicitly denies that household formation occurs in two, temporally ordered stages:

> "The split is not made because people necessarily go through these two stages, but it is made for analytic reasons. It is difficult to develop tractable models yielding predictions of economic effects when the two levels of analysis are combined.... Thus, for analytical convenience the analysis is broken down into these two levels" (pp. 28-29).

A signal advantage of the Ermisch scheme is that it is applicable across-the-board, that is, to all age, sex, and marital status groups, whereas many other behavioural models in the literature are customized for one particular sub-group (for example, unmarried young adults, formerly married older women, etc.).

Ermisch has estimated the model with some success for several sets of data for Britain. But the model has not guided empirical work of other investigators, nor have others tried to develop it theoretically. As suggested earlier, cumulative work on a shared body of theory is not the demographic way.

Ermisch's distinction between "levels of analysis" and "stages" is interesting, insofar as it is a way of trying to finesse the problem of temporal sequence. Whether the attempt is successful is another question, since, despite his disclaimers, the implication of temporal ordering is hard to avoid: first come the demographic choices, then the residential.

4.3.2. Household Formation of Young Adults: Some Recent Econometric Modelling

In several papers modelling the household formation of young adults, Haurin, Hendershott, and Kim (1990, 1992a, 1992b) tackle the problem of temporal ordering of choices and events more directly.

Their models deal with four interrelated decisions: to leave the parental home or not; to marry or not; to have children or not; if not married, to live alone or with others. The authors make a careful distinction between choices and actions, noting that the sequencing of the two may not be the same. For example, one may decide to marry and then decide to leave the parental home before marriage. But the act of leaving the parental home will precede the act of marriage. They also emphasize that data relevant to these choices and events will be 'truncated'; for example, the tendency to live alone is unobserved if a person has not in fact left the parental home. And in general, it is the behaviour or action, not the choice itself, that is observed.

The authors' empirical analysis focusses on the causal role of "...the price of independent living and the individual's ability to pay this price" (1990, p. 1). Their 1990 paper, on which we focus here, uses data from the 1987 wave of the U.S. National Longitudinal Survey of Youth.

The statistical model consists of a series of interrelated equations, in order to deal with various issues of temporal ordering, endogeneity/exogeneity, simultaneity, and truncation.

With regard to income, for example, the authors note:

> "(it) is the product of wage rates and the quantity of labor supplied, and the latter is a decision made jointly with household formation. In a statistical sense, income is endogenous, and its use in estimates will lead to biased coefficients. Similarly, current wages do not accurately represent earning capacity because either the respondent ... may not work ... or the current job may be transitory or part time. The correct concept for the household formation decision is the wage that could be earned should the respondent take on the responsibility of independent living" (1990, p. 3).

The relevant wage is estimated using an incidental equation.

Similar considerations with regard to other variables and their proper measurement and interrelationships lead to a total model consisting of several separate equations, two central equations relating to living outside the parental household and to separate living, and incidental equations relating to income, marriage, and fertility.

The model specification is impressive and the empirical results sensible, even though they are based on what ultimately are arbitrary assumptions about temporal and causal ordering. At times, the authors seem to forget their distinction between choice and behaviour; at times they admit that their data measures behaviour when what they would like to measure is the choice itself. Their use of *estimated* values of marriage probability and of fertility as predictor variables would not pass muster with many demographers. But the work is an important reminder that much social demographic work lags behind that of economists in taking seriously the methodological implications of the processual character of household formation (see below, Section 4.4.4).

4.3.3. Whom Shall I Live With? Demographic Choice as Secondary

In their conceptualization of factors affecting household formation, Burch and Matthews (1987) turn Ermisch on his head. The choice of living arrangements becomes primary, demographic choices and behaviours somewhat instrumental and secondary. Like Ermisch and unlike Haurin and his colleagues, they try to finesse the issue of time rather than to face it head-on.

Building on earlier work by Lam, J. Goodman, and Ermisch, they begin by developing a list of ten 'household goods,' the things people seek through living arrangements. A classification of the goods is essayed in terms of the standard microeconomic notions of public/private, complements/substitutes, and normal/inferior goods. Household decision-making is seen as a matter of tradeoffs and compromises, since not all household goods can be maximized simultaneously.

Moving beyond economics, they note that 'payment' for household goods received may be in terms of money or household labour, but also in terms of not strictly economic commodities such as companionship, deference, and respect.

Against this background, several hypotheses are formulated to explain the rise of separate living in developed societies. These focus on the roles of income, kin numbers, changing preferences for privacy, age-gender role change in respect to household crowding and the quality of household services, and the role of modern communications and transportation technology in transforming the meaning of 'living alone.'

The authors suggest that a finite program of research would eliminate some of the nine hypotheses. But such empirical work has never been carried out. Nor has the preliminary theoretical scheme been further developed.

4.3.4. Michigan Models and Supply Factors

In connection with the now-famous Panel Study of Income Dynamics, a group at Michigan developed a general framework for studying co-residence decisions with attention both to demand and to supply factors (Duncan & Morgan 1976). A representative application (Hill & Hill 1976) deals with home-leaving of adult children.

On the demand side, the framework is typical, with emphasis on household services but also on privacy and autonomy. What distinguishes the model from much other work is serious attention to the supply side:

> "... the chance that the child will remain in the household during any given period is a function of both his willingness to stay (the demand for current arrangement) and the willingness of other family members to have him stay (the supply of current arrangement). Each of these relationships would be affected, in opposite directions,

by a 'price' reflecting the extent to which the child carries his own weight financially and around the house" (p. 136).

The child's willingness to stay is seen also as a function of the standard of living in other available households, and of employment opportunities in the area of the parental household.

Supply and demand are equilibrated through adjustments of what is termed the "net transfer variable," the extent to which the child carries his own weight in the household:

"... if at some level of net transfer, the child's desire to remain in the parental family is greater than the desire of other family members to have him do so, he should be able to rectify the situation by increasing his net transfer position by working more either in the home or in the marketplace" (p. 137).

Empirical estimates of the model for PSID data yielded fairly low R^2's and some specific results at odds with predictions — for example, child's income decreased rather than increased parental willingness to have him remain at home. It was suggested by Burch (1981) that these outcomes may have been due to inadequate consideration of social psychological factors, but neither he nor others have tried systematically to build on the excellent foundation supplied by Hill and Hill.

4.4. FAMILY DEMOGRAPHY: SHORTCOMINGS AND CHALLENGES

Each of the above formulations has its strong points. But even taken together they do not provide a well-rounded theory of household formation. Four areas of deficiency in present household research and theory stand out, some of them shared with contemporary demography generally, others more proper to family demography.

4.4.1. Supply Side Factors

As is apparent, only one of the formulations reviewed above gives serious attention to supply-side determinants of household status, despite the fact that only one household status decision, the decision to live alone, is relatively free from bargaining and joint decision-making.[8] Co-residence with another or

[8] Even in this decision, of course, there are impersonal supply factors relating to the provision of housing. The focus here is on the personal market for co-residence. And others may have legitimate interests in one's individual decision, as in the case of a spouse, a dependent parent, etc. But in the final analysis, one is not absolutely dependent on the assent and

others requires their choice as well as mine. The proper paradigm for the behavioural study of household status choices is similar to that for the study of mate selection rather than that for the study of, say, the demand for children.

As in marriage, social norms help define a pool of eligibles for co-residence, and, at least in the past, have defined preferred matches (for example, a dependent parent with an adult son or daughter).

Pitkin and Masnick (1987) have opened up leads to the aggregate analysis of supply factors, but their work seems to have lain fallow. A long tradition of research begun by Kobrin has documented the importance of kin availability (specifically adult children) as a determinant of household status of older, formerly married women (e.g. Wolf 1983). And research on the household status of the young has suggested that age at home-leaving is related both to number of siblings (perhaps affecting the quality of co-residence), and to the existence of an alternate residence with the non-custodial parent following divorce (Mitchell *et al.* 1989). But these only begin to get at the behavioural processes underlying joint decision-making.

A major barrier to empirical research (but not necessarily theorizing) on these issues is the absence of data sets with information on all the relevant potential decision makers. In the typical case, detailed information is collected on respondent, with little if any detail on prospective co-residents, largely but not exclusively kin.[9] As a start, respondents might be questioned more closely on objective aspects of their supply situation and on perceived alternatives to their current living arrangements.

4.4.2. Subjective and Other 'Soft' Variables

Like microeconomics generally, the study of household formation behaviour exhibits the dilemma that at its theoretical core it deals with subjective decision processes aimed at maximum subjective satisfaction, yet much theory and almost all research deal with objective behaviours, objective constraints, and objective household goods.

Considerable progress has been made in elaborating on the various household goods which people are thought to pursue in their decisions, but hardly any empirical data exist on what economists would call 'preference mappings' with respect to these goods. Recent trends in household status are

---

continuing co-operation of another in order to enter and maintain this household status.

9 To the extent that past social preferences for working out co-residential arrangements with kin weaken, the data problems become more severe in the sense that a potential pool of co-residents becomes more difficult to define. On the other hand, the pool is larger so that sheer availability problems are apt to diminish: for example, if an elderly person is willing to co-reside with any suitable roommate, not just with an adult child.

seen as a manifestation of individualism (seen as a largely Western cultural trait), and an overweening pursuit of 'privacy and autonomy,' at great economic cost (both individual and societal), and often at the presumed cost of companionship and support in daily living.

The collection of good data on these matters is not easy and that helps explain their rarity. The subjective realm needs to be differentiated (values, perceived norms, emotions, time horizons, etc. — see Burch 1991). Scales must be carefully constructed (contrast the casual use of proxies or hastily derived instruments), and measurements must precede the behaviours they are to be used to explain. The latter requirement points to surveys with a longitudinal design, since all agree that retrospective measures of subjective factors are of questionable value.[10]

In the absence of better direct measures of the subjective realm, we shall continue to assume, with the free-market economists, that people vote with their behaviour, and that if people live alone or apart from their parents or adult children, for example, then ultimately that is their preference. Theoretical elaboration will tend to be curtailed, seemingly too academic and sterile in the absence of more than the occasional opportunity for rigorous empirical testing.[11]

4.4.3. Networks and Other Household Contexts

In the study of one-person households, the phase *living alone* became something of a technical term, despite connotations that may have claimed too much. In a census context, the meaning was clear enough. But what did the fact of 'living alone' tell us of the person's social life? Strictly speaking, very little. But the connotation was there, and often 'living alone' stretched into 'isolated,' 'lonely,' 'bereft,' etc., and one-person households were seen as a social problem.[12]

The anecdote underlines an old complaint against household data, namely, that the four walls of the housing unit used to define it are too small and too arbitrary to capture the social, economic, and psychological realities. The complaint continues to be made, taking on added force as we learn more about the realities of non-Western household and family systems, and as the systems

---

[10] This clearly is so in the case of medium to large-scale surveys. Whether valid data could be collected through skilful intensive interviewing is worth further investigation.

[11] This is not what happened in the case of economics, of course, but it seems the most likely outcome in demographic circles where theory is less-valued for its own sake.

[12] These developments were in many ways similar to those with respect to the so-called 'isolated nuclear family' in family sociology. So strong was the image of isolation in some circles, that scientific reputations were made by the 'discovery' of moderate degrees of involvement with extended kin in modern societies.

in our own societies become more fluid and fuzzy (cf. the residence of the child in joint custody of divorced parents). A compelling statement of this theme as it relates to female well-being in developing countries is provided by Bruce and Lloyd (1992).

With respect to household decision-making, recognition of household boundaries as in part arbitrary leads to a consideration of supply-side factors described in section 4.4.1 above. But it also leads to the consideration of growing institutional competition with the household for the provision of 'household' goods. In different language, the functions of the household are changing, with loss of some functions long considered pre-eminently domestic.

From time immemorial, for example, the preparation and sharing of meals (what anthropologists have called the *commensal* dimension) has been a defining characteristic of the domestic group. But this household function has been sharply attenuated, with the decline of full-time homemakers, increased eating out, intrusion of TV into meal times, and individually prepared microwaved convenience foods. Similarly, the storage of household property has now been shared with commercial storage firms, as housing units become somewhat smaller, or at least are designed without attics, basements, and other 'extra' space.

The choice of household statuses thus poses new opportunities, constraints, and tradeoffs, as former household goods are no longer provided in the household, or are better provided for elsewhere. As Burch and Matthews (1987) ask "Why should a young adult stay at home if dinner consists of carry-out food, and if no one does his or her laundry?" There may be good reasons, but they are not as strong as in the past.

In some ways, theoretical research on these issues gets back to that on subjective factors discussed above in section 4.4.2, since perceived goals and perceived means to achieve them presumably drive decisions.

4.4.4. Taking Time Seriously

Of the four frameworks reviewed above in section 4.3, only one, that of Haurin, Hendershott, and Kim makes a serious effort to deal with household decision making as a *process*, something that takes place over time, with important temporal orderings among choices and observable events. Ermisch speaks of two stages, but emphasizes that these are distinguished mainly for analytic convenience. Burch and Matthews subordinate everything causally, and therefore temporally, to some more or less permanent vision of one's desired living arrangement, analogous to the microeconomist's lifetime utility. Hill and Hill work in the timeless time of standard economic analysis, modelling relationships at equilibrium. These theories could almost have been chosen as illustrations of McNicoll's (1992) comment that demographers tend to focus on "steady states and stable equilibriums" (p. 24).

At the empirical level, considerable progress has been made in the charting of household and family formation processes. As noted earlier, household and family demography has been one of the principal beneficiaries of the new emphasis on the collection of event histories and of the development of multi-state techniques for their analysis. In time, better documentation of typical household status paths or careers will pose clearer tasks of behavioural explanation, and in some cases, the addition of co-variates to multistate models will provide direct tests of behavioural hypotheses regarding specific transitions.

But the prospect of truly dynamic behavioural theory seems far off, as contemporary social science barely begins to rouse from the slumbers of static, linear, and continuous analysis. Perhaps only in the next century will household demography be written in the language of chaos and fuzzy sets.

Demography, in particular, may have to overcome an 'accounting mentality,' if more progress is to be made in the study of dynamics. In a recent study of home-leaving, F. Goldscheider and DaVanzo (1985) comment:

> "We recognize that decisions about living arrangements may be made simultaneously with other life-course decisions. ... However, we assume here that these other roles are predetermined with respect to decisions about living arrangements in order to simplify 'accounting' for their contributions in explaining variations in residential dependence" (p. 552).

The urge to simplify is found also in a recent paper by Bongaarts (1992), where the interpretation is in terms of determinants of fertility transition processes, but the analysis is a decomposition and a simple regression of cross-sectional data.

Apart from issues of model misspecification and biased coefficients, one can question, as does McNicoll (1992), the value of orderly accounting and 'explained variance' for behavioural understanding of causal processes and sound policy advice.

4.5. THE PROSPECTS FOR HOUSEHOLD THEORY

Increasing criticism is heard of contemporary demography for an alleged neglect of theory and of broader substantive issues. Livi-Bacci, for example, has spoken of the danger of demography becoming "... more a technique than a science" (1984, p. iii). In a recent critique, McNicoll (1992) argues that contemporary demography, with its emphasis on technique, measurement and description, has abdicated its role as a behavioural science and a source of substantive policy guidance, despite some comparative advantages in this regard with respect to other, narrower disciplines. Burch (1993) speaks of the existence of "two demographies," one a branch of applied statistics, showing little concern for behavioural theory, the other a behavioural science, in which at least in principle, theory is central.

Critical voices from outside demography also are heard. Olsen (1988), in a review of a review article on demographic models of marriage and fertility, finds much of this work trivial and lacking in behavioural content, and speaks of one large part of mainstream demography as a "... refined variant of actuarial science" (p. 577).

The last phrase is a reminder of the sensitivities involved in ranking disciplines in terms of intellectual value. Not all demographers and hardly any actuarial scientists would agree that being a "refined variant of actuarial science" is such a bad thing, and would point to the rich intellectual traditions behind both disciplines or sub-disciplines. Some, for example Ryder (1992) might argue that demography's tendency to stick to well-measured facts and necessary relationships and its restraint vis-à-vis theorizing are both signs of strength rather than weakness. Demography is often compared favourably in this regard with sociology, in which theoretical speculation has run rampant, or with economics, in which abstract theory has been refined *ad infinitum,* accepted as gospel rather than tested against empirical data.

It may be possible to view the current state of demography as a reflection of mature specialization in a growing scientific and policy analysis enterprise dealing with population, in short, a healthy manifestation of scientific division of labour. But all specialization tends inevitably to stratification, and, as McNicoll (1992), Keyfitz (1984), and others have argued, demography's focus of effort may have led it to occupy a more lowly and less influential place than it is capable of or deserves.[13]

To focus on broad theory, for a discipline or an individual, however, is to make an act of faith that fruitful theory is possible, a point on which there is by no means consensus. A brief word is in order, then, on the prospects for theories of household formation.

In household and family demography, virtually all commentators agree that the great barrier to progress is *complexity*. There are many different units involved: the individual; the couple; the nuclear family (as both a kinship and a residence group); the wider kinship group or network; the household (as a

---

[13] Some have argued (e.g. Notestein 1982) that at least in some contexts, purely descriptive social and demographic science can have more rather than less policy relevance and influence. He was confident that the long-term population control agenda had been favoured by a purely descriptive focus in the early work of the U.N. Population Division, a focus that forestalled deadly opposition from international political forces such as the Roman Catholic Church and the Soviet Bloc. It is not clear that such an argument is relevant today.

domestic group, with or without reference to kinship). And the formation and modification of these units involve several different demographic behaviours: marriage and remarriage; cohabitation; divorce; birth; death; home-leaving of adult children; other forms of entry and exit.

Corner (1987), in an attempt to model cohort headship rates, speaks of the "compound" nature of the headship curve, in effect a function of most of the events just mentioned, but with different sets of events empirically relevant at different ages. McNicoll (1992) speaks of "apparently simple events" whose explanation, however, requires almost limitless depths of behavioural analysis and understanding (p. 11).

Additional complications arise from the multiple possibilities of analyzing collectivities such as household and families as units in their own right (with attendant problems of definition, delimitation, and continuity, and problems of "fission" and "fusion," (Ryder 1992), or viewing household and family development with reference to one or more individual members (so-called 'marker individuals'). Most analysts to date have shied away from the former approach, as being beyond the conceptual and technical capabilities of contemporary social science.

In discussing complexity as a barrier to theoretical development, I am reminded of a favourite axiom of Marion J. Levy, Jr.: that the *apparent* complexity of a subject matter is in inverse proportion to the state of its theory. In short, good theory makes things look simple, or at least simpler. That is the precise function of theory. But this too is an act of faith, and one can't be certain until after the fact — which in terms of behavioural theory may be a very long time indeed.

Recent methodological discussions (e.g. Turner 1987) have called such faith into question, resurrecting an earlier tradition of scientific methodology associated with Karl Pearson, which sees little prospect for the cumulative development of verified social theory, due to an unbridgeable gap between theory and data ("underdetermination"). And McNicoll (1992) identifies the dangers of theoretical work on "big questions" as "diffuseness and 'softness'..." (p. 25), the very hallmarks of much sociological and demographic theory, and frequently cited reasons for its dismissal or neglect.

Thus the prospects are uncertain. But if there's any hope at all for the development of cumulative verified theory in our field, then two things seem especially necessary.

The first is a willingness to state our theories with sufficient rigour and formalization that at least some of our many competing ideas can be falsified and dropped from the canon. Cumulative theory may not be possible at all, but it certainly is not possible if we continue to theorize in such loose, verbal forms (Lenski 1988).[14]

The second is a willingness to allow theory some time and space to develop, giving it some temporary autonomy. McNicoll describes the pursuit of Hammel's agenda for cultural analysis as an "...empirically hopeless task....," but nonetheless one that "covers the right theoretical bases" (p. 13). For many demographers, an empirically hopeless task is not a proper task at all; a model that can't be directly empirically estimated is no scientific model.

Of course, theory must ultimately be kept on the leash of empirical verification if it is to escape unbridled speculation. But the leash need not be constantly short. Theory needs to go its own way in order to develop, to be pulled back only from time to time for empirical testing.[15]

ACKNOWLEDGEMENTS

This paper was prepared while the author was Visiting Professor, Dipartimento de Scienze Demografiche, Università degli Studi di Roma 'La Sapienza,' on sabbatical leave from the University of Western Ontario. It is based on a research program on family demography supported by the Canadian Social Sciences and Humanities Research Council [#T120C2] and by the UWO Academic Development Fund.

REFERENCES

BENOIT, D., P. LEVI, & P. VIMARD (1983), Structures des Ménages dans les Populations Rurales du Sud Togo, *Cahiers Orstom Série Sciences Humaines*, Vol. 19(13).

BERTINO, S., A. PINNELLI, & M. VICHI (1988), Two Models for Microsimulation of Family Life Cycle and Family Structure. *Genus*, Vol. 44, pp. 1-23.

BERQUÒ, E., & P. XENOS (eds.) (1992), *Family Systems and Cultural Change*, Clarendon Press: Oxford.

---

[14] The appropriate language for such formalization, beyond more careful verbal formulations, is not at all clear. Formal logic has typically proven a dead-end, although it is not clear whether attempts in demography or closely related fields have been competently carried out. Hanneman (1988) makes a strong case for user-friendly dynamic simulation packages. An obvious starting point for household formation might be the use of one or more of the household simulation models described in this volume, with input of fertility, marriage, etc. described in terms of behavioural equations rather than empirically derived distributions. Some pedagogical exercises of this sort have been carried out at Berkeley by Wachter and Hammel using SOCSIM.

[15] To pursue the analogy, one might say that sociological theory was never on a leash, and that economic theory got so 'strong' that it broke free.

BONGAARTS, J. (1983), The Formal Demography of Families and Households: an Overview, *IUSSP Newsletter*, No. 17, pp. 27-42.

BONGAARTS, J. (1987), The Projection of Family Composition over the Life Course with Family Status Life Tables. *In*: BONGAARTS, BURCH & WACHTER, pp. 189-212.

BONGAARTS, J. (1992), The Supply-Demand Framework for the Determinants of Fertility: an Alternative Implementation. Working Paper 44, Population Council Research Division: New York.

BONGAARTS, J., T.K. BURCH, & K.W. WACHTER (eds.) (1987), *Family Demography: Methods and Their Applications*, Clarendon Press: Oxford.

BRASS, W. (1983), The Formal Demography of the Family: an Overview of Proximate Determinants. *In*: *The Family*, British Society for Population Studies Occasional Paper 31, Office of Population Censuses and Surveys: London, pp. 37-49.

BRUCE, J., & C.B. LLOYD (1992), Finding the Ties that Bind: beyond Headship and Household. Working Paper 41, Population Council Research Division: New York.

BULATAO, R.A., & R.D. LEE (eds.) (1983), *Determinants of Fertility in Developing Countries*, Academic Press: New York.

BURCH, T.K. (1967), The Size and Structure of Families: a Comparative Analysis of Census Data, *American Sociological Review*, Vol. 32, pp. 347-363.

BURCH, T.K. (1970), Some Demographic Determinants of Average Household Size: an Analytic Approach, *Demography*, Vol. 7, pp. 61-69.

BURCH, T.K. (1979), Household and Family Demography: a Bibliographic Essay, *Population Index*, Vol. 45, pp. 173-95.

BURCH, T.K. (1980), The Index of Overall Headship: a Simple Measure of Household Complexity Standardized for Age and Sex, *Demography*, Vol. 17, pp. 93-107.

BURCH, T.K. (1981), Interactive Decision Making in the Determination of Residence Patterns and Family Relations. *In*: *Proceedings, International Population Conference, Manila, 1981*, IUSSP: Liège, Vol. 1, pp. 451-463.

BURCH, T.K. (1985), Changing Age-Sex Roles and Household Crowding: a Theoretical Note. *In*: *Proceedings, International Population Conference, Florence, 1985*, IUSSP: Liège, Vol. 3, pp. 253-261.

BURCH, T.K. (1989), Improving Statistics and Indicators on Families for Social Policy Purposes, Discussion Paper 89-3, Population Studies Centre, University of Western Ontario.

BURCH, T.K. (1991), De Gustibus Confusi Sumus? *In*: J.J. SIEGERS, J. DE JONG GIERVELD & E. VAN IMHOFF (eds.) *Female Labour Market Behaviour and Fertility: A Rational-Choice Approach*, Springer-Verlag: Berlin, pp. 61-73.

BURCH, T.K. (1993), Theory, Computers and the Parameterization of Demographic Behaviour. Contributed paper, Session #36, 'The Foundations of Demographic Models: Theories or Empirical Exercises?' IUSSP Montreal.

BURCH, T.K., S. HALLI, A. MADAN, K. THOMAS, & L. WAI (1987), Measures of Household Composition and Headship Based on Aggregate Routine Census Data. *In*: BONGAARTS, BURCH & WACHTER, pp. 19-39.

BURCH, T.K., & B.J. MATTHEWS (1987), Household Formation in Developed Societies, *Population and Development Review*, Vol. 13(3), pp. 495-511.

COALE, A.J. (1965), Appendix: Estimates of Average Size of Households. *In*: COALE, FALLERS, LEVY & SCHNEIDER, pp. 64-69.

COALE, A.J., L. FALLERS, M.J. LEVY JR., & D.M. SCHNEIDER (1965), *Aspects of the Analysis of Family Structure*, Princeton University Press: Princeton.

COALE, A.J., & D. MCNEIL (1972), The Distribution by Age of the Frequency of First Marriages in a Female Cohort, *Journal of the American Statistical Association*, Vol. 67, pp. 743-749.

CORNER, I. (1987), Household Projection Methods, *Journal of Forecasting*, Vol. 6, pp. 271-284.

DEVOS, S., & A. PALLONI (1989), Formal Models and Methods for the Analysis of Kinship and Household Organization, *Population Index*, Vol. 55, pp. 174-198.

DUNCAN, G.J., & J.N. MORGAN (eds.) (1976a), *Five Thousand American Families — Patterns of Economic Progress. Volume IV: Family Composition Change*, Institute for Social Research: Ann Arbor.

DUNCAN, G.J., & J.N. MORGAN (1976b), Introduction, Summary and Conclusions. *In*: DUNCAN & MORGAN, pp. 1-22.

ERMISCH, J. (1988), An Economic Perspective on Household Modelling. *In*: KEILMAN, KUIJSTEN & VOSSEN, pp. 23-40.

GOLDSCHEIDER, F.K., & J. DAVANZO (1985), Living Arrangements and the Transition to Adulthood, *Demography*, Vol. 22, pp. 545-562.

GOODMAN, L.A., N. KEYFITZ, & T. PULLUM (1974), Family Formation and the Frequency of Various Kinship Relationships, *Theoretical Population Biology*, Vol. 5, pp. 1-27.

GREBENIK, E., C. HÖHN, C., & R. MACKENSEN (eds.) (1989), *Later Phases of the Family Life Cycle*, Clarendon Press: Oxford.

HAMMEL, E.A. (1990), SOCSIM II, Working Paper 29, Graduate Group in Demography, University of California: Berkeley.

HANNEMAN, R.A. (1988), *Computer-Assisted Theory Building: Modeling Dynamic Social Systems*, Sage: Newbury Park, Ca.

HAURIN, D.R., P.H. HENDERSHOTT, & D. KIM (1990), Real Rents and Household Formation: the Effect of the 1986 Tax Reform Act. NBER Working Paper 3309, National Bureau of Economic Research: Cambridge, Ma.

HAURIN, D.R., P.H. HENDERSHOTT, & D. KIM (1992a), The Impact of Real Rents and Wages on Household Formation, WPS 92-6. College of Business, Ohio State University: Columbus.

HAURIN, D.R., P.H. HENDERSHOTT, & D. KIM (1992b), Housing Decisions of American Youth, WPS 92-16. College of Business, Ohio State University: Columbus.

HERNES, G. (1972), The Process of Entry into First Marriage, *American Sociological Review*, Vol. 37, pp. 173-182.

HILL, D., & M. HILL (1976), Older Children and Splitting off. In: Duncan & Morgan, pp. 117-154.

KEILMAN, N., A. KUIJSTEN, & A. VOSSEN (eds.) (1988), *Modelling Household Formation and Dissolution*, Clarendon Press: Oxford.

KEYFITZ, N. (ed.) (1985), *Population and Biology*, Ordina Editions: Liège.

KEYFITZ, N. (1985), *Applied Mathematical Demography*, 2nd edition, Springer-Verlag: New York.

KRISHNAN, P., & A.K. KAYANI (1976), Model Divorce Tables, *Genus*, Vol. 32, pp. 106-126.

KUZNETS, S. (1978), Size and Age Structure of Family Households: Exploratory Comparisons, *Population and Development Review*, Vol. 4, pp. 187-223.

LASLETT, P., & R. WALL (eds.) (1972), *Household and Family in Past Time*, Cambridge University Press: Cambridge.

LEE, R. (1981), From Rome to Manila: How Demography has Changed in Three Decades. *In*: *Proceedings, International Population Conference, Manila, 1981*, Volume 5, IUSSP: Liège, pp. 505-516.

LENSKI, G. (1988), Rethinking Macrosociological Theory, *American Sociological Review*, Vol. 53, pp. 163-171.

LEVY JR., M.J. (1965), Aspects of the Analysis of Family Structure. *In*: COALE, FALLERS, LEVY & SCHNEIDER, pp. 1-63.

LIVI-BACCI, M. (1984), Foreword to N. KEYFITZ (ed.), *Population and Biology*, Ordina Editions: Liège.

MCNICOLL, G. (1992), The Agenda of Population Studies: a Commentary and Complaint, Working Paper 42, Population Council Research Division: New York.

MASON, A. (1987), HOMES: A Household Model for Economic and Social Studies, Paper of the EWPI 106, East-West Population Institute: Honolulu.

MITCHELL, B.A., A.V. WISTER, & T.K. BURCH (1989), The Family Environment and Leaving the Parental Home, *Journal of Marriage and the Family*, Vol. 51, pp. 605-613.

NOTESTEIN, F.W. (1982), Demography in the United States: a Partial Account of the Development of the Field, *Population and Development Review*, Vol. 8, pp. 651-687.

OLSEN, R.J. (1988), Review of M. MONTGOMERY & J. TRUSSELL, Models of Marital Status and Childbearing, *Journal of Human Resources*, Vol. 23, pp. 577-583.

PINNELLI, A., & M. VICHI (1988), *FAMily micro-SIMulation System: Instructions for the PC Computer Program*, Materiali di Studi e di Ricerche nr. 16e, Università degli Studi di Roma 'La Sapienza': Rome.

PITKIN, J., &. G. MASNICK (1987), The Relation between Heads and Non-Heads in the Household Population: an Extension of the Headship Rate Method. *In*: BONGAARTS, BURCH & WACHTER, pp. 309-326.

PRESTON, S.H. (1987), Estimation of Certain Measures in Family Demography Based upon Generalized Stable Population Relations. *In*: BONGAARTS, BURCH & WACHTER, pp. 40-62.

RAVANERA, Z.R., R. FERNANDO, & T.K. BURCH (1993), From Home-Leaving to Nest-Emptying: a Cohort Analysis of Life Courses of Canadian Men and Women. Contributed paper, Session #20, 'Women's and Men's Life Strategies in Developed Societies', IUSSP Montreal.

ROGERS, A., & L.J. CASTRO (1981), Model Migration Schedules. Research Report 81-30, International Institute of Applied Systems Analysis: Laxenburg.

RYDER, N. (1987), Discussion. *In*: BONGAARTS, BURCH & WACHTER, pp. 345-356.

RYDER, N. (1992), The Centrality of Time in the Study of the Family. In: BERQUÒ & XENOS, pp. 161-175.

SWEET, J. (1977), Demography and the Family, *Annual Review of Sociology*, Vol. 3, pp. 363-405.

TODD, E. (1985), *The Explanation of Ideology: Family Structure and Social Systems*, translated by D. Garrioch, Basil Blackwell: London.

TURNER, S.P. (1987), Underdetermination and the Promise of Statistical Sociology, *Sociological Theory*, Vol. 5, pp. 172-184.

VANCE, R.B. (1952), Is Theory for Demographers?, *Social Forces*, Vol. 31, pp. 9-13.

WACHTER, K.W. (1987), Microsimulation of Household Cycles. *In*: BONGAARTS, BURCH & WACHTER, pp. 215-227.

WALL, R., J. ROBIN, & P. LASLETT (eds.) (1983), *Family Forms in Historic Europe*, Cambridge University Press: Cambridge.

WATKINS, S.C., J. MENKEN, & J. BONGAARTS (1987), Demographic Foundations of Family Change, *American Sociological Review*, Vol. 52, pp. 346-358.

WOLF, D.A. (1983), Kinship and the Living Arrangements of Older Americans, Project Report, The Urban Institute: Washington D.C.

WRIGHT, R.E. (ed.) (1992), *CEPHID Modelling Project: Canada's Elderly — Projecting Health, Income and Demography*, Institute of Research on Public Policy: Montreal.

YANAGISAKO, S.J. (1979), Families and Households: the Analysis of Domestic Groups, *Annual Review of Anthropology*, Vol. 8, pp. 161-205.

ZENG YI (1991), *Family Dynamics in China: A Life Table Analysis*, University of Wisconsin Press: Madison.

Part II

Data and Analysis

5. HOUSEHOLD CONCEPTS AND HOUSEHOLD DEFINITIONS IN WESTERN EUROPE: Different levels but similar trends in household developments

Nico Keilman

Statistics Norway
Division for Social and Demographic Research
P.O. Box 8131 Dep.
N-0033 Oslo
Norway

Abstract. The purpose of the present chapter is to investigate the consequences of differences in definitions of households and household members for household and family data, and for the interpretation of observed trends in this field. We will largely concentrate on information obtained from censuses. The focus is on Western Europe, but the findings of this chapter apply equally well to other industrialized countries with comparable data collection systems and household and family trends. A brief review is given of definitions and concepts of the household, of household types "consensual union", and "one-parent family", and of the notion of "child", as these are practised by the statistical agencies in European countries for their censuses and a few special surveys. Next, the consequences of different definitions are analysed for these concepts and notions. In order to trace the consequences of different definitions, we mainly rearranged observed and simulated numbers of various types of households.

5.1. PROBLEMS

Several reviews have revealed consistent trends in household and family developments in Europe during the last few decades: decreasing average household and family sizes, growing numbers of one-person households and of consensual unions, more one-parent families originating from divorce and less from widowhood, growing proportions of childless couples, and, quite recently, a higher age at which young adults leave the parental home (Hall 1986; Roussel 1986; Keilman 1988; Schwarz 1988; Gonnot & Vukovich 1989; Linke

Household Demography and Household Modeling
Edited by E. van Imhoff *et al.*, Plenum Press, New York, 1995

et al. 1990; Höpflinger 1991). At the same time, many authors noted problems of definition regarding household and family data. These problems apply not only to the household as such, but also to specific *types of households and families*, for instance, one-parent families and consensual unions (Linke *et al.* 1990; Duchêne 1990; Festy 1990). And finally, various *household members* are defined in different ways: we mention in particular the concept of child (Linke *et al.* 1990; Duchêne 1990). Given these problems of definitions regarding household and family data, it is not surprising that some authors doubt whether trends in household and family composition in Europe point, roughly speaking, into the same direction: "...La complexité des familles et des ménages ne cesse de s'accroître, et les définitions utilisées par les pays lors de leurs recensements ne permettent pas toujours de cerner correctement les convergences et les divergences des tendences ..." (Duchêne 1990, p. 116).

The purpose of the present chapter is to investigate the consequences of differences in definitions of households and household members for household and family data, and for the interpretation of observed trends in this field. We will largely concentrate on information obtained from censuses. Sample surveys are frequently used as well to trace household developments. However, comparisons of household trends between countries and/or over time on the basis of sample survey data are hampered not only by problems of definitions (as is the case for censuses), but also by problems caused by variations in sample design, in formulation of the relevant questions, and by selective non-response. The focus is on Western Europe, but the findings of this chapter apply equally well to other industrialized countries with comparable data collection systems and household and family trends.

Below, we first give a brief review of definitions and concepts of the household, of a few household types, and of some categories of household members, as these are practised by the statistical agencies in European countries for their censuses and a few special surveys. Because extensive inventories regarding these problems have been compiled in the past (CES 1983; UN 1989; Duchêne 1990; Festy 1990; Linke *et al.* 1990, summarized in Linke 1991), this review does not include any new findings.

In section 5.2, we consider the consequences of differences in definition for international comparisons. There are, in principle, three approaches to show the possible impact of alternative definitions on comparative household analyses between European countries. The first one is to look for data from countries that have collected information according to various definitions, and hence have compiled two or more time series for the same variable for some time. From the remainder of this chapter, it will become clear that such data are available for a few cases only. The second approach is to identify a shift in definitions, and investigate whether any shift in observed trends can be noted in parallel. This is much less convincing than the first approach, but still useful. Another possibility within this second approach is to *construct* a second series, according to a variant definition. Analysing such parallel series of observed counts assumes that individuals, although they fall in a different category according to the variant definition, still exhibit the behaviour (expressed by occurrence-exposure rates) of the persons who remained in the original category. For instance, in some countries, the census definition implies that a cohabiting couple without children is treated as two single persons. However, the marriage propensities of this couple will be much higher than those of persons living alone. By explicitly *simulating household formation and dissolution* (third approach), the assumption of equal behaviour can be relaxed. In an earlier paper (Keilman 1991), we used the LIPRO model (see Van Imhoff, chapter 12 in this volume) to simulate household dynamics and to investigate variant definitions concerning children, consensual unions, and one-parent families. Because the findings based on the latter simulation exercise by and large were in agreement with those stemming from the analysis of parallel series and rearranged data, the simulation results will not be reported here.

5.1.1. The Household Definition

The United Nations recommended to use the following definition of a household for the 1980 round of censuses: a household is "... either (a) a one-person household, that is, a person who makes provision for his or her own food and other essentials for living without combining with any other person to form a part of a multi-person household or (b) a multi-person household, that is, a group of two or more persons living together who make common provision for food and other essentials for living ..." (e.g. UN 1989, p. 4). These recommendations are considered to be valid as a guide for census-taking in the 1990 round as well (UN 1990). The definition formulated above recognizes two important concepts of the household: "housekeeping", i.e. sharing resources to provide household members with food and other essentials for living, and "dwelling unit" or "housing unit". The household concept is closely linked to the specific type of housing of the household members: "...Households usually occupy the whole, part of or more than one housing unit but they may also be found living in camps, boarding houses or hotels or as administrative personnel in institutions, or they may be homeless...". A distinction is made between *private* (or *domestic*) households and *institutional* households: institutional households are comprised of persons "... living in military installations, correctional and penal institutions, dormitories of schools and universities, religious institutions, hospitals and so forth...". Institutional households are often characterized by the fact that its members are subject to common rules and/or have common objectives. Persons living in hotels and boarding houses do not belong in this category, and they should be counted as members of private (i.e. non-institutional) households.

The United Nations recommended to use both the housekeeping concept and the dwelling unit concept in the household definition for the 1980 round of censuses. A definition based on the dwelling unit concept alone would probably be sufficient for many studies in the field of housing, but when studying socioeconomic aspects of the household, the housekeeping concept should be included in the definition as well. To what extent did countries follow the UN-recommendation? We found three reviews of the practice of countries in Europe, carried out by the United Nations, by Linke *et al.* for the Council of Europe, and by the Conference of European Statisticians (CES).

Table 5.1 indicates that the majority of the countries follow the UN-recommendation and use both the housekeeping concept and the dwelling unit concept in their household definition. A definition on the basis of the dwelling unit only is being used by four Nordic countries (Denmark, Finland, Norway, and Sweden), by France, and by Spain (only families). Since we have only one source, we can state that this definition is *probably* also used by the Færoer Islands (only families), by Iceland, and by the USSR (only families). For Norway (and possibly for the other countries as well), there are at least two reasons to define the household on the basis of the dwelling unit only. The first one is the fact that it is relatively easy to check and to explain to the census respondents (Johansen 1990, p. 45). The second reason is connected to the population register that is used, amongst others, for census purposes in many of the countries that employ the dwelling unit definition. Sometimes the census questionnaire contains preprinted register information (Norway is a case in point), and in other instances the "census" is a mere register count (this is the case in Denmark). A dwelling unit definition of a household is easier to handle in a register than the UN definition. As Linke *et al.* state in their report, Austria collects household statistics according to *two* definitions: the one recommended by the UN and the one based on the dwelling-unit concept only. This facilitates a comparison between the number of households evaluated according to these two different definitions, as will be illustrated in section 5.2.[1] Three countries out of the 35 listed in Table 5.1 show inconsistencies in the sources: Luxembourg, Portugal, and Switzerland.

---

[1] Richard Gisser of the Demographic Institute in Vienna pointed out to me that dual information can most probably be collected for many more countries in Table 5.1, just by appropriate processing of the household information collected in the census. The fact that, in the report by Linke *et al.*, only Austria appears to collect household information according to a dual definition may be due to a misinterpretation (by the other countries) of the relevant question in the questionnaire which was used by Linke *et al.*

Table 5.1. Household definitions in the 1980-round of population censuses, various countries in Europe

	(1)	(2)	(3)
Austria	B	B	B
Belgium	B	B[1]	B
Bulgaria	B	-	B
Channel Islands	B	-	-
CSSR	B	-	B
Cyprus	-	B	B
Denmark	D	D	D
Færoer Islands	D[1]	-	-
Finland	D	-	D
France	-	D	D
FRG	-	B	B
GDR	B	-	-
Greece	-	B	-
Hungary	B	-	B
Iceland	-	D	-
Ireland	B	B	-
Isle of Man	B	-	-
Italy	B[1]	B[1]	-
Liechtenstein	B	B	-
Luxembourg	D	B	-
Malta	-	B	-
Netherlands[2]	-	B	B
Northern Ireland	B	-	-
Norway	D	D	-
Poland	B	-	B
Portugal	D[1]	B[1]	-
Romania	-	-	B
Scotland	B	-	-
Spain	D[1]	D[1]	D[1]
Sweden	D	D	D
Switzerland	D	B	B
Turkey	-	B	B
USSR	-	-	D[1]
UK	B	B	B
Yugoslavia	B	-	-

Meaning of codes:
D: dwelling unit concept; B: both housekeeping and dwelling unit concept; -: no information available.
Sources: column 1: UN (1989); column 2: adapted from Linke *et al.* (1990, pp. 19-20); column 3: adapted from CES (1983).
[1] Families only.
[2] Household concept used in Labour Force Surveys; most recent Census taken in 1971.

The data in Table 5.1 suggest that one should be careful when comparing, for instance, household trends in Nordic countries with those in other European countries (except for France). All other things being equal, the dwelling unit definition gives a lower number of (small) households than the UN definition, because two or more households (according to the UN definition) which provide for their own housekeeping but which live in the same dwelling are counted as one household in the dwelling unit definition. In section 5.2, we will try to assess how large the difference is in number of households according to the two definitions.

An important aspect of the household definition is the issue how to determine whether an individual lives in a certain dwelling unit or not. Many European countries have some registration system in which persons are linked to addresses. These systems are maintained for administrative purposes, and very seldom are they geared towards demographic research. Therefore it is not surprising to note cases in which a person actually lives (for instance, stays four nights a week) at one address, whereas he or she is registered at some other address. As an example, consider Norway. The Norwegian Population Register has an extensive set of rules for registration, of which we mention two particular cases (e.g. SSB 1985, p. 4).[2] (1) A never-married person who resides outside the home of the parents because of education or military service is registered as living at the parents' home. (2) A married person who resides outside the partner's dwelling because of labour, education, military service, etc. is registered at the partner's house. Consequently, any demographer who is interested in the *actual* household situation of the population, but who relies on the *formal* figures following from the Population Register, will be confronted with a downward bias in, for example, numbers of young adults living single or living in a consensual union (due to the first registration rule), as well as a downward bias in numbers of spouses who live separated from each other before they divorce formally (due to the second rule).

The examples given here demonstrate that the concept of "living in the same dwelling unit" is far from unambiguous. Unless census information records the actual dwelling situation of household members, and not the formal, strong biases may be expected to occur in household figures, in particular for non-traditional households. In section 5.2.2, we shall give some numerical examples of the underestimation of consensual unions.

---

2 To a large extent, these rules are the same in other Nordic countries.

5.1.2. Household Members and Household Types

The scoring of household members has important implications for the breakdown of the overall set of households into households of various types. For instance, when persons over 18 years of age are no longer considered as "children", irrespective of the type of household they reside in or the dependency structure with other household members, we will see a decrease in the number of households with children. In this section, we first discuss the concept of "child", and next two types of households: consensual unions and one-parent families. This selection of issues does not imply that other types of households and household members do not pose any problems of definition. Consider, for example, the scoring of relatives (parents, uncles and aunts, etc.) in a family household. Various possibilities exist here, e.g. three-generation family and family with other adult persons. However, we have chosen children, consensual unions, and one-parent families because for these groups we have a long time series of data from the Netherlands, which facilitates looking at the consequences of various definitions and categorizations (see section 5.2).

Children. Defining the concept of "child" in the context of household composition involves not only tracing co-residing offspring (either by blood or adoption) of adult household members. One should also consider dependency structures in the household, both from an emotional and an economical point of view. But such structures are difficult to operationalize in a census. Therefore, there is considerable variation in the definition of "children". In some cases one takes any direct descendants, whatever their age — in other cases they are counted as part of the family (-household) only if still under 16, or in full-time education, or still unmarried (Eversley 1984, p. 15). Other age limits are possible as well, of course. The UN recommendation is to restrict children to those who are never-married, irrespective of age (UN 1980, p. 72). This may lead to peculiar situations. For instance, a never-married woman aged 50 living in the same household as, and taking care of, her aged mother will be denoted as "child". Thus it makes much sense to include an upper age limit as well, for instance, 18, or 20, or 25 years. Any child staying in the parental household after reaching this maximum age will from then on be counted as "person not member of the nuclear family" (UN 1987, p. 38). (The notion of "family" will be defined below.) The consequences of various age limits will be investigated in section 5.2.1.

Consensual unions. To determine whether a pair of adults form a cohabiting couple is far from easy. Two conditions should be met (Festy 1990, p. 80): the persons should live in the same household, and they should live "as husband and wife" without being married (*to each other*). Problems connected with the first condition were discussed more generally in section 5.1.1. Regarding the second condition, no valid objective criteria can be used (see Trost 1988, for a review of issues connected to defining cohabitation). Thus, in practice (for instance for Norway, Sweden, France, England — see Festy 1990, pp. 80-82; and also for the Netherlands at the occasion of the 1985/1986 Housing Demand Survey carried out by Statistics Netherlands), it is left to the respondents to determine whether they are cohabiting or not. This is what Trost calls the "phenomenological definition". Sometimes the partners are just asked whether they live as a couple, and combination with marital status results in a separation between married couples and cohabiting couples. Using a phenomenological definition will introduce a bias for various reasons (Trost 1988, p. 4): social acceptance, or tax avoidance, or due to differences in perception between the partners (one considers the relationship as marriage-like, the other one looks upon it as more casual). As a result, two unmarried adults living in the same household may be recorded as a consensual union at one occasion, and as two non-related adults at the other. In case there are dependent children as well in the household, the alternative registration may be a one-parent family, and a non-related adult living in the same household.

Consensual unions also have implications for international comparisons regarding statistics on families. According to UN recommendations, a family (or family nucleus, to be more precise) is comprised of persons living in the same household (either private or institutional) who are related as husband and wife or as parent(s) and never-married children by blood or adoption. Couples living in consensual unions should be regarded as married couples (UN 1980, p. 72). However, among the 19 European countries for which Linke *et al.* reviewed the family concept, it turned out that Belgium and the Netherlands do not take consensual unions into account by their family concept, whereas the same was reported for Denmark regarding couples living in a consensual union and having no common children (Linke *et al.* 1990, p. 13).[3] Other things being equal, this implies an underestimation of families in these three countries, as compared with other countries.

One-parent families. A one-parent family consists of a single parent and one or more co-residing dependent children. Except for the concept of "co-residing child", which was discussed more generally above, this definition is straightforward, but one is confronted with problems in cases where one or more other adults live in the same household, too. Unless such an other adult is a relative of the lone parent, for instance, his or her father or mother (-in-law), it may very well be the lone parent's partner. This situation mirrors the

---

[3] It should be noted, however, that the category "living in consensual union" is present in Denmark's family statistics.

one described above with respect to consensual unions with children. Höpflinger (1991, p. 321) refers to a study written by Neubauer, and published in 1988, which shows that in the Federal Republic of Germany, 10 per cent of women who declared themselves as "lone mother" were, in fact, cohabiting — for "lone fathers" the figure was even 28 per cent. In such cases it is not clear whether one should speak of a one-parent family which includes another adult in the same household, or of a consensual union with one or more children. A household consisting of a widowed mother, her child from the previous marriage, and her new partner may be counted as a one-parent family in case the man does not accept parental responsibility for the child (he then cannot be defined as "father"). In fact, there is only one "parent-child unit", and two adults forming a "conjugal unit" (Trost 1990, p. 29). A consensual union with a child would imply a conjugal unit and *two* parent-child units. As Duchêne (1990, p. 119) points out, in Belgium (and in many more countries as well, she contends), an unmarried cohabiting couple with one or more children is counted as one-parent family. In section 5.2.3, we discuss some numerical consequences.

5.2. CONSEQUENCES

Below we give two examples of countries for which we have numbers of households according to both the UN definition and the dwelling-unit definition.

In Austria in 1981, 2.764 million households were observed according to the UN definition. By using the dwelling-unit definition, one arrives at 80 thousand households less — a difference of 2.9 per cent.[4] In Norway in 1980, the difference between the number of households according to the UN definition and that on the basis of the dwelling-unit definition may be estimated as roughly 30 thousand (see Ås 1990, p. 57), or some two per cent of the total number of households (according to the UN definition). Given the fact that the average growth in the number of Norwegian households was 1.62 per cent *per annum* during the years 1970-1980, the difference between the number of households according to the two definitions may be considered small, see Table 5.2. The current dwelling-unit definition was introduced at the occasion of the census of 1970. Before 1970, the UN definition was used (Ås 1990, p. 54).

"Housekeeping unit" or "dwelling unit" are not the only two alternatives — other possibilities fall somewhere in between these two, as will be demonstrated by the following example. Todd and Griffiths (1986) carried out a study into the effects of a household definition change upon numbers of households in England. In many major household surveys of the Office of

---

[4] The help of Richard Gisser of the Demographic Institute in Vienna in providing these numbers is gratefully acknowledged.

Table 5.2. Households in Norway[1]

1930	1950	1960	1970	1980	1990
		in thousands			
653	964	1139	1296	1524	1769[2]

[1] 1930-1960: both housekeeping and dwelling-unit concept.
1970-1990: dwelling-unit concept only.
[2] Private households only; provisional figure.
Sources: SSB (1991b); Ås (1990).

Population Censuses and Surveys up to 1981, the housekeeping unit concept was used to define a household. But in 1981 it was considered more appropriate to bring the instructions for interviewers in line with Census practice, in which persons who shared a livingroom were also included in the household, even if they did not share catering arrangements. The effect of this change in definition was that the overall estimate of the number of households was reduced by 108,000, or 0.6 per cent.

The three examples for Austria, Norway, and England might suggest that historical and international comparisons concerning the total number of households are not obscured by differences in defining the concept of a household. That this does *not* hold more generally is demonstrated by findings for (former) Czechoslovakia: the difference in the number of households implied by the dwelling unit definition, as compared to the UN definition, was eight per cent in the Czech republic, both in 1970 and 1980; in the Slovak republic the difference was nine per cent in 1970 and eleven per cent in 1980 (Kalibova 1991). The reason for this relatively large difference is unknown.

Most of the examples in the subsequent sections, where we have rearranged observed data according to various definitions, apply to a single point in time. However, for the Netherlands, we are able to rearrange data in a long time series of household statistics, spanning some 65 years. The data are not observed, but they are the results of a projection carried out with the help of the household projection model LIPRO. The reason that we use projected data as if they were observed is that the data are extremely rich in detail: seven household types are compiled on the basis of information for individuals broken down by five-year age group, sex, and eleven different household positions, see below. In this section, we report on several rearrangements of this data set according to different definitions and concepts concerning children, consensual unions, and one-parent families.

LIPRO (LIfestyle PROjections) is a multidimensional model (and corresponding computer program) developed at the Netherlands Interdisciplinary Demographic Institute for the projection of a population broken down by age, sex, and an additional third characteristic. Van Imhoff gives a brief account of the model in chapter 12 of this volume, whereas a complete reference is Van Imhoff & Keilman (1991). In LIPRO applications so far, household position and marital status have been used for the third characteristic, but the model is flexible enough to facilitate other choices, for instance, region of residence, labour market status, or educational level. In this chapter, we shall use the model's variant which focusses on household position.

The model projects events that individuals experience as they move between household positions: not only events due to household formation and household dissolution, but also birth, death, emigration, and immigration. Interactions between individuals who (will) belong to the same household (e.g. marriage, start of consensual unions, divorce, leaving the parental home) are included in the model. At each point in time, numbers of households of various types are derived from numbers of persons broken down by household position.

The following set of private household positions was used for individuals for each combination of age and sex:

1. CMAR child in family with married parents
2. CCOH child in family with cohabiting parents
3. C1PA child in one-parent family
4. SING single (one-person household)
5. MAR0 married, living with spouse, but without children
6. MAR+ married, living with spouse, and one or more children
7. COH0 cohabiting, no children present
8. COH+ cohabiting, with one or more children
9. H1PA head of one-parent family
10. NFRA non family-related adult (i.e. an adult living with family of types 5 to 9)
11. OTHR other position in private household (member of a multiple family household; multiple single adults living in the same household)

The eleven household positions which individuals may occupy at any point in time result in the following seven types of private households:

A. SING one-person household
B. MAR0 a married couple without dependent children, but possibly with other adults

C.	MAR+	a married couple with one or more dependent children, and possibly with other adults
D.	COH0	a cohabiting couple without dependent children, but possibly with other adults
E.	COH+	a cohabiting couple with one or more dependent children, and possibly with other adults
F.	1PAF	a one-parent family, possibly with other adults (but no partner to the single parent!)
G.	OTHR	other household, such as multiple family household, or co-resident adults without partner relation

Number of households of types A-F follow straightforwardly from number of adults in households positions 4-9. The number of "other" households (type G) is found as the number of individuals in household position "other" (position 11), divided by the average size of this household type (2.82 in the Netherlands in 1985).

In Table 5.5 (to be discussed later), we give some summary results of a projection, for which most of the occurrence-exposure rates for household events and vital events (broken down by sex and five year age group) were estimated from the 1985 Housing Demand Survey of the Netherlands. This survey contains information on current and past household status of some 47,000 private households. The parameters were kept constant for the entire projection period 1985-2050.[5] (The LIPRO program gives the user the possibility to employ time-varying rates, but for the purposes of the present chapter this is not necessary.)

5.2.1. Children

A variable upper age limit for the definition of a child can easily be applied when we have data about children in households broken down by age of the child (provided the oldest age group is not too low).

For the case of Belgium in 1981, Duchêne (1990, p. 116) reports a drop in the proportion of one-parent families (among all families with children) from 14.7 per cent to 10.7 per cent as a result of introducing an upper age of 21 years for children who belong to a family, instead of the current practice of using no upper age limit (see also Duchêne 1991). However, in Denmark, where one-parent families are more frequent, and perhaps more similar to two-parent families, than in Belgium, shifting the upper age of children has a somewhat

---

[5] We applied the so-called linear model with the harmonic mean consistency algorithm. The unit projection interval was five years (equal to the width of the age brackets) and the scenario type was "fixed".

more limited effect: in 1988, 21.0 per cent of the families with one or more children under 26 are a one-parent family (some of these one-parent families are, in fact, consensual unions with children, see the discussion of Table 5.6 below), and still 18.0 per cent of the families with one or more children under 18 (Danmarks Statistik 1988, p. 6). Other countries in which the share of one-parent families as a percentage of all families goes down by a few percentage points as a result of a lower maximum age of the children, are the Netherlands (1985, from eight per cent to six per cent when a maximum age of 20 years is introduced, see Table 5.5), Sweden (1985, from 17 per cent to 13 per cent when the age goes down from 18 to 16 years), Spain (1981, from 11 per cent to five per cent when a maximum age of 16 years is introduced), and Italy (1983, from 12 per cent to 6 per cent when a maximum age of 18 years is introduced), see Duchêne (1991) for the latter three examples. For the case of France, Table 5.3 illustrates very clearly sudden shifts in indicators of family composition as a result of an upper age limit for children of 25 years, instead of 17 years. The proportion of families without children shows a slow but steady rise, but it drops by more than ten percentage points when elderly children are included. On the other hand, the declining proportion of families with four or more children (from 6.5 to 2.7 per cent between 1968 and 1982) suddenly goes up by 2.3 percentage points. For Norway we observe an increase of a few percentage points in the proportions of families with children (both married couples and lone parents), when no age limit for children is taken into account, as compared to the situation in which children can be no older than 20 years of age, see Table 5.4. However, the trends in the various proportions for the period 1974-1989 are not affected by introducing an age limit: proportions of married couples (both with and without children) go down and proportions of single parent families grow.

In panel 1 of Table 5.5, children are defined without taking any upper age limit into consideration. This follows the practice for the definition of a child as used in the Housing Demand Survey. The implication is that the age distribution of children has a relatively long, but very flat tail at high ages: 3 % for the age group 25-29, 0.9 % for 30-34, 0.5 % for 35-39, 0.2 % for both 40-44 and 45-49, and finally 0.1 % for age groups 50-54 and 55-59.

The trends indicate a steady increase in the number of private households which ends around 2030, and a slight decrease thereafter. When considering households by type, we observe a relatively strong and continuous growth in the share of one-person households, a decline in the proportion of households consisting of a married couple and one or more co-residing children, and for one-parent families first a rise in their relative number, followed by a modest fall in the first decade of the next century. However, as total numbers of households increase until around 2030, the absolute number of one-parent families does not fall until the year 2015, and in 2050 the number of one-parent

Table 5.3. Distribution of families by number of children, France

	1968[1]	1975[1]	1982[1]	1982[2]
				per cent
0	48.7	49.2	51.7	41.0
1	20.9	21.9	21.4	23.0
2	15.8	16.8	17.7	21.8
3	8.0	7.3	6.6	9.3
4	3.5	2.7	1.7	3.0
5	1.6	1.1	0.6	1.1
6 and over	1.4	0.9	0.4	0.9

[1] Children under 17 years.
[2] Children under 25 years.
Source: Rallu *et al.* (1993), table 5.

Table 5.4. Families by type, Norway

	no age limit for children				age limit of 20 years for children				
	MAR0	MAR+	SMOT	SFAT	MAR0	MAR+	SMOT	SFAT	TOTAL[1] (=100%)
	per cent								in thousands
1974	18.9	39.8	5.5	1.2	24.0	34.6	3.5	0.6	1,590
1977	18.7	38.9	6.0	1.1	24.0	33.6	3.9	0.6	1,629
1980	18.3	37.1	6.6	1.2	23.6	31.8	4.5	0.6	1,684
1982	18.0	35.6	7.0	1.3	23.2	30.4	5.0	0.7	1,737
1984	17.6	34.1	7.4	1.3	22.8	28.8	5.4	0.8	1,784
1987	16.7	31.7	8.1	1.6	22.1	26.2	6.1	1.0	1,858
1989	16.2	29.7	8.8	1.6	21.7	24.2	6.8	1.0	1,930

Meaning of codes: MAR0 married couple, no children.
MAR+ married couple with children.
SMOT one-parent family headed by a single mother.
SFAT one-parent family headed by a single father.

[1] Including other family types, for example cohabiting couples with common children, and one-person families.
Source: SSB (1991c, tables 1 and 2).

Table 5.5. Private households by type, the Netherlands; rearrangement of data

	SING	MAR0	MAR+	COH0	COH+	1PAF	OTHR	TOTAL (=100%)
				per cent				in thousands
1. Benchmark variant								
1985	26.7	22.0	39.3	4.0	0.8	5.6	1.6	5,567
2000	34.3	22.9	30.1	2.8	0.7	7.9	1.3	6,620
2015	40.5	23.7	23.8	2.6	0.7	7.5	1.3	7,213
2030	45.7	22.7	20.8	2.4	0.6	6.5	1.3	7,413
2050	47.1	21.7	20.5	2.5	0.6	6.2	1.3	6,809
2. Children's upper age limit 25								
1985	26.7	22.8	38.5	4.0	0.8	4.8	2.4	5,567
2000	34.3	24.3	28.7	2.8	0.7	5.5	3.7	6,620
2015	40.5	24.9	22.6	2.6	0.7	4.7	4.1	7,213
2030	45.7	23.7	19.8	2.4	0.6	4.0	3.8	7,413
2050	47.1	22.7	19.5	2.5	0.6	3.8	3.7	6,809
3. Children's upper age limit 20								
1985	26.7	25.3	35.9	4.1	0.7	3.9	3.3	5,567
2000	34.3	26.1	26.9	2.8	0.7	4.0	5.2	6,620
2015	40.5	26.5	21.0	2.6	0.6	3.3	5.5	7,213
2030	45.7	25.0	18.5	2.5	0.6	2.8	4.9	7,413
2050	47.1	24.1	18.2	2.5	0.6	2.7	4.8	6,809
4. Consensual unions doubled, recruitment from SING and 1PAF								
1985	18.8	23.1	41.3	8.4	1.7	5.0	1.7	5,300
2000	29.0	23.7	31.2	5.8	1.5	7.5	1.3	6,388
2015	35.7	24.5	24.6	5.4	1.4	7.0	1.3	6,982
2030	41.5	23.4	21.4	4.9	1.2	6.1	1.3	7,191
2050	42.9	22.4	21.2	5.2	1.2	5.8	1.3	6,591
5. Consensual unions doubled, recruitment from SING (50%) and from MAR (50%)								
1985	26.7	18.0	38.5	8.0	1.6	5.6	1.6	5,567
2000	34.3	20.1	29.4	5.6	1.4	7.9	1.3	6,620
2015	40.5	21.1	23.1	5.2	1.4	7.5	1.3	7,213
2030	45.7	20.3	20.2	4.8	1.2	6.5	1.3	7,413
2050	47.1	19.2	19.9	5.0	1.2	6.2	1.3	6,809
6. Ten per cent of 1PAF regarded as COH+								
1985	26.7	22.0	39.3	4.0	1.4	5.0	1.5	5,561
2000	34.3	22.9	30.1	2.8	1.5	7.1	1.2	6,611
2015	40.5	23.7	23.8	2.6	1.5	6.8	1.2	7,211
2030	45.8	22.7	20.8	2.4	1.3	5.9	1.2	7,404
2050	47.2	21.7	20.5	2.5	1.2	5.6	1.2	6,795

Meaning of codes:
SING one-person household.
MAR0 married couple, no children.
MAR+ married couple with children.
COH0 cohabiting couple, no children.
COH+ cohabiting couple with children.
1PAF one-parent family.
OTHR other private household (multiple family household, or alternatively two or more adult persons who have no partner relation or parent-child relation to each other, for example several students living in one dwelling, or a brother and a sister who stayed behind after the death of their parents).

families (425,000) is considerably higher than that observed in the mid-1980s (311,000). When we look at the proportion of households with children (columns headed by MAR+, COH+, 1PAF) we note a fall in the proportion from 46 per cent in 1985 to 31 per cent in 2030. This decrease by 15 percentage points over a period of 30 years is a direct continuation of the trend observed during the years 1960-1985, namely a fall by 17 percentage points (Kuijsten 1990, p. 44; Van Imhoff & Keilman 1990, p. 69). These developments are largely consistent, at least qualitatively, with those observed for other European countries, see section 5.1.

To investigate the consequences of variable age limits for co-residing children, we removed, for each projection year, numbers of children who are aged 25 or over, and who live in one of the following three household types: two married adults and co-residing child(ren) (MAR+), two cohabiting adults and co-residing child(ren) (COH+), and one-parent family (1PAF). Numbers of the latter three types of households were reduced accordingly. Some of these old children are the only child in, say, a MAR+ household. For each of the young adults who is no longer considered a child, the number of MAR+ households is reduced by one, and the number of MAR0 households goes up by one. However, some of these old children are the eldest child in a multi-children household, and removing them should leave the number of MAR+ households unchanged (except for the few multi-children households in which *all* children are 25 or older). As a first approximation, the number of MAR+ households were reduced by the number of children aged 25 or older, divided by the mean number of children in a MAR+ household.[6] The number of MAR0 households was increased by the same amount. Similar operations were carried out for children in COH+ households (which led to an increase in the number of COH0 households), and for one-parent families (implying a growth in the number of OTHR households, because a lone parent living together with her or his adult child is considered as an OTHR household here). The second panel of Table 5.5 shows results of such a rearrangement of data in case children cannot be older than 25, and in panel 3 the age limit is set to 20.

What do these variable age limits for children imply for major household trends? Total numbers of households remain unchanged, because children and adults just get different labels. One-person households remain unchanged, too. Comparison of panels 1-3 in Table 5.5 indicate that the general trend for households with a married couple and one or more children is barely affected. The largest implications can be observed for one-parent families and households of type "other". The initial upward trend in the former has almost disappeared, in particular when the age limit is set at 20 years — since the number of one-

[6] A more accurate correction should take household composition by age of all children into account. Such detailed information is not available, however.

parent families plus the number of households of type "other" remains the same, the latter type displays a relatively strong rise during the first 20 years. Measured in percentage points, the differences between the three variants for the proportion one-parent families are similar to those for married couples with or without children — however, since the percentage of one-parent families is rather low, the latter household type is more easily affected than the former three. Indeed, in contrast with the Benchmark variant, the absolute number of one-parent families at the end of the projection period is *lower* than at the beginning: 183,000 with an age limit of 20 years, and 262,000 with one of 25 years. There are two reasons why a change in the maximum age of co-residing children has larger consequences for one-parent families than for two-parent families. First, on average, there are only 1.71 children living in a one-parent family in the Netherlands in 1985, whereas the mean number of co-residing children in a two-parent family is 1.93. Second, children in one-parent families are relatively old: 31 per cent are over 20 years of age, 14 per cent over 25, and still eight per cent are over 30. For two-parent families, the corresponding proportions are much lower: 17, 4, and 1 per cent, respectively. The relatively high mean age of children in one-parent families in the Netherlands is, no doubt, a consequence of the fact that the 1985 Housing Demand Survey contained no clear instructions as to how to define a child. As a result, many elderly lone parents, who were dependent of a co-residing adult child for reasons of health or economy, will have been reported as head of a one-parent family.

A general conclusion of this section is that trends in most household types are relatively insensitive to a particular choice for the highest age at which young adults can still be regarded as children belonging to the household. The consequences for one-parent families, however, can be substantial, in particular when families of this type have relatively few co-residing children, and/or when these children are relatively old. To avoid a bias in international comparisons, one should control for the maximum age of these children.

5.2.2. Consensual Unions

To investigate issues of definition connected to consensual unions, two analyses will be carried out on the basis of rearrangement of data. First, we look at the consequences of the practice followed by Belgium and the Netherlands, where consensual unions are not included in the definition of the family. Second, the impact of an underestimation of consensual unions caused by registration of the formal dwelling situation for each of the partners instead of the actual one (see sections 5.1.1 and 5.1.2) is analysed. The issue of whether a household consisting of two adults and one or more children should be counted as a consensual union with children, or rather as a one-parent family plus an additional non-related adult, will be taken up in section 5.2.3 when we investigate one-parent families.

In case consensual unions are not included in the family definition, trends in numbers of families of various types are only affected when numbers of consensual unions (with or without children) are of substantial importance, compared to numbers of other families. This is the case in Sweden and Denmark — for instance, for the latter country, the number of cohabiting couples has risen quickly from about one-eighth of the number of married couples in the mid-1970s to nearly one-fifth in 1985 (Manniche 1990, p. 88). In other countries, consensual unions are less frequent, and the consequence for family developments of disregarding consensual unions from families are only small. Consider, for instance, the case of the Netherlands in the first panel in Table 5.5. When we add the proportion of consensual unions with children to that for one-parent families (and the proportion of consensual unions without children to that for "other" households), the shift is modest and hence main trends change only very little. But for Denmark, where the proportions of one-parent families and consensual unions (with or without children) are much larger, the shifts are considerable, see Table 5.6. When consensual unions are considered as families, two-adult families make up 89 per cent of all families. When consensual unions are not treated as two-adult families, the proportion drops to 74 per cent.

In section 5.1.1, we noted that there may be a downward bias in the number of consensual unions recorded in population censuses as a result of problems with registration of the actual dwelling place of the respondents. This underestimation is illustrated with figures for Norway, France, and England and Wales.

Table 5.6. Families by type, Denmark, 1985

MAR0 (1)	MAR+ (2)	1PAF (3)	COH0 (4)	COH+ (5)	TOTAL (6)
		in thousands			
472.4	588.3	163.6	139.4	78.4	1,442.0

Meaning of codes: See table 5.5.
Sources: Columns 1-3: Danmarks Statistik (1988, p. 3); columns 4 and 5: Danmarks Statistik (1988, p. 3) gives 50.1 thousand families consisting of a cohabiting couple with *joint* children. According to Manniche (1990, p. 90) these couples form 64 per cent of *all* cohabiting couples with children, whereas there are 1.8 as many COH0 couples as there are COH+ couples.

Sample surveys carried out in Norway in 1977 and 1984 revealed that among all women aged 18-44, 5 and 11 per cent, respectively, lived in a consensual union in Norway (Østby & Strøm Bull 1986, pp. 142-143). In 1988, the figure was 18 per cent (SSB 1991a, p. 42). For the age group 20-24, the figure rose from 12 per cent in 1977 to 19 per cent in 1984 (Festy 1990, p. 84), and for women aged 22 the trend continued to 34 per cent in 1988 (SSB 1991a, p. 42). These developments can be explained quite well on the basis of what we know about consensual unions in Norway. In spite of the increase between 1977 and 1984, the 1980 census resulted in the *same* figure for women aged 18-44 (5 per cent) as that in 1977 — for the age group 20-24 it was even *lower*: only 7 per cent. An important reason for the underregistration in Norway is the fact that only persons registered at the same address *according to the Population Register* are to be recorded as belonging to the same household. A second reason may be the relatively good "rapport" between interviewer and respondent in the surveys, as a result of which census information regarding specific issues is somewhat less reliable than survey information. Østby and Strøm Bull suggest, on the basis of certain adjustments in the raw data from the census, that the underestimation of consensual unions (for all ages) amounted to *at least* 30 per cent, and probably more. This figure is largely consistent with what one would expect on the basis of a linear interpolation between the figures for 1977 and 1984. It should be noted, however, that (an unknown) part of the bias may be the result of the sampling method, possibly selective non-response, and/or different wording of the relevant questions in the sample surveys. Festy (1990, pp. 84-85) finds that the number of consensual unions in a French survey in 1982 is twice as high as that found by an entirely comparable analysis on the basis of census data. For England and Wales, Penhale (1989, p. 12) reports an underestimation of roughly 40 per cent in the proportion of women who are cohabiting according to the 1981 Census. Noack and Keilman (1993) find 150,000 consensual unions in Norway in 1990 on the basis of survey data, whereas the census results in 100,000 couples only.

To look at the consequences of such a bias, we assumed numbers of consensual unions (both with and without children) twice as high (in other words, an underregistration of 50 per cent) as the ones simulated with LIPRO.[7] Panel 4 in Table 5.5 displays the results of calculations in which it was supposed that each additional consensual union without children was erroneously recorded as two one-person households, and that each consensual union with children

---

[7] Because the data used for the LIPRO simulations stem from a survey, and not from a census, numbers of consensual unions are to be considered quite reliable (for reasons explained earlier in section 5.1.2). Thus, these numbers should, in fact be reduced by 50 per cent. Instead we decided to make them twice as high, in order to give a clear picture of the consequences of an underregistration. The real effect will be much less for the Netherlands.

is found by combining a one parent family with a one-person household. In panel 5 we assumed that one half of the additional households stem from one-person households, and the other half from married couples with or without children. For each married person who lives, in fact, in a consensual union with a partner (not the formal spouse), his or her spouse should be registered as living alone, and in case the household consists of a married couple with children, it becomes a one-parent family. The assumptions underlying panels 4 and 5 are quite bold — they are certainly not the most plausible ones possible, but on the other hand they do not reflect an entirely impossible situation either. The way we proceed here facilitates the investigation of the boundaries of *possible* trajectories for household trends, but not *the most probable* one.

In spite of the rather drastic assumptions, the trends displayed by panels 4 and 5 are the same as or parallel to those noted in the Benchmark calculations. Total numbers of households are a little lower in panel 4 (because two one-person households are combined into one consensual union without children, and each consensual union with children is formed out of a one-person household and a one-parent family), and they remain the same in panel 5 (for each new consensual union without children household, a one-person household disappears, and a married couple without children is split in two, and similarly for new consensual unions with children — in fact, a partner in a formal union is merely replaced by one in a consensual union). Because numbers of households consisting of a married couple (with or without children), and "other" households are not affected by the rearrangement in panel 4, their shares in all households are a little higher than in the Benchmark. The proportion of one-person households is relatively low in panel 4, but also here it rises strongly. As a direct consequence of the assumptions we used, numbers of consensual unions are much higher than in the Benchmark variant, but their development over time remains unchanged.

5.2.3. One-Parent Families

Table 5.7, compiled by Höpflinger (1991, p. 323), shows that on average, roughly 80 per cent of the one-parent families in the countries concerned is comprised of only the lone parent together with one or more children. On the basis of this average percentage we assumed that 10 per cent of the heads of one-parent families, or, in other words, half of the lone parents with other adults in the household should, in fact, be regarded as cohabiting with a partner. Thus, in panel 1 of Table 5.5 for the year 1985, 31.1 thousand households were changed from one-parent family to consensual union with children. Where did we find the partners to these quasi lone parents? In the Netherlands in 1985, 60.4 thousand persons were recorded as a non-related adult living with a family,

Table 5.7. One-parent families

Country	year	(1)	(2)	(3)	(4)	(5)	(6)
				per cent			
Austria	1987	10.2	8.9	12.7	1.6	1.2	25.0
Belgium	1981	7.7	6.7	13.0	1.9	1.5	21.0
Denmark	1981	9.9	6.6	33.3	1.5	1.2	20.0
France	1982	5.2	4.4	15.4	0.9	0.7	22.2
FRG	1981	7.9	7.0	11.4	1.5	1.3	13.3
Ireland	1981	10.1	8.2	18.8	2.6	2.1	19.2
Italy	1981	7.5	6.1	18.7	3.0	1.6	46.7
Luxembourg	1980	7.7	5.9	23.4	1.6	1.1	31.3
Netherlands	1981	6.8	6.0	11.8	1.3	1.1	15.4
Spain	1981	6.7	5.4	19.4	1.5	1.1	26.7
UK	1981	9.0	7.3	18.9	2.5	1.5	40.0

Meaning of codes:
1 All female headed one-parent families, as a percentage of all family households.
2 Female headed one-parent families without others, as a percentage of all family households.
3 Difference between columns 1 and 2, as a percentage of column 1.
4 All male headed one-parent families, as a percentage of all family households.
5 Male headed one-parent families without others, as a percentage of all family households.
6 Difference between columns 4 and 5, as a percentage of column 4.
Source: Höpflinger (1991, p. 323).

for instance, a lodger, a distant relative, etc. (Van Imhoff & Keilman 1990, p. 69). Many of them will not be the partner of a lone parent. For instance, 52 per cent of these adults are aged over 60, whereas for lone parents the percentage is only 18. Thus it is reasonable to assume that many of these adults are an elderly father or mother of (one of) the adult(s) in the household. We assumed that half of the 31.1 thousand required partners were erroneously recorded as a non-related other adult living in a one-parent family (and even this proportion is probably quite high), while the remaining 15.6 thousand partners are to be found in "other" households (in which some 250 thousand individuals reside). On the basis of these rearrangements of individual household positions and household types, panel 6 of Table 5.5 was constructed. The results demonstrate that the proportion of consensual unions with one or more children doubles, but that it remains of minor significance only. The proportion of one-parent families is reduced somewhat, as compared with panel 1. For all household types, the development over time is left unchanged.

5.3. CONCLUSIONS AND DISCUSSION

In this chapter we have looked at the international practice regarding definitions for household and family concepts, and we have investigated possible consequences for international comparisons of the variety in definitions that are actually employed. The focus was on information obtained from population censuses, and we analysed the following issues: (i) the concept of the household (dwelling-unit and housekeeping unit); (ii) a maximum age for co-residing children; (iii) consensual unions; and (iv) one-parent families. The international practice with respect to issue (i) was reviewed on the basis of earlier inventories carried out in the recent past. In order to trace the consequences of different definitions regarding issues (ii)-(iv), we rearranged observed and simulated numbers of households of various types. Our conclusions may be summarized as follows.

The overall development in European countries during the past few decades towards small average household and family sizes, large numbers of one-person households and of consensual unions, many one-parent families originating in divorce and few in widowhood, and high proportions of childless couples, is little distorted by different definitions regarding the household, a co-residing child, a consensual union, or a one-parent family. Some trends may be accelerated or retarded, but their direction remains the same. Moreover, defining the household according to the dwelling-unit concept or to the housekeeping concept (or both, as the UN recommends) has a very limited impact on the total number of households. On the other hand, when the focus is on a specific type of household, or of household members, large differences may arise, when variant definitions are applied. In case one lowers the maximum age for co-residing children, the number of one-parent families diminishes rather strongly, in particular when families of this type have relatively few co-residing children and/or when these children are relatively old. At the same time, the upward trend in one-person households which is already present when no age limit is practised becomes even steeper. Finally, a large proportion of consensual unions is erroneously recorded as two one-person households, or as a one-parent family and a one-person household. Correcting for the underregistration of consensual unions (underregistrations of 50 per cent are not unlikely) may reduce the number of one-person households by up to one-third.

The general conclusions formulated here must be regarded as tentative. The present analysis is only partial, for two reasons. First, the effects of variant definitions for only a limited number of types of households and of household members were simulated: children, consensual unions, and one-parent families. Second, the calculations were only applied to the case of the Netherlands, and this country is certainly not representative of the general household situation in Europe (nor is any other single country). Fitting the model to data from more than just one country would possibly give a much firmer ground to our conclusions, or perhaps refute them.

Meanwhile, the analysis in this chapter has shown that different definitions have little impact on overall household trends, but levels may be shifted upwards or downwards. This makes a cross-sectional analysis hazardous, unless it is combined with a time series analysis. In other words, many European countries move into the same direction, but in order to distinguish forerunner countries from those starting later, we need data collected according to the same definitions.

The effects of different definitions are better understood the better knowledge one has of underlying household behaviour. One example is the mean number of children in a one-parent family and their age distribution (section 5.2.1). But that closes the circle: in order to increase our understanding, we need accurate data and unambiguously defined household indicators!

ACKNOWLEDGEMENTS

Most of the material contained in the present chapter is taken from an unpublished paper presented at the 1991 Quetelet Seminar "The collection and comparability of demographic and social data in Europe", Gembloux, Belgium, September 17-20, 1991, organized by the Institute of Demography, Catholic University of Louvain.

REFERENCES

ÅS, D. (1990), Det er Husholdninger som bor i Boligene: Om å Telle og Beskrive Husholdninger ("Dwellings are Occupied by Households: About the Enumeration and Description of Households".). *In*: GULBRANDSEN & MOEN, p. 51.

CONFERENCE OF EUROPEAN STATISTICIANS (1983), Sources of data and definitions of households and families in countries in the ECE-region, Working Paper 2 prepared by the Secretariat of the Conference of European Statisticians, January 1983.

DANMARKS STATISTIK (1988), Familie- og Husstandsstatistik 1. Januar 1988 ("Family and household statistics, 1 January 1988"), *Statistiske Efterretninger*, Vol. 6(9), pp. 1-15.

DUCHÊNE, J. (1990), Les Familles Monoparentales et Recomposées: Quelles Données pour une Mesure de leur Incidence? *In*: PRIOUX, pp. 115-134.

DUCHÊNE, J. (1991), Ménages et Familles dans les Pays Industrialisés au Cours de la Décennie 1980, Paper for the 1991 Quetelet Seminar 'The collection and comparability of demographic and social data in Europe', Gembloux, September 1991.

EVERSLEY, D. (1984), *Changes in the Composition of Households and the Cycle of Family Life*, Population Studies no. 11. Council of Europe: Strasbourg.

FESTY, P. (1990), Fréquence et Durée de la Cohabitation: Analyse et Collecte des Données. *In*: PRIOUX, pp. 72-86.

GONNOT, J.-P. & G. VUKOVICH (1989), Recent Trends in Living Arrangements in Fourteen Industrialized Countries, Working Paper WP-89-34, International Institute for Applied Systems Analysis: Laxenburg.

GULBRANDSEN, O., & B. MOEN (eds.) (1990), Husholdninger: Data og Definisjoner ("Households: Data and Definitions"), Prosjektrapport 63, Norwegian Institute for Building Research: Oslo.

HALL, R. (1986), Household Trends within Western Europe 1970-1980. *In*: A. FINDLAY & P. WHITE (eds.) *West European Population Change*, Croom Helm: London, pp. 19-34.

HÖPFLINGER, F. (1991), The Future of Household and Family Structures in Europe. *In*: *Seminar on Present Demographic Trends and Lifestyles in Europe*, Council of Europe: Strasbourg, pp. 291-338.

JOHANSEN, S. (1990), Husholdningsetablering i Folke- og Boligtellingen 1990 ("Household Definitions in the 1990 Population and Housing Census"). *In*: GULBRANDSEN & MOEN, pp. 45-50.

KALIBOVA, K. (1991), Typology of Households and Families: Czechoslovak Experience, Paper for the 1991 Quetelet Seminar 'The collection and comparability of demographic and social data in Europe', Gembloux, September 1991.

KEILMAN, N. (1988), Recent Trends in Family and Household Composition in Europe, *European Journal of Population*, Vol. 3, pp. 297-325.

KEILMAN, N. (1991), Household Statistics in Europe: Consequences of Different Definitions, Paper for the 1991 Quetelet Seminar 'The collection and comparability of demographic and social data in Europe', Gembloux, September 1991.

KUIJSTEN, A. (1990), Facteurs d'Evolution de la Structure des Familles Nucléaires. *In*: PRIOUX, pp. 41-59.

LINKE, W. (1991), Statistics on Households and Families in the Member Countries of the Council of Europe: Definitions, Methods and Sources, *INSEE Méthodes*, Vol. 8, pp. 117-124.

LINKE, W., M. DE SABOULIN, G. BALDURSSON, & A. KUIJSTEN (1990), HOUSEHOLD STRUCTURES IN EUROPE: REPORT OF THE SELECT COMMITTEE OF EXPERTS ON HOUSEHOLD STRUCTURES, Population Studies no. 22, Council of Europe: Strasbourg.

MANNICHE, E. (1990), Quelques Aspects de la Cohabitation au Danemark. *In*: PRIOUX, pp. 87-95.

NOACK, T., & N. KEILMAN (1993), Familie og Husholdning ("Family and household"), forthcoming in *Sosialt Utsyn 1993* ("Social Survey 1993"), Norwegian Central Bureau of Statistics: Oslo/Kongsvinger.

ØSTBY, L., & K. STRØM BULL (1986), Omfang og Utbredelse av Samliv uten Vigsel ("Extent and Development of Cohabitation outside Marriage"), *Tidsskrift for Rettsvitenskap*, Vol. 99, pp. 140-166.

PENHALE, B. (1989), Living Arrangements of Young Adults in France and England and Wales, CEPR Conference 'Beyond national statistics: Household and family patterns in comparative perspective', London, April 1989.

PRIOUX, F. (ed.) (1990), *La Famille dans les Pays Développés: Permanences et Changements* (Actes du Séminaire sur les Nouvelles Formes de Vie Familiale, Vaucresson, Octobre 1987), Congrès et Colloques no. 4, INED: Paris.

RALLU, J.-L., J. GAYMU, & A. PARANT (1993), Households Trends, Care for the Elderly, and Social Security in France. *In*: N. VAN NIMWEGEN, J.C. CHESNAIS & P. DYKSTRA (eds.), *Coping with Sustained Low Fertility in France and the Netherlands*, NIDI CBGS Publications 27, Swets & Zeitlinger: Amsterdam/Lisse.

ROUSSEL, L. (1986), Evolution Récente de la Structure des Ménages dans quelques Pays Industriels, *Population*, Vol. 41(6), pp. 913-934.

SCHWARZ, K. (1988), Household Trends in Europe after World War II. *In*: N. KEILMAN, A. KUIJSTEN & A. VOSSEN (eds.) *Modelling Household Formation and Dissolution*, Clarendon Press: Oxford, pp. 67-83.

STATISTISK SENTRALBYRÅ (1985), Forslag til Standarder for Kjennemerker Knyttet til Familier og Husholdninger ("Proposal for Standards of Characteristics of Families and Households"), Interne Notater no. 85/31, SSB: Oslo/Kongsvinger.

STATISTISK SENTRALBYRÅ (1991a), Familie- og Yrkesundersøkelsen 1988 ("Family and Occupation Survey 1988"), Norges Offisielle Statistikk B959, SSB: Oslo/Kongsvinger.

STATISTISK SENTRALBYRÅ (1991b), Folke- og Boligtelling 1990: Foreløpige Hovedtall ("Population and Housing Census 1990: Provisional Key Figures"), Norges Offisielle Statistikk B961, SSB: Oslo/Kongsvinger.

STATISTISK SENTRALBYRÅ (1991c), Familiestatistikk 1. januar 1991 ("Family Statistics 1 January 1991"), *Statistisk Ukehefte*, Vol. 24/91, pp. 1-10.

TODD, J., & D. GRIFFITHS (1986), CHANGING THE DEFINITION OF A HOUSEHOLD. Office of Population Censuses and Surveys, Social Survey Division, HMSO: London.

TROST, J. (1988), Cohabitation and Marriage: Transitional Pattern, Different Lifestyle, or just Another Legal Form. *In*: H. MOORS & J. SCHOORL (eds.) *Lifestyles, Contraception and Parenthood*, NIDI CBGS Publications 17, Netherlands Interdisciplinary Demographic Institute: The Hague.

TROST, J. (1990), Stabilité et Transformation de la Famille. *In*: PRIOUX, pp. 25-39.

UNITED NATIONS (1980), *Principals and Recommendations for Population and Housing Censuses* (ST/ESA/STAT/SER.M/67), UN Statistical Office: New York.

UNITED NATIONS (1987), *Recommendations for the 1990 Censuses of Population and Housing in the ECE Region: Regional Variant of the World Recommendations for the 1990 Round of Population and Housing Censuses*, United Nations Statistical Commission, and Economic Commission for Europe, Conference of European Statisticians, Statistical Standards and Studies no. 40, United Nations: New York.

UNITED NATIONS (1989), *Demographic Yearbook: 39th Issue*, United Nations: New York.

UNITED NATIONS (1990), Supplementary Principles and Recommendations for Population and Housing Censuses (ST/ESA/STAT/SER.M/67/Add.1), UN Statistical Office: New York.

VAN IMHOFF, E., & N. KEILMAN (1990), *Huishoudens en Uitkeringen in de 21e eeuw: De Gevolgen van Veranderende Huishoudenssamenstelling voor de Sociale Zekerheid* ("Households and Social Security Benefits in the 21st Century: The Consequences of Household Dynamics for Social Security"), NIDI Report 18, Netherlands Interdisciplinary Demographic Institute: The Hague.

VAN IMHOFF, E., & N. KEILMAN (1991), *LIPRO 2.0: An Application of a Dynamic Demographic Projection Model to Household Structure in the Netherlands*, Swets & Zeitlinger: Amsterdam/Lisse.

6. THE COLLECTION OF SURVEY DATA ON HOUSEHOLD STRUCTURES

Jenny de Jong Gierveld

Netherlands Interdisciplinary Demographic Institute (NIDI)
P.O. Box 11650
2502 AR The Hague
The Netherlands

Abstract. Nowadays, survey research is one of the main tools for investigating characteristics of the size and composition of households. Some of the methodological aspects of surveys are dealt with, such as the definitions and operationalizations of key household concepts, the possibilities for using complex routing procedures, and the choice between cross-sectional surveys using retrospective questions and panel designs to collect dynamic household data. Two Dutch examples of recent survey research into households are presented and compared on the possibilities to observe 'new', still infrequent types of living arrangements: Living Apart Together (LAT) relationships among the elderly.

6.1. INTRODUCTION

Selecting the household as a common focus for demographic research and analysis has both theoretical and practical justifications:

> "The household is a fundamental social unit. Households are more than groups of dyadic pairs. They have an emergent character that makes them more than the sum of their parts. They are a primary arena for the expression of age and sex roles, kinship, socialization, and economic cooperation where the very stuff of culture is mediated and transformed into action...." (McNetting *et al.* 1984, pp. xxi-xxii).

This quotation indicates the importance of households for understanding social life in general, and, more specifically, demographic developments. Indirectly, this quotation also underlines some of the problems involved in analysing households: the mediation of the rapidly changing 'stuff of culture' confronts household researchers with questions of the up-to-date conceptual-

Household Demography and Household Modeling
Edited by E. van Imhoff *et al.*, Plenum Press, New York, 1995

ization and operationalization of the household concept. Do we use the optimal definitions and suitable instruments (e.g. household schemes) to investigate the sometimes very complex household structures?

In his introduction to the book "Family Forms in Historic Europe" (Wall *et al.* 1983), when talking about the past, Wall comes to one of the main points we have to deal with:

> "... Assumptions about the households may either be explicit or implicit. In 'Household and family in past time' the household was defined in terms of residence (sleeping together) and in terms of consumption (at least one main meal together) (Laslett & Wall 1972). It was implicitly assumed that in order to eat together, all working members of the household would pool their income, except for servants who shared the household in the residential sense but had no 'right' to its communal income. (...) In the censuses, and before that in the unofficial listings of inhabitants, there appear blocks of names that look like households. The better listings and censuses refer to the sorts of people we would expect to find in households: parents, children, relatives, a few non-relatives (not identified as inmates), and servants. But in exactly what sense were these groups of people households?......"

For the present Dutch situation, Du Bois-Reymond (1992) gives an interesting overview of the complexity of new types of household structures. These new types of household compositions are currently only recognizable in the more individualized sectors of society. But the percentage of households consisting of a nuclear family per se is declining very rapidly. For example, in the Netherlands, the percentage of households consisting of a married couple and their children declined from 48.9% in 1960 to 35.0% in 1989. For the situation in Great Britain, Clarke (1992) provides an overview of the households with children aged 0-18 with *eleven* distinct categories. The figures prove that, of the children aged 16-18, only 62% is currently living with both natural parents.

What we see here is that where research on household structures was already very complex, new events in the last decades didn't make it easier for researchers to get a grip on developments in households. A description of the current household situation has to take new features into account, and must be properly equipped to register new or still infrequent types of households.

This chapter will accordingly focus on the following research question: is survey research[1] a suitable research design for providing data on household patterns and changes therein? Concentrating on *the methodological aspects* of surveys, we formulated the following more specific research questions:

---

[1] Sample surveys consist of relatively systematic, standardized approaches to the collection of information on individuals, households, or larger organized entities through the questioning of systematically identified samples of individuals (Rossi *et al.* 1983).

- What types of definitions and which types of operationalizations of households are promoted (and what will be the expected outcomes of these strategies)?
- What types of research designs are considered to be suitable in investigating households?
- What types of additional methodological rules are put forward to reach the goal of providing data on *changes* in patterns of households?

Next, this chapter will focus on *illustrating* the outcomes of the methodological rules as advocated for use in household survey research at the empirical level. We will investigate some recent examples of Dutch household surveys, each using somewhat different concepts, operationalizations, and survey designs, to find out if these household surveys are properly equipped to register households, including new or still infrequent types of households.

In section 6.2, we will firstly discuss opinions of leading demographers on the concepts and instruments to be used in research into household types and structures. In section 6.3, after comparing surveys with censuses, we will present some more specific characteristics of survey research, followed in section 6.4 by methodological issues concerning the collection of dynamic household data.

Next, we will deal with surveys as *realized* and will describe the characteristics of Dutch household surveys: section 6.5 presents a general overview of Dutch datasets on households. Section 6.6 examines the characteristics of the household concepts that are used. The typical characteristics of two example surveys are looked at in section 6.7, with an in-depth look at the routing. Section 6.8 deals with the longitudinal aspects of both surveys. A case study is set out in section 6.9 by way of example, using both surveys. Finally, section 6.10 will present a summary of the results.

6.2. KEEPING IN TOUCH WITH THE GROWING VARIETY OF HOUSEHOLD TYPES

6.2.1. Alternative Options: Households and Individuals

The first option that comes to mind when developing a research strategy for describing and analyzing household types is to adhere to classical ideas and conceptual instruments in the research designs, and to extend or update these where possible and necessary.

It is in line with this option that Höhn makes a plea for integrating family life cycle concepts *and* life course analysis for studying non-conventional as well as conventional family patterns and household types (Höhn 1987, p. 75). Her objective is to reach a certain typology that allows for differentiation according to marriage and consensual unions, to first and remarriages, and to the status of never-married. Each set of life courses will be defined in terms of ages at events: age at first marriage/consensual union, age at marital/union dissolution, age at remarriage/second union, age at death of spouse/partner, etc. Children from the current as well as from previous unions are considered separately in the new typology, also according to age at first birth prior to remarriage/age at first birth in remarriage, etc. The typology suggested by Höhn is based on a large set of biographical data, and includes at least 25 types of life courses. For each type, a set of indicators has to be collected for each generation of females and males, so that changes between generations can be shown. In addition, an overview must be provided of the percentage of females and males following a certain type of life course. Such a weighting of life courses reveals major changes in life styles, providing data on changes within life course cycles, as well as changes in the prevalence of different types of life courses.

The appealing and positive points in this development of a new typology are that Höhn provides ample opportunity to register conventional as well as non-conventional life courses; marriage as well as consensual unions, one-parent families after divorce, etc. The use of such a typology can overcome a lot of the shortcomings that are usually connected to more traditional registrations of marital status and family life cycles.

In order to use this typology as an instrument for describing and analysing household structures, one must also be aware of its negative aspects: e.g. its strict link with family stages. The leaving home processes of young adult children and sometimes their return home, the inclusion in the household of 'friends' of young adult children and their leaving the household, the inclusion of old-old parents or other relatives and their leaving the household are examples of the more complex and unstable situations that household demography is confronted with.

A second option is to forget the classical ideas and concepts about households as connected to family life cycles, and immediately direct the description and analysis to household types.

An advocate of this option is Keilman (1988). He states: "The traditional role of the family has weakened: members of married couples often live apart, cohabiting persons are not necessarily legally married, and cohabiting married persons are not always married to each other". Instead of studying marital status and nuptiality processes, the concept of a household should be used. (Keilman 1988, p. 123).

Van Imhoff and Keilman (1991, p. 9) give the following description of a household:

> "The household is one of the many operationalizations of the concept of a living arrangement. Other examples are marital status and family type. The trichotomy

marital status, family type, and household type runs parallel to Ryder's distinction between the conjugal dimension, the consanguineal dimension, and the co-residence dimension of family demography (Ryder 1985). In this order, these alternative operationalizations describe living arrangements ranging from a less to a more complex type of structure. First, the conjugal and the marital status perspective explore the formation and dissolution of marital unions. Second, the consanguineal and the family relationship explore links between parents and children. (...) Finally, we consider the household. (...) Households are the most complex type of primary units, embracing all the aspects of the less complex definitions of the above classification. (...) A household is a co-resident group regardless of consanguineal or affinal ties."

This *co-residence concept of households*, also called '*household-dwelling*' concept describes the (private) household as the aggregate number of persons occupying a housing unit. This description is a rather broad one; only one criterion is used to indicate the household and the household members, and this is the co-residence criterion.

A second type of household concept is more restrictive: combining the co-residence indicator with a micro-economic view of the household as a producer of goods and services (Burch & Matthews 1987, p. 498; Ermisch 1988, p. 23; Van Imhoff & Keilman 1991, p. 10). These productional functions of the household are seen as composite goods or bundles of goods, including physical shelter, storage of common and personal property, domestic duties (meals, laundry, repair of clothing, etc.), personal care for dependent household members, companionship, recreation, privacy, and economies of scale in consumption of goods. The definition that incorporates both indicators is the so-called '*housekeeping unit*' concept (see Keilman, chapter 5 in this volume).

Van Imhoff and Keilman (1991, p. 15) advocate concentrating on the description of (events experienced by) individuals (broken down by household position). Modern dynamic household models require information on the numerous types of events which individuals may experience. Such information provides household data that allow for multi-level analysis. The first level concentrates on the individual members of the household, providing insight into the development of different household positions of the individuals, and the household events they experience. The second is the level of the household, providing insight into the development of types and numbers of households.

The appeal of this research approach is the new attention for (events experienced by) *individuals*, broken down by household position. Empirical research oriented towards the investigation of characteristics of individuals (broken down by household position) can combine objective characteristics of the individual's position and changes in position with subjective indicators of the cultural backgrounds of the individuals (norms, standards, ideals, stereotypes, etc.). Consequently, a broader array of relevant characteristics will be available for the explanation of demographic events experienced by the individual members of the household. Furthermore, this will enable a more thorough description of events and their consequences, as well as multivariate analyses of the factors 'causing' certain events.

In summary, modern household demography, eliciting data from the individual members of a household, can provide a more realistic picture of the household situation that new generations are confronted with.

6.2.2. Implications for Data Collection

Selecting either option (1) or (2) has different consequences for data collection. In the case of option (1), concentrating on a typology of (conventional and non-conventional) life courses, the researcher needs a research design that takes into account a description of the marital and parental life course or the family stage that the individual has been confronted with. Assumptions are required on the degree of correspondence between family stages and household structures. A description of household structures via the intermediate family stages allows the researcher to only interview the head of the family (or the spouse of the head). This head of the family is supposed to provide the details required for the description of the household. It is generally typical of census-taking to interview only that person — either the head of the household or the spouse of the head of the household — and to ask this person to provide the data for herself/himself and for the other family members. The procedure is rather straightforward: facts about the family can be provided rather easily, particularly when these facts are not too complicated. Analysis will, in general, concentrate on the aggregate level of the household or family.

When selecting option (2), data collection procedures firstly and mainly concentrate on (interviewing) an individual member of the household in order to elicit several individual characteristics 'behind' and connected with the events that this individual has experienced. Additionally, the researcher asks about the composition and other characteristics of the household. Depending on the specific research questions of the project, inquiries into aspects of the productive functions of the household can be included: questions about the division of domestic duties, who is mainly responsible for the personal care of dependent household members, which recreational activities are provided within the realm of the household, and so on.

Additionally, as mentioned already, the analysis procedures suitable for this type of data are twofold: firstly, at the level of the individual (investigating the interrelationship between sets of characteristics of the individuals); and secondly, at the aggregate level of the household, providing insight into the (development of) types and numbers of households. The possibility of collecting detailed, valid, and reliable data on several aspects of the individuals' *and* household situation is more generally guaranteed via this procedure.

6.3. POSSIBILITIES AND RESTRICTIONS OF SURVEY DESIGNS IN THE INVESTIGATION OF CHARACTERISTICS OF HOUSEHOLDS AND INDIVIDUAL HOUSEHOLD MEMBERS

The differences in survey research design, especially between large- and small-scale survey research, is basically a matter of gradual differences. Even (the most frequently used type of) census-taking can be considered to belong to this continuum of survey research, situated at the large-scale, fully structured end of the continuum. As a matter of fact, census-taking is normally used in special contexts and with specific characteristics; characteristics that, as a rule, differ significantly from the guidelines and characteristics, the standards and norms of, and the methods and techniques used in survey research (see also Klijzing 1988). In principle, each of these research designs will include interchangeable sets of questions and sets of items. The modules (sets of internally-related questions) that will be involved in either census-taking or survey research are selected by the researchers, and guided by the intrinsic goal of the study.

One must remain aware of the fact that the design[2] of a survey is directly connected to:

- the central (substantive) research questions;
- the theoretical ideas that are involved in describing and explaining the research findings; and
- the practical/financial scope of the research project.

Consequently, when comparing household data derived from either census or survey research, one has to keep in mind that, in most cases, the practical and financial possibilities of the two methods differ enormously, as do the central aims and options of the two methods. A census will, in principle, cover all households (and individuals) in a country, whereas surveys are based on samples of households or individuals. Census-taking is normally supported by law which stimulates individuals to participate, whereas survey researchers are regularly confronted with (high percentages of) non-response, especially in those countries whose citizens are highly individualized and directed toward protecting their 'privacy'.

On the other hand, the costs of interviewing all the households in a country are tremendous, resulting in relatively short and manipulative questionnaires, that concentrate on the most important characteristics of households, families,

---

[2] A research design is the arrangement of conditions for the collection and analysis of data in a manner that aims to combine relevance to the research purpose with economy in procedure (Selltiz *et al.* 1961)

and individuals. Interviewing a sample of all households provides the opportunity of increasing the number of questions and enhancing the complexity of the questionnaire procedure.

In most countries of the world, census-taking is a regular activity. (Exceptions are the Netherlands and Denmark; year of last census in the Netherlands is 1971, in Denmark 1981). In many countries, the ten-year period between two successive rounds of census-taking is used for large-scale surveys (Linke 1991).

An important issue concerns the comparability of the data between the various countries (Linke 1991, p. 120). This comparability is directly related to the definitions of households and families, as used in each of these censuses and surveys.

Given a solid data base of accurate and up-to-date national demographic indicators, elicited either via census-taking or population registers, supplementary data sets concentrating on households can be provided by survey research. A survey has some interesting features that together offer an outstanding opportunity for investigating households and families.

These features include:

- *Flexibility*. The survey research design is flexible and can easily be adjusted to the requirements of differing research purposes, and can be applied to both exploratory and explanatory studies. The survey could consist of strictly structured questions with closed answer categories, but also (a number of) open questions without prepared answer categories. The survey can be concluded via face-to-face interviews, by mail, or by telephone interviewing. Modules of the questionnaire can be aimed at facts and opinions, as well as at investigating values, beliefs, and attitudes. The time frame of the survey could be the present and the past, but also the future.
  Thus, it gives the researcher a wide range of opportunities to more closely examine specific research questions of the project.
- *Complexity*. The survey research design will benefit from a clear and straightforward research purpose, but a more complicated purpose can be handled. Life histories can be investigated; special routings through a questionnaire are possible and frequently used. Face-to-face interviews can be combined with the use of laptop-computers (see also next section); telephone interviewing is possible via CATI-systems (computer assisted tele-interviewing), etc.
- *Individual and aggregate levels of data collection are possible*. The survey design has displayed its true value for research questions that concentrate on individual data, but also for research questions that are directed towards the investigation of data on larger entities. For

instance, questions can be asked about the person her/himself, but also — if necessary in the same questionnaire — about other people, about broader entities like households, families, firms, etc. The survey research design is to a certain degree (see footnote 2) suitable for research strategies that directly concentrate on family characteristics at the aggregate level (option 1 in part 2), but is outstanding for those strategies that initiate the description and analysis of households and families via characteristics of the individual (option 2 in part 2).

- From the other positive characteristics of the survey research design, we select here the possibility to organize the observational situation of face-to-face interviews in such a way that *'rapport'*[3] *between interviewee and interviewer* will be reached. The rapport effect is valuable in those data collection situations that are more or less threatening or fatiguing for the respondent. Rapport can more easily be built up during the face-to-face surveys than in the shorter period of time available in census-taking. This includes the advantage that the registration of informal data, in particular, threatening and/or socially undesirable data, will be more reliable than in other designs.

6.4. ISSUES IN THE COLLECTION OF DYNAMIC HOUSEHOLD DATA

The study of household *transitions* requires longitudinal data. Caldwell and Hill (1988) have pointed out the importance of longitudinal data collection, especially during periods of stress or crises, to gain greater insight into the systems and processes governing what is taking place. Moreover, longitudinal data should include macro- as well as micro-level variables. Macro-level data, that is, statistics about the long-term transitions at the national level, can be seen to indicate changes in institutional and structural processes, and the evolution of societal values, standards, and norms.

Macro-level data on their own, however, provide only a partial explanation of the transitions to be studied (Wall 1989). Meso-level and micro-level data are needed to study the transitions in individual life histories and their consequences for the household composition of the individual, taking into account the specific values, norms, and behavioural rules of the groups in which individuals find themselves, as well as individual perceptions and evaluations of different options.

---

[3] A rapport effect is an effect of the interviewers activities in establishing a good relationship with the respondent in order to prevent breaking off the interview early and/or to prevent item non-response.

In case of major fluctuations and changes, the 'snapshot' provided by cross-sectional data will not be a good picture of the situation because the analysis will depend upon the specific conditions prevailing at the time of survey.

In principle, longitudinal data can be provided by retrospective or prospective research designs. An example of a retrospective design is a survey characterized by a strict investigation of life histories as experienced by individuals in their educational or fertility career, etc. An example of a prospective design is the panel study: the same individuals are investigated at several points in time.

For practical purposes, such as the cheaper price and covering a life history by interviewing only once, a cross-sectional study using retrospective questions is sometimes performed to gather data about two (or more) moments instead of a panel study. A panel study requires investments over time, needs complex procedures to stay in contact with the respondents, and it avoids sample attrition. However, the quality of retrospective questions is disputable. It is directly related to the chance of response bias in the answers to retrospective questions.[4]

Either as a result of recall lapse or the deliberate twisting of one's answers, a discrepancy can be observed between actual facts and the representation of the facts in a questionnaire. Two types of recall lapse can be distinguished (Sikkel 1985). The first one is referred to as 'memory effect' and is related to the (social-)cognitive operation of memory. This memory effect generally results in systematic underreporting of events due to gaps in a person's memory. The second type of recall lapse is referred to as 'mistake effect'; the respondent remembers the event, but misreports the real date of occurrence either by telescoping or receding (Keilman 1985). Telescoping is said to take place when the respondent reports the event to have occurred more recently than it has in actual fact. Receding is the opposite, that is, when the event is reported to be of an older date than the actual one.

As yet, little research has been done on the reliability of retrospectively collected data on household transitions. Freedman and her colleagues (Freedman *et al.* 1988) interviewed young adults both in 1980 and 1985 on family, educational, and occupational statuses. Very minor differences in the reports

---

[4] Researchers must remain aware of the bias that will be introduced in the response distribution when retrospective questions and questions about other persons are included in the questionnaire. The chance of occurrence of memory lapses and other faults in providing these facts is directly related to the length of time between the moment of occurrence and the moment of inquiring, the 'saliency' of the research theme for the respondent, the 'threatening' aspects of the topic versus the social desirability of certain answer categories, and to the 'distance' between the interviewed person/head of the household and the family member whose characteristics have to be elicited (Sudmon & Bradburn 1982).

concerning marriage and childbirth were found. The differences in the reported statuses concerning education and occupation were somewhat larger.

Further evidence that questions regarding the transitions of marriage and childbirth are among those most reliably remembered is provided by Courgeau and Lelièvre (1989).

In addition to the problem of recall lapse in retrospective studies, there is the problem of possible deliberate twisting of the answers to certain research questions. As years go on, one tends to prefer to 'forget' and 'underreport' unfavourable or socially undesirable events or facts (Meijer 1990).

Panel and life history research explicitly take into account change and the dynamics of empirical phenomena. Ideally, life history research should be coupled with a panel design, where the same respondents are interrogated on two or more occasions. That way one obtains the kind of data required to test hypotheses concerning the determinants and consequences of household transitions (cf. Ott, chapter 7 in this volume).

6.5. DUTCH EXAMPLES OF RESEARCH INTO HOUSEHOLDS; GENERAL CHARACTERISTICS

In the following, we will present two examples of survey research oriented towards the investigation of households. We will give an overview of the operationalization of the concept 'household'. In addition, we will present the schemes (or questionnaire boxes) to elicit the household composition, followed by an illustration of the specific survey characteristics. In this context, the routing procedures, among others, will be presented.

Attention will also be paid here to the longitudinal aspects of both types of surveys. To what extent is it possible to measure the dynamics in life courses?

6.5.1. The Housing Demand Survey

Since the final census of 1971 in the Netherlands, researchers of households use different data sets. Since the first of January, 1987, a register count has been held in 393 municipalities using a computerized population register, from which statistics on families and on persons not living in a family relationship have been derived. This register count is administrative, at the municipal level, from which demographic information is collected on an annual basis. Besides these register counts, a large-scale data base on households is built up, also on a more or less regular time basis: the Housing Demand Survey (HDS). Information is collected on the housing situation of households, related to the demographic *and* socio-economic features of these households. For the purpose of research on current household structures, the most important survey is this large-scale, cross-sectional HDS, representative of the Netherlands.

The first wave of this survey dates back to 1977, the second was realized in 1981, the third in 1985, and the most recent one in 1989/1990. This survey is more specifically aimed at collecting information on the housing situation, expenditures for housing, and realized and desired migrations of households. Statistical data are provided concerning the description of the housing supply and the housing conditions of households. The 1989/90 sample size contains 76,358 persons, 18 years and over, and is based on the Municipal Registers. The survey results are primarily used by the Ministry of Housing, but are also available to other users, e.g. demographers.

6.5.2. The NESTOR-LSN Survey

The second example includes the first wave of a survey of the *NESTOR Program, Living arrangements and social networks of older adults*, a medium-sized survey of a (partially) longitudinal nature. The first wave of this survey was conducted in the period January to July 1992. The objective of this program is to provide insight into the older adult's well-being. Determinants of older adults' well-being include: the composition of the household they are in, and their kin and non-kin networks. Furthermore, it aims to provide insight into the outcomes of older adults' households and their kin and non-kin networks in terms of the availability of social support which is essential for daily functioning, for coping with the problems associated with life events, maintaining well-being, and to specify the assumptions essential in constructing models of future trends.[5]

Face-to-face interviews conducted among 4500 older adults form the basis for the main study. A stratified sample of older persons, males and females between 55 and 89 years, has been taken from the population registers of various municipalities in three regions in the Netherlands. The sample is stratified according to sex and year of birth of the older adults.

---

[5] The research program has been developed at the request of the Netherlands Steering Committee Program for Research on Ageing, instituted by the Ministry of Welfare, Health and Cultural Affairs, and by the Ministry of Education and Science. Researchers from three academic institutes are involved in this program: the Department of Sociology and the Department of Social Research Methodology at the Vrije Universiteit in Amsterdam, and the Netherlands Interdisciplinary Demographic Institute (NIDI) in The Hague.

6.6. CHARACTERISTICS OF THE HOUSEHOLD CONCEPTS AND HOUSEHOLD QUESTIONS AS USED IN BOTH SURVEYS

6.6.1. The Housing Demand Survey

The following definitions are used in this survey:

Households: A group of two or more persons who live together in a household and share the household tasks, as well as a person who runs a household independently. The first case is called a multi-person household, the second a one-person household. Living together in a household and sharing the household tasks means that the members of the household live in (a part of) the same accommodation, share the main room in the home, regularly share meals, and share the communal expenses.

Multi-person households are subdivided accordingly:

- households with one family (one-family households);
- households with two or more families (multi-family households), and
- households without families (non-family households).

In the case of family households, persons not belonging to the family could be a part of the household.

Head of the Household: The person who rents or is owner of the accommodation. In the case of shared rent or ownership, the main breadwinner or the oldest person is considered to be the head. Persons in one-person households are, by definition, head of the household. These definitions show that the 'housekeeping unit' concept of households is used in the HDS.

There is currently an important debate in the Netherlands on how questions on household structures should be formulated. The use of terms by interviewers such as 'head of household' and 'family head' can sometimes be confusing, and could lead to the respondent becoming irritated. The concept 'main breadwinner' has the same objections. However, the choice of one member of the household as a reference person seems to be necessary in order to be able to determine the composition of the household and the respondent's position therein. In order to prevent irritations and ambiguities, a workgroup of the 'Foundation of Research Institutes' proposes the use of the term 'reference person', defined as follows:

> "The person who owns the accommodation or is responsible for paying the rent. If this is more than one person, then this is the male. If there are more males, then the oldest male. If there is no male, then the oldest female".

It must be stated that although the reference person within the household is given another name, there are strong similarities with the HDS definition of 'head of the household'. More innovative is the Statistics Netherlands (NCBS) approach in the new standard household box as used in the CAPI and CATI procedures[6]. In this household box, neither 'head of the household' nor 'reference person' is used, but 'household core' which is defined as follows:

> "The household core is formed by the person or persons managing the household, who regulates the affairs in the household".

This means that it is irrelevant who pays the rent or who owns the home. It is up to the respondents to state how they perceive the positions within the family. Thus, contrary to previous methods, there is no need for a diligent search for one 'head of the household' or reference person. When we examine the routing within the household module, we see an essential difference (see Figures 6.1 and 6.2).

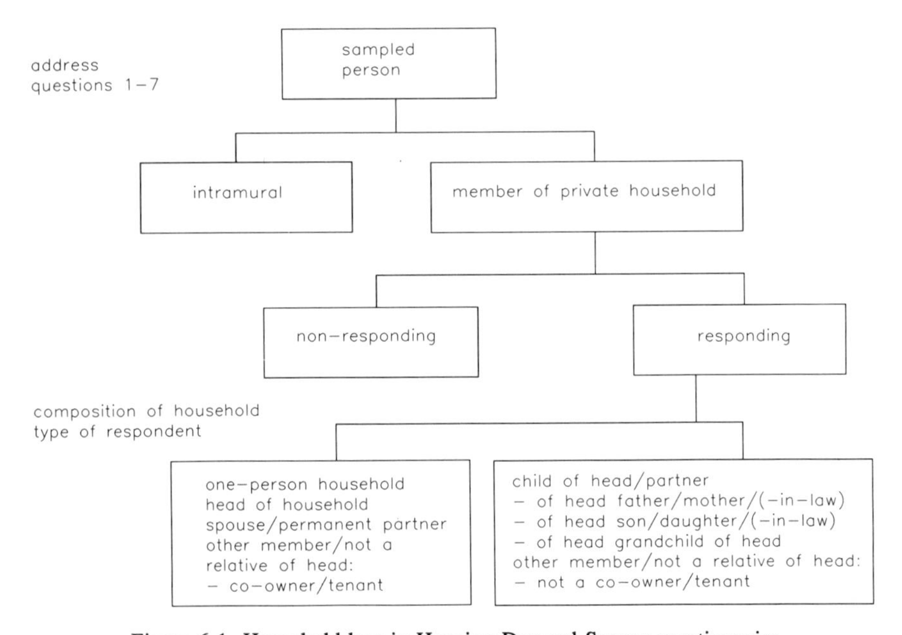

Figure 6.1. Household box in Housing Demand Survey questionnaire

[6] CAPI = Computer-Aided Personal Interviewing.
CATI = Computer-Aided Telephone Interviewing.

In Figure 6.1, the first set of questions (of the HDS) are directed in such a way to indicate the position of the respondent as either head of the household, partner of the head of the household or other positions. The first questions as provided in the household module of Figure 6.2, of the new NCBS approach are primarily meant to indicate the household *core*; *both* partners of the household core are treated equally.

6.6.2. The NESTOR-LSN Survey

In the NESTOR-LSN survey, respondents are interviewed about all kinds of demographic and socio-economic variables. Personality characteristics and values, attitudes, etc. are also included in the interview design. Asking questions about the composition of the interviewee's household is always ego-centred in this survey, in other words: the interviewee is the 'reference' person.

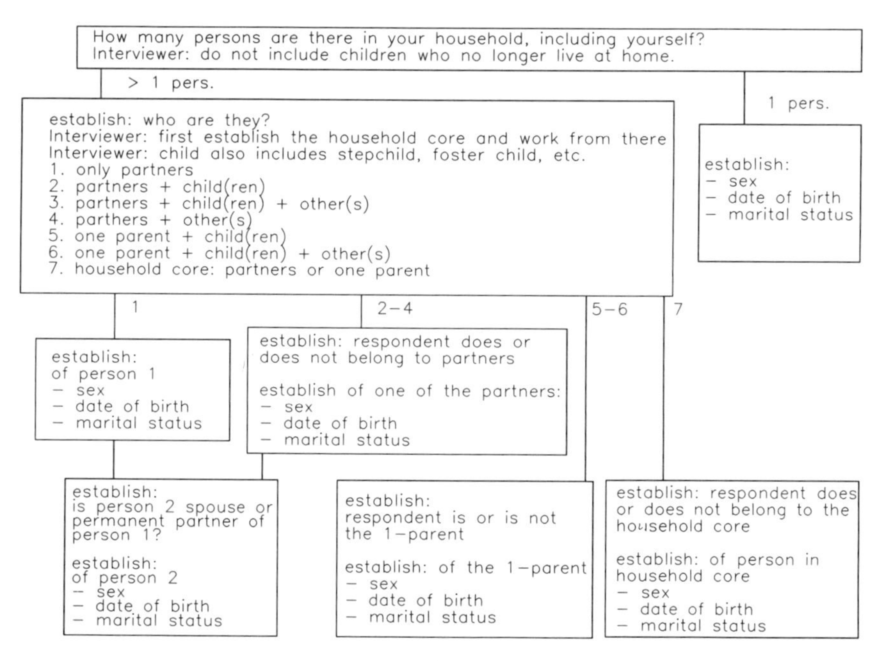

Figure 6.2. Standard household box of Statistics Netherlands

The module on the relationship between the respondent and — if available — other household members includes five questions on type of partner relationships with a total of six answer categories; 15 other relationship categories have been included.

We conclude that both surveys, the HDS as well as the NESTOR-LSN, provide an optimal opportunity to investigate the composition of the household. The NESTOR-LSN allows for more details as far as the compositions are concerned; the HDS allows a representative overview of Dutch households.

6.7. SOME SPECIAL CHARACTERISTICS OF THE TWO DUTCH HOUSEHOLD SURVEYS: ROUTING PROCEDURES

As mentioned before, the flexibility allowed by survey research designs provides the opportunity to use more complex routings in the questionnaire. In censuses, but also in mail surveys, the researchers must develop questionnaire layouts that preferably present each question in a fixed order, without going back to another page and without skipping questions because of being 'not applicable'. In the case of face-to-face and telephone interviewing, a more complicated routing is feasible, but only to a certain extent. In those designs where time is available to give intensive instruction to interviewers — mostly restricted to small-scale designs of scientists — the researchers can devise special modules and special sets of questions for separate subcategories of the sample. Those designs that concentrate on using laptop computers or CATI procedures have more opportunities for special routings than paper-and-pencil variants.

6.7.1. The Housing Demand Survey

In the HDS, nearly every page has some minor routing options; but there is one major routing procedure that discriminates between, on the one hand, the head of the household, the (married or unmarried) partner of the head, and on the other hand, all other household members. If the respondent belongs to the first subgroup, she/he has to answer a larger set of questions on the composition and other characteristics of the household; members of the second subgroup are confronted with fewer questions. For example, people in a one-person household, head of a household, and the (married or unmarried) partner of the head of the household will be interviewed fairly intensively to register the size and composition of the household. If the respondent is a child in the household, or a father/mother (-in-law) living in with the family, or a live-in domestic, the set of questions to be answered about the household is rather moderate; the size and composition of the household is not registered. Modules that must be answered by each respondent include questions on migration in the recent past, the intention to move within the next two years, and a description of the financial status of the respondent. The interviewer is needed to take care of the routing and other instructions. When constructing the layout of the questionnaire, special attention has been given to supporting the interviewer in finding the way through this paper-and-pencil undertaking.

6.7.2. The Nestor-LSN Survey

The NESTOR-LSN program board decided to use laptop computers in all stages of the data processing of the survey. In view of all the information that needed to be collected, the only way to accomplish the task was to use a computerized questionnaire. A properly functioning computerized questionnaire takes on the interviewer's tasks by routing the interview conversation; the program describes under which conditions and in which order the questions have to be asked. The system is able to handle complex data structures, while concealing the complexity from the user (Croft 1992). Examples of broadly-used systems for interviewing on microcomputers can be found in Bethlehem & Hundepool (1992) and in Pageau *et al.* (1992).

In addition to providing suitable routing, the computer program takes care of controlling the internal consistency of the given answers. An inconsistency is immediately shown on the computer screen; control and correction of the answers is realized within the realm of the interview setting. At the end of the interview, all the answers are coded and registered. Analysis can in principle start immediately after the end of the interview.

In summary, the smaller scale surveys have more opportunities to integrate complex routing procedures and specific modules in the questionnaire than larger scale surveys (or censuses). Face-to-face interviewing broadens the scope of specific surveys; the HDS is a good example of in-depth interviewing of specific subsamples of respondents. The extra possibilities of optimal questionnaire design in computer-assisted, face-to-face interviewing are illustrated by the NESTOR survey.

6.8. LONGITUDINAL DATA

The use of longitudinal datasets and retrospective questions in surveys have been mentioned above. Both surveys discussed here provide opportunities to describe and analyze dynamics in individual life courses, in one way or another. The following will briefly explain how various issues were handled.

6.8.1. The Housing Demand Survey

In the HDS, retrospective questions — mainly restricted to changes in the housing and in the composition of the household during the past year — are included in the questionnaire. For each member of the household, it is asked whether (s)he belonged to the household twelve months ago as well. In the case of a denial, further questions are asked. If the respondent already lived in the Netherlands, then more details are requested of the previous household situation. In addition, the respondent is asked about the date when a change took place in marital status. Finally, the respondent is asked when (s)he first went to live on his/her own and whether (s)he went to live alone/with a spouse/with a permanent partner, or with others.

6.8.2. The NESTOR-LSN Survey

The NESTOR survey pays particular attention to changes in personal life events over a longer period. All respondents are sent checklists every six months, whereby particularly the changes in household situation are recorded. If the change in the household situation is an important one from the point of view of the researchers, the respondents will be re-interviewed about the specific situations they are in now. For example, widowed persons — whose permanent partners deceased after the time of the first interview are re-interviewed several times about their adjustment to widowhood. Questions are asked about their network changes, their activities, and how they deal with their loss, etc.

6.9. A CASE STUDY OF HOUSEHOLD TYPES AS REGISTERED IN THE HDS AND NESTOR-LSN SURVEYS

In this section, we will try to describe the survey as a *method for observing new trends in household compositions*. As mentioned before, our goal is to identify infrequent, 'new' types of living arrangements as perhaps realized by an innovative group of elderly. We must keep in mind that the percentage of elderly who live according to these 'new', more individualized values and norms will perhaps be very low. The household type we are firstly interested in is cohabitation.

In general, social changes develop gradually. Society is confronted with new ideas and new behaviour in a process that develops in several stages. In the first stage, the so-called forerunners, precursors — that is a small group of open-minded people: 'the innovators' — start to change their behaviour. After some time, the new ideas and behaviour will be accepted by the majority of society: the moment that the original view on the behaviour changes from elite behaviour to rank-and-file behaviour. The new behaviour as 'Gesunkenes Kulturgut' reaches nearly everyone in society, except the small group of those who lag behind. It is in the third and final stage that those who lag behind — perhaps whether they like it or not — swing round (see Rogers & Shoemaker 1971; Rogers & Kincaid 1981; Kaufmann & Schmidt 1976).

As far as changes in social values and norms concerning the family and demographic behaviour in general are concerned, De Feijter (1991) and Lesthaeghe (1991) have proved that innovators are concentrated among younger, highly-educated individuals, those without religious bonds, and in general living in the larger urban centres. This innovative behaviour has been described for the start and introduction of consensual unions, and for the start of independently living alone among adults in the age group 18 to 55 years (De Feijter 1991; De Jong Gierveld *et al.* 1991).

Up until now, consensual unions of the elderly population have not been well recorded. We are interested in finding out whether these 'new' living arrangements can be found among the elderly. What percentage of the population aged 55 years and over live in a consensual union with a partner of the opposite sex? Additionally, we would like to know what percentage of the population aged 55 years and over realize a so-called LAT relationship, that is, the two partners in a LAT union live independently, and neither 'de facto' nor 'de jure' at the same address.

We also want to find out about the individual characteristics that are related to the realization of these living arrangements. The elderly cannot be viewed as a homogeneous group (Dooghe 1992). At least a differentiation in several age groups is necessary.

Besides innovative motives and intentions that could lead to the innovative behaviour, we have to include the restrictions and opportunities that the elderly are confronted with on their path to the realization of innovative behaviour. Objective restrictions and/or subjectively perceived restrictions could form a barrier to the realization of intentions. Among the (perceived) restrictions, the following can be recognized: the ideas/values and attitudes of significant others, e.g. children and neighbours that obstruct one's plans to start a consensual union or LAT relationship. So, we expect that due to differences in social control of neighbours and family members, the realization of cohabitation among the elderly will differ according to characteristics of the region one lives in. One factor favouring LAT relationships of the elderly is related to the rules of the social security system of the country that in general reduce old age or widows' pensions in case of remarriage or consensual unions.

This analysis will concentrate on that category of elderly who live in private households. Those elderly who live in institutions are excluded from this analysis.

Table 6.1 provides an overview of the population aged 55 years and over, according to their position in the household. The data source is the Housing Demand Survey 1989/1990 (CBS 1992).

Table 6.1. The elderly population of the Netherlands according to household types, in percentages (CBS, HDS survey: 1989/1990; n = 24.456; n minus married persons = 4.408). (Excluded: persons in institutions)

Age group	married persons	cohabitation	one-person household	one-parent household	parent in household of children	other	total minus married persons	total
55-64	79.2	2.2	14.6	3.2	0.2	0.5		100
65-74	67.0	2.1	26.9	2.3	0.9	0.8		100
75-84	45.5	1.4	45.8	2.9	2.4	2.1		100
55-64		10.4	70.4	15.4	1.2	2.6	100	
65-74		6.5	81.3	6.9	2.8	2.5	100	
75-84		2.6	83.8	5.3	4.3	3.8	100	

Table 6.2. The elderly population of the Netherlands according to household types, in percentages (NESTOR-LSN survey: 1992; n = 3.696; n minus married persons = 1.215). (Excluded: persons in institutions)

Age group	married persons	cohabitation	one-person household	one-parent household	other	total minus married persons	total
55-64	77.5	2.1	16.3	2.4	1.7		100
65-74	66.6	1.8	26.5	2.6	2.5		100
75-84	41.9	1.8	50.8	2.9	2.7		100
55-64		9.3	72.6	10.6	7.4	100	
65-74		5.3	79.5	7.7	7.4	100	
75-84		3.0	87.5	4.9	4.6	100	

Table 6.1 points out that living alone (in a one-person household) is the predominant living arrangement among unmarried people aged 55 and over. Cohabitation is also important: ranging from about ten per cent among the younger unmarried elderly towards 2.6 per cent among the older categories.

As can be seen in Table 6.2, the percentage of people registered as living in a consensual unit is about the same in the Dutch Nestor survey as in the HDS study. The same tendency is recognizable in both surveys.

Included in the NESTOR-LSN questionnaire is an additional set of questions intended to identify intimate relationships in the form of LAT relationships: Living Apart Together. In most regions of the Netherlands, LAT relationships have not yet been fully accepted. This will be especially true for the elderly population who have been brought up with more traditional ideas of family life and responsibilities. Thus, we expect that the elderly interviewee will in general be somewhat hesitant in disclosing, in 'confessing' their involvement in a LAT relationship. This will be the more so in case the elderly are living in rural areas characterized by more social control, e.g. in the eastern and southern regions, and in traditional or religious groups of society. Special attention has been paid to instructing the interviewers in an open, social interview style in order to ensure optimal openness of both persons involved in the interview.

In Figure 6.3, the living arrangements are differentiated so that LAT relationships are included in the overview. Figure 6.3 demonstrates that LAT relationships are indeed a recognizable phenomenon among elderly in the Netherlands. In the younger age groups, about ten per cent of the unmarried respondents mentioned a LAT relationship, but it is also known among the very old! In my opinion it is worthwhile to describe this starting trend and the characteristics of the innovators. In the near future we will concentrate on that, but a first view on this interesting development will be given now.

As mentioned before, it is expected that cohabitation and LAT relationships are not evenly distributed over the elderly population, but are concentrated in specific categories. The following significant relationships have been found in the NESTOR-LSN dataset:

- cohabitation and LAT relationships are more common among the younger age groups of the elderly than among the (very) old;
- cohabitation and LAT relationships are more common among those elderly living in Amsterdam and commuter areas than among elderly living in the eastern or southern regions of the Netherlands;
- cohabitation and LAT relationships are more common among those elderly who have no religious affiliation than among church members.

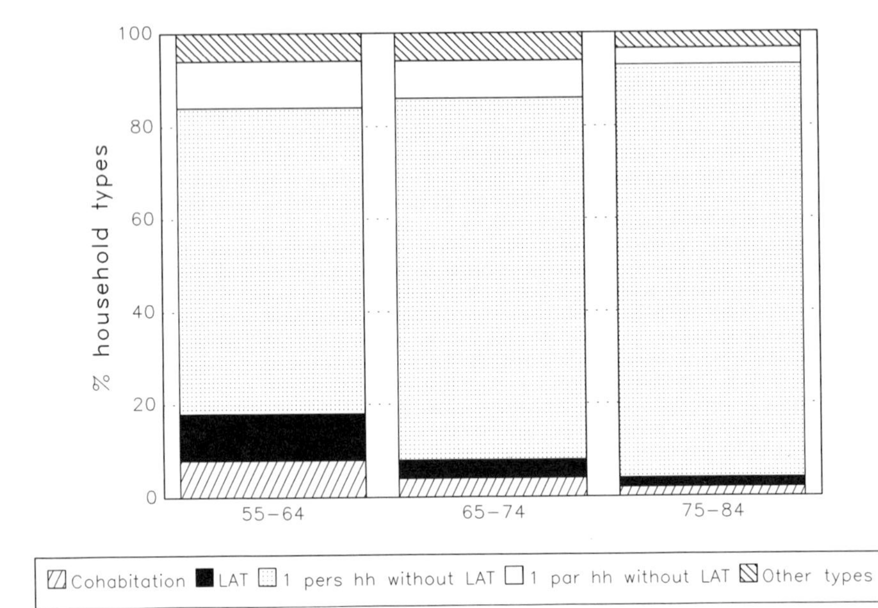

Figure 6.3. Household types/living arrangements of unmarried elderly NESTOR-LSN data 1992

A summary of these main findings is presented in Figure 6.4. We constructed some contrast groups:

- elderly, 55 to 64 years of age, in the western regions (Amsterdam and commuter area) characterized by absence of a religious affiliation, in contrast to
- elderly aged 65 years and over in the eastern and southern regions of the country, and characterized by church membership.

Figure 6.4 clearly points out sharp discrepancies in the distribution of 'new' types of households and living arrangements over subcategories of elderly. It is in line with ideas mentioned about 'innovators' for describing the situation of younger categories of elderly living in Amsterdam or surroundings and without church affiliation as the innovators among the elderly. About 13.5 per cent of this category is involved in unmarried cohabitation, another 13 per cent is characterized by a LAT relationship, together embracing 26.5 per cent of the younger elderly.

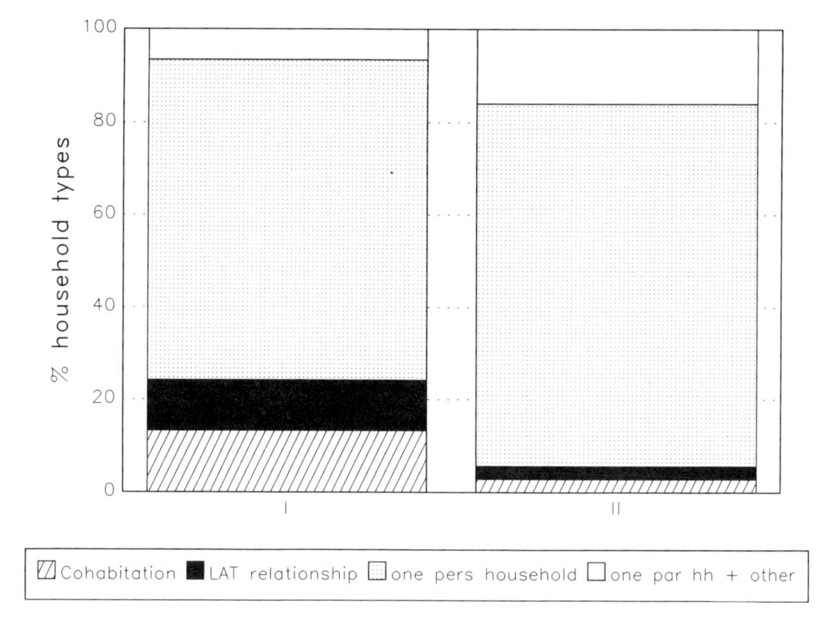

Figure 6.4. Household types/living arrangements of unmarried elderly NESTOR-LSN data 1992; contrast groups

I = Elderly 55-64 years of age, in Amsterdam (or commuter area, no church affiliation (n=86))
II = Elderly 65 years and over, in Eastern & Southern regions of the Netherlands, church members (n=367)

Out of the 65 and over category living in eastern and southern regions and characterized by church membership, only 2.4 per cent is cohabiting and 2.3 per cent is involved in a LAT relationship. We hypothesize that different norms and values and differences in social control from the surrounding society members will be of crucial importance in the explanation of this discrepancy.

6.10. SUMMARY

This chapter presents an overview of some methodological differences in the characteristics of empirical research into households. We differentiated between censuses and surveys (both large- and small-scale), their positive and negative points in investigating the numbers, composition, and changes of households. Attention was given to definitions of households, plus several options for questionnaire modules investigating the characteristics of households and household members. Furthermore, routing procedures and the possibilities of retrospective and prospective longitudinal research was investigated. In the second section, a comparison of two Dutch surveys that explicitly concentrate on households was put forward:

- the Dutch Housing Demand Survey, a large-scale survey, representative of the country, and
- the Dutch NESTOR Living Arrangements Survey, of smaller scale and based on three regional subsamples.

We illustrated the high quality of both Dutch surveys.

The HDS provides a large dataset, useful for all kind of researchers into households, allowing for regional and other break-downs into subcategorical studies. The NESTOR survey, with its 4500 respondents, is less well-equipped for regional and country level studies, but is characterized by a nuanced set of question modules that allow in-depth studies of several household and household connected themes.

ACKNOWLEDGEMENTS

The author would like to thank Erik Beekink for his assistance in preparing this chapter.

REFERENCES

BETHLEHEM, J.G., & A.J. HUNDEPOOL (1992), Integrated Statistical Information Processing on Microcomputers, Paper for the IUSSP/NIDI Expert Meeting on Demographic Software and Computing, NIDI, The Hague, 29 June-3 July 1992.

BURCH, T.K., & B.J. MATTHEWS (1987), Household Formation in Developed Societies, *Population and Development Review*, Vol. 13 (3), pp. 495-511.

CALDWELL, J.C., & A.G. HILL (1988), Recent developments using micro-approaches to demographic research. *In*: J.C. CALDWELL & V.J. HULL (eds.) *Micro-approaches to Demographic Research*, Kegan Paul International: London.

CLARKE, L. (1992), Children's Family Circumstances: Recent Trends in Great Britain, *European Journal of Population*, Vol. 8(4), pp. 309-340.

COURGEAU, D., & E. LELIÈVRE (1989), *Analyse Démographique des Biographies*, Institut National d'Etudes Démographiques: Paris.

CBS (1992), *Huishoudens 1989: Sociaal-Demografische Cijfers*, Resultaten uit het Woningbehoeftenonderzoek 1989/1990, CBS: Voorburg.

CROFT, T. (1992), The Integrated System for Survey Analysis (ISSA): Design Principles behind the Development of ISSA, Paper prepared for the IUSSP/NIDI Expert Meeting on Demographic Software and Computing, NIDI, The Hague, 29 June-3 July 1992.

DE FEIJTER, H. (1991), Voorlopers bij Demografische Veranderingen, NIDI Report 22, NIDI: The Hague.

DE JONG GIERVELD, J., A.C. LIEFBROER & E. BEEKINK (1991), The Effect of Parental Resources on Patterns of Leaving Home among Young Adults in the Netherlands, *European Sociological Review*, Vol. 7 (1), pp. 55-71.

DOOGHE, G. (1992), De Bevolkingsveroudering binnen de Europese Gemeenschap, *De Gids op Maatschappelijk Gebied*, Vol. 83(1), pp. 3-17.

DU BOIS-REYMOND, M. (1992), Pluraliseringstendensen en Onderhandelingsculturen in het Gezin, Paper for the Sociaal-Wetenschappelijke Studiedagen 1992, 'Culturen en Identiteiten', Amsterdam.

ERMISCH, J. (1988), An Economic Perspective on Household Modelling. *In*: N.W. KEILMAN, A. KUIJSTEN & A. VOSSEN (eds.) *Modelling Household Formation and Dissolution*, Clarendon Press: Oxford, pp. 23-40.

FREEDMAN, D., A. THORNTON, D. CAMBURN, D. ALWIN & L. YOUNG-DE MARCO (1988), The Life History Calendar: a Technique for Collecting Retrospective Data. *In*: C.C. CLOGG (ed.) *Sociological Methodology 1988*, American Sociological Association: Washington D.C.

HÖHN, C. (1987), The Family Life Cycle: Needed Extensions of the Concept. *In*: J. BONGAARTS, T.K. BURCH & K.W. WACHTER (eds.), *Family Demography: methods and their application*, Clarendon Press: Oxford, pp. 65-80.

KAUFMANN, K., & P. SCHMIDT (1976), Theoretische Integration der Hypothesen zur Erklärung der Diffusion von Innovationen durch Anwendung einer allgemeinen kognitiv-hedonistischen Verhaltenstheorie. *In*: P. SCHMIDT (ed.) *Innovation*, Hamburg, pp. 313-386.

KEILMAN, N. (1985), De Opzet van een Longitudinale Relatie- en Gezinsvormingsurvey, Internal Report 39, NIDI: The Hague.

KEILMAN, N.W. (1988), Dynamic Household Models. *In*: N.W. KEILMAN, A. KUIJSTEN & A. VOSSEN (eds.) *Modelling Household Formation and Dissolution*, Clarendon Press: Oxford, pp. 123-138.

KLIJZING, F. (1988), Household Data from Surveys Containing Information for Individuals. *In*: N.W. KEILMAN, A. KUIJSTEN & A. VOSSEN (eds.) *Modelling Household Formation and Dissolution*, Clarendon Press: Oxford, pp. 43-55.

LASLETT, P., & R. WALL (eds.) (1972), *Household and Family in Past Time*, Cambridge University Press: Cambridge.

LESTHAEGHE, R. (1991), The Second Demographic Transition in Western Countries: an Interpretation, IPD Working Paper, Free University of Brussels.

LINKE, W. (1991), Statistics on Households and Families in the Member Countries of the Council of Europe: Definitions, Methods and Sources, *INSEE Méthodes*, Vol. 8, pp. 117-124.

MCNETTING, R., R.R. WILK & E.J. ARNOULD (eds.) (1984), *Households: Comparative and Historical Studies of the Domestic Group*, University of California Press.

MEIJER, L. (1990), Studiesnelheid van Universitaire Studenten: een Empirisch Onderzoek naar de Invloed van Individuele en Sociale Factoren op de Studiesnelheid van Universitaire Studenten, Unpublished Master's Thesis, Faculteit Bestuurskunde, Universiteit Twente: Enschede.

PAGEAU, M., P. GERLAND & N. MCGIRR (1992), Data Processing Strategies: Hardware and Software for Data Entry Editing and Tabulating. Paper for the IUSSP/NIDI Expert Meeting on Demographic Software and Micro-computing, NIDI, The Hague, 29 June-3 July 1992.

ROGERS, E.E., & F.F. SHOEMAKER (1971), *Communication of Innovations*, New York.

ROGERS, E.E., & D.L. KINCAID (1981), *Communication Networks*, New York.

ROSSI, P.H., J.D. WRIGHT & A.B. ANDERSON (1983), Sample Surveys: History, Current Practice and Future Prospects. *In*: P.H. ROSSI, J.D. WRIGHT & A.B. ANDERSON (eds.) *Handbook of Survey Research*, Academic Press: London, pp. 1-20.

RYDER, N.B. (1985), Recent Developments in the Formal Demography of the Family. *In*: *International Population Conference, Florence, 1985*, Vol. 3, IUSSP: Liège, pp. 207-220.

SELLTIZ, C., M. JAHODA, M. DEUTSCH & S.W. COOK (1961), *Research Methods in Social Relations*, Holt Rinehart and Winston.

SIKKEL, D. (1985), Models for Memory Effects, *Journal of the American Statistical Association*, Vol. 80, pp. 835-841.

SUDMON, S., & N.M. BRADBURN (1982), *Asking Questions*, Jossey-Bass: San Francisco.

VAN IMHOFF, E., & N.W. KEILMAN (1991), *LIPRO 2.0: An Application of a Dynamic Demographic Projection Model to Household Structure in the Netherlands*, NIDI CBGS Publications 23, Swets & Zeitlinger: Amsterdam/Lisse.

WALL, R., J. ROBIN & P. LASLETT (1983), *Family Forms in Historic Europe*, Cambridge University Press: Cambridge.

WALL, R. (1989), Leaving Home and Living Alone: an Historical Perspective, *Population Studies*, Vol. 43, pp. 369-389.

WALL, R. (1991), European Family and Household Systems, *Historiens et Populations*, Société Belge de Démographie, pp. 617-636.

7. THE USE OF PANEL DATA IN THE ANALYSIS OF HOUSEHOLD STRUCTURES

Notburga Ott

Johann Wolfgang Goethe University
Department of Economics
Senckenberganlage 31
D-60054 Frankfurt am Main
Germany

Abstract. Longitudinal data are necessary for the analysis of household dynamics. In comparison to retrospective data, data from panel surveys have some advantages in this type of analysis. On the other hand, some specific problems arise with panel data. Both options and problems are illustrated in the German Socio-Economic Panel. To support different types of analysis with panel data, a special data management in a relational database with time tables is helpful. Finally, the 'fuzziness' of household definitions is shown with some selected examples.

7.1. INTRODUCTION

The development of household structure is an important issue for policy-makers as well as for private enterprises. The reason is that the household is usually seen as the decision unit which determines the welfare level for its members by deciding labour supply and consumption demand. Therefore, the analysis of household dynamics is important, particularly for forecasting the future household structure. Transition probabilities between different household types are often calculated on the basis of aggregate data.[1] However, such a procedure can only evaluate the trend for a limited number of attributes. For more complex dynamic household models, longitudinal data on the micro level are necessary.

---

[1] Some examples of such macro models are presented by Keilman (1988).

Household Demography and Household Modeling
Edited by E. van Imhoff *et al.*, Plenum Press, New York, 1995

This chapter deals with the use of *panel data* in such dynamic modelling. The differences between a panel and a retrospective design in data collection are illustrated in Section 7.2. Section 7.3 describes the design of the German Socio-Economic Panel, which allows a wide range of analyses of household dynamics. In Section 7.4, a type of data management is presented which supports different kinds of analysis. Some particular problems in the use of panel data in household analysis are discussed in Section 7.5 and 7.6. Finally, some results about the relationship of household dynamics and family events show the 'fuzziness' of the household concept.

7.2. COLLECTING LONGITUDINAL DATA: RETROSPECTIVE AND PANEL DESIGN

In principle, there are two ways to collect longitudinal data: by retrospective questions within a cross-sectional survey or by using a panel design. Collecting retrospective data is a quicker method to obtain longitudinal information, but it has an important disadvantage which implies considerable problems for many questions, especially for slow processes with infrequent events like demographic processes. As only those persons who live at the time of the interview can answer the retrospective questions, the sample is biased because of the survival process (Tuma & Hannan 1984, p. 129). If this survival process is not independent of the processes of interest, there is a selection bias in the sample of collected histories, even for a representative cross-sectional sample of the population. For example, if single persons die earlier than persons ever married, the probability of marriage cannot be estimated correctly from retrospective data without additional information on the survival process. This type of selection bias does not occur in a panel design, because the biographies which are observed from the beginning constitute an unbiased sample of the panel population. However, other sample selection problems arise in a panel design, due to attrition in the course of the panel, if patterns of nonresponse are not independent of the characteristics of interest. But this type of selection bias is easier to handle in the analysis (see Rendtel 1991).

Further problems arise due to the different kinds of information which can be collected by these two methods. In a retrospective design, the dates of events are collected, whereas a panel design gives the status information at the time of each interview. This is shown in Figure 7.1 with a hypothetical biography of a person, including schooling, employment, and family situation.

The information gathered with a retrospective design (symbol ○) are the characteristics at the time of interview and the dates of previous events. In Figure 7.1, at the time of interview, the respondent is employed, married, and has two children. This is the same information as collected in a simple cross-

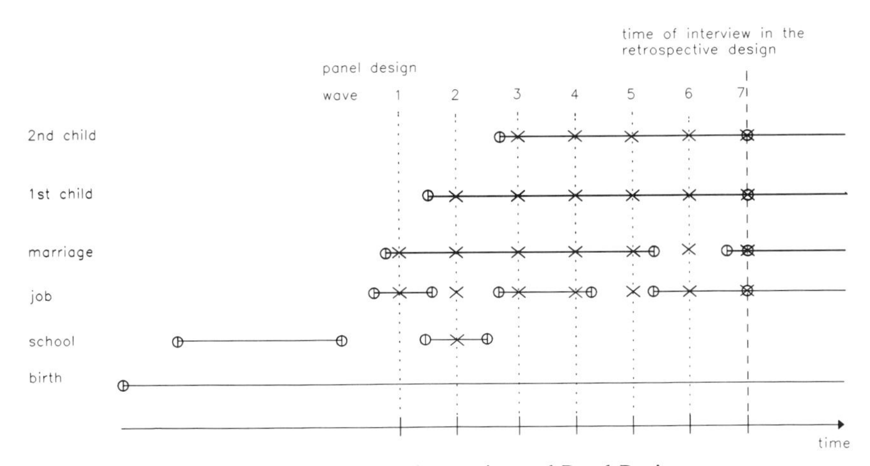

Figure 7.1. Retrospective Design and Panel Design

sectional survey. With the additional retrospective questions, the information is gathered on previous events, such as the dates of the start and end of all previous jobs and the characteristics of the job at these dates.

In a panel design, data are collected from the same person at different points in time. The data record contains the characteristics of the person at each interview (symbol ×). A change in one of these status variables indicates an event in the time interval between the two interviews.

Both kinds of information have typical advantages and disadvantages. Usually, dynamic processes are analyzed with respect to specific events. Therefore, exact dates of the events are essential information. On the other hand, data on characteristics at different points in time allow the analysis of slight changes in slow processes. However, there are also some problems with both designs, especially specific measurement problems. Because people do not always remember all information correctly, recall errors are a special problem in retrospective data (Bailar 1989, pp. 10-16). Panel data, on the other hand, may be biased by the so-called time-in-sample error, i.e. that the responses are not independent of the participation time of the respondent in the survey (Bailar 1989, pp. 16-21, Kalton *et al.* 1989, pp. 254-256).

Hence, to get complete life histories, both kinds of information seem to be necessary. Therefore, many studies combine both approaches by introducing retrospective questions in a panel design. This is usually done for the time before the first wave and for the time intervals between the interviews. Then the data record for each person contains not only the status variables at the time of the interviews, but also the exact dates of events.

7.3. THE GERMAN SOCIO-ECONOMIC PANEL

The Socio-Economic Panel (SOEP) is an example of this type of combined panel-retrospective survey. It is a representative longitudinal survey of private households in Germany introduced in 1984 (Wagner *et al.* 1991). The initial sample contains about 4,500 German households and about 1,500 households with a foreign head. In 1990, a third subsample of about 2,000 households from the former GDR was added. Similar to other household panel studies, the data are gathered annually, both at the household level and at the level of persons living in the household. In each household, all adult household members aged 16 years and over are interviewed. Information about the younger household members and the household as a whole is collected from the 'head of the household'. The questionnaire covers the topics income, work and education, time use, health, personal satisfaction, biography, and family.

7.3.1. The Follow-Up Concept

A 'household' is not a stable, well-defined unit in the course of time because persons move in and out. Therefore, the follow-up concept must deal with the household members. The SOEP follows all persons of the interviewed households, independent of their household membership in the next year. If a member moves out to start a new household or joins an already existing household which was not previously included in the panel, the corresponding household becomes a new panel household and all persons living in it become new respondents. The same applies to all persons who move into a household.

Figure 7.2 shows this follow-up concept for a hypothetical household. At the time of the first interview, the household contains a married couple with two children. In wave 2, the second child has left and lives in a one-person household. Before the third wave, this child has started a consensual union with another person, who becomes an additional respondent. Between wave 3 and 4, the child still living in the parental household also begins a partnership, and the partner moves into the household as a new respondent. In the next wave, the partnership of the second child has been dissolved and both live in their own household. The parents have divorced in the meantime and the husband has left the household. Before wave 6, he has found a new partner and lives in a consensual union. During the same time, the first child marries and the second child returns to the mother's household. During the 6 waves, four households have originated from the initial one, and three new respondents are in the panel. The initial household, a family (married couple with children), has changed to a 'one-parent family with a non-relative' and finally to a 'couple with relatives'. During this time, the formation of three new households and the dissolution of one household are observed. Therefore, all types of household dynamics

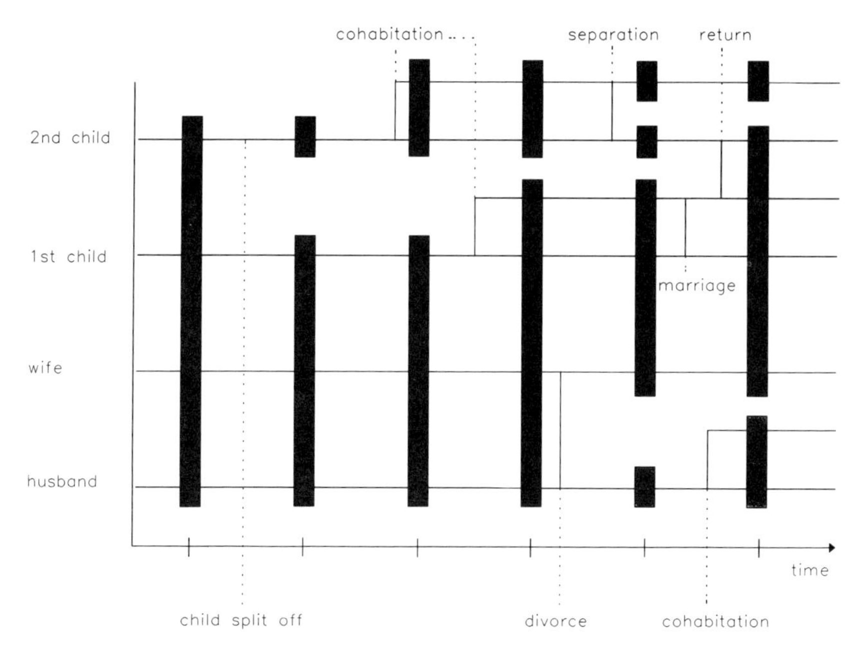

Figure 7.2. The follow-up Concept

— changes in household composition, changes in the relationship between household members, household formation, and dissolution — are covered by this follow-up concept.

7.3.2. Demographic Information in the SOEP

A focal point of the SOEP is providing data for the analysis of demographic development (Hanefeld 1987, pp. 39ff). Therefore, demographic information is gathered in detail (see Table 7.1). There are three types of information. At each interview, the head of the household gives information on household composition in terms of the relations of all household members to the head. In addition, the demographic status variables of each household member are collected at each interview. The third type of information contains retrospective data on events since the beginning of the last year. If, for instance, the interview takes place in the middle of 1986, all demographic events since January 1 of 1985 are to be reported. This design guarantees that the complete time interval since the last interview is included.

Normally unvariable characteristics, like gender, year of birth, and nationality, are collected in each wave and are used in controlling longitudinal

Table 7.1. Demographic Information in the Socio-Economic Panel

Person		
- gender		
- date of birth		
- nationality		
- marital status:	- marital status at each interview	
	- events between the waves	
	- all marriages before 1984	
- unmarried couples:	- status at each interview	
	- events between the waves	
	- year of starting a common household	
- age and gender of child	- for all children in the household	
	- for women	- date of birth of all children
		- year of leaving home
- second household		
- childhood:	- residential surrounding	
	- time of leaving the parental household	
- for foreigners:	- native country	
	- year of immigration	
	- residence of the spouse living abroad	
	- number and age of the children living abroad	
Household		
- number of household members		
- moves in and out between two waves		
- relationship to the head of household		
- number, age, and gender of children younger than 16 years		
- change of residence between two waves		

links. All other variables vary over time and describe changes in the personal status. At the same time, they describe developments at the household and family level as well, because events in the family biography of one person usually correspond with those of other members of the same household.

7.4. DATA MANAGEMENT

What would be a proper way to organize this type of longitudinal data in order to support different types of analysis? In the SOEP, data are collected on both the household and person level. Usual ways of combining such information are either to add the household information to each household member (then household information is seen as a personal characteristic), or to put all information of the members on the household record. Both procedures only support one kind of analysis. In the first case it is easy to analyze persons, but difficult to analyze households when the information of all household members is needed. In the other case the opposite is true.

A better way is to store the data separately in a relational data base.[2] In such a data base, different tables are set up for the personal and the household information. The columns in the table are defined by the variables, and the information of one case is stored in a row. The information of different tables is combined by key variables. In the case of different tables for persons and for households, two key variables are necessary for each wave (see Figure 7.3). An identifier for each person is required in order to find and link the data of this person in different waves. In addition, an identifier is required for each household to link all members to the household and to each other. It is now easy to link all information of households and other household members to the data of a person, even if he has changed the household. For example, at the first interview, in Figure 7.3, the person with ID-number 1903 lives in household number 19 with two other persons. All information of the first wave can be linked by the household ID. At the second interview, the person has left the household and lives in a new household with two other persons who are new respondents in the panel. To link the information of wave 2, the new household ID must be used, whereas the information over time can be found through the person ID. In addition, pointer variables for partners, as well as for children and their parents, are helpful in combining the information of couples and families.

Such a data base structure supports every kind of analysis which refers to the status variables at the time of the interviews. However, the second type of information which is gathered in the SOEP, the dates of various events which occur between the interviews, cannot be managed well within this concept. If the data of the events are stored in the different tables for the different waves, a very complicated retrieval system is needed to recover the information of interest. One example is the duration of a marriage. Supposing the person marries between wave 1 and 2, and is divorced between wave 5 and 6. If the date of the marriage is stored in the person table of wave 2, and the date of divorce in that of wave 6, a search through the person tables of all waves is necessary to find the events of interest. Therefore, for the data management of events, it is much better to choose a concept which is independent of the waves, and to store the information in an additional time table. This can be done with a table referring to events or referring to spells.

A spell table contains the starting time, the ending time, and the final status for each episode (e.g. for each marriage). This concept is appropriate for well-defined episodes with a small number of destination states, for example, spells

---

[2] Solenberger *et al.* (1989) discuss the organization of household panel data in a hierarchical structure, and David (1989) points out the advantages of a relational data base.

WAVE 1

Person Table '84

Person ID	Household ID	other Variables
1901	**19**	....
1902	**19**	....
1903	**19**	....
2701	27	....
2702	27	....
2703	27	....
39601	396	....
39602	396	....
39603	396	....

Household Table '84

Household ID	other Variables
19	....
27	....
396	....

WAVE 2

Person Table '85

Person ID	Household ID	other Variables
1901	19	....
1902	19	....
1903	***6013***	....
601302	***6013***	....
601303	***6013***	....
2701	27	....
2702	27	....
2703	6225	....
39601	390	....
39602	390	....
39603	390	....

Household Table '85

Household ID	other Variables
19	....
6013	....
27	....
6225	....
396	....

Figure 7.3. Data Tables in a Relational Data Base

of unemployment. In demographic analyses, slightly different definitions for similar phenomena are often used. For example, in analyses of partnerships, episodes starting with a marriage or with the formation of a common household may be regarded. Then a concept with spell tables would contain redundant information. For such cases, event tables seem to be more appropriate. All information on dates of events, as well as on the states at each interview, are stored in a table as shown in Table 7.2. New information gathered in the next wave is added to the same table.

Table 7.2. Event (and Status) Table for Biographical Data

Person ID	Year	Month	Day	Event/Status at time of interview*
211	1926	-	-	year of birth, woman
211	1953	-	-	1st married
211	1956	-	-	1st child born, daughter
211	1960	-	-	2nd child born, son
211	1965	-	-	divorce
211	1984	3	-	other
211	1984	4	9	* divorced, interview '84
211	1984	12	-	cohabitation
211	1986	4	2-	* divorced, interview '86
211	1986	11	-	separation
211	1987	3	13	* divorced, interview '87
1242	1961	-	-	year of birth, woman
1242	1984	-	-	1st child born, son
1242	1984	3	-	birth
1242	1984	4	6	* single '84
1242	1984	4	6	* cohabitation '84
1242	1984	11	-	marriage
1242	1985	3	15	* married, interview '85
1242	1986	5	23	* married, interview '86
1242	1986	12	-	birth
1242	1987	3	17	married, interview '87
3487	1911	-	-	year of birth, woman
3478	1931	-	-	1st marriage
3478	1934	-	-	1st child born, daugther
3478	1938	-	-	2nd child born, son
3478	1942	-	-	2nd child leaves home
3478	1942	-	-	3rd child born, daugther
3478	1947	-	-	4th child born, son
3478	1958	-	-	1st child leaves home
3478	1968	-	-	3rd child leaves home
3478	1972	-	-	end of 1st marriage by death
3478	1975	-	-	4th child leaves home
3478	1984	6	7	* widowed, interview '84
3478	1986	4	15	* widowed, interview '86
3478	1987	3	2	* widowed, interview '87

With such a concept, it is very simple to find out the sequence of events of interest during the course of time, which is an essential requirement for longitudinal analysis. Missing events (e.g. a marriage between the events cohabitation and divorce) or inconsistent dates can also be discovered very easily. A similar time table can be generated for household processes as well.

However, episodes on the household level are always delimited by events on the person level. Therefore, one only needs to link the data records of the involved persons, for instance, those of spouses and their children. To manage this properly, pointer variables are necessary between spouses or between parents and their children. Then the dates of different household members can be compared and matched in a simple way. This type of data management can handle different types of household analysis.

7.5. OPTIONS AND PROBLEMS WITH PANEL DATA IN HOUSEHOLD ANALYSIS

The information for each household in a follow-up design as described provides many options for demographic analyses. Besides individual biographies, changes in household composition like moves into and out of the household, changes in internal relationship, and the formation and dissolution of households can be analyzed.[3] Furthermore, family subgroups can be followed simultaneously, even if they live in different households. Examples are the follow-up of both partners after a divorce in order to compare their economic situation, or adult children leaving the parental household. The relation of demographic to other socio-economic processes can be analyzed for all of these questions. Topics of interest here may be the economic situation and time allocation in the household during the course of a marriage, economic changes after the birth of a child, or the duration of consensual unions, or the transition to marriage depending on the economic situation.[4]

Although, in principle, many questions can be analyzed with this panel design, not enough events have been observed for some of them until now. This points to a large problem in panel surveys. For slow processes with few events, a long observation time is necessary for sufficiently long histories. Demographic events are typically infrequent: less than 100 cases per wave are observed in the SOEP for each of the demographic events. This leads to a high proportion of incomplete spells in the sense that either the beginning (left censoring) or the end (right censoring) is not observed.

---

[3] Examples of such analyses are Pischner & Witte (1988) for biographies, Witte (1987 and 1988) for changes in household composition, Giesecke (1989) for divorces, Ott (1990) for consensual unions, and Witte & Lahmann (1988a and 1988b) for formation, dissolution, and mobility of one-person households.

[4] See for example Berntsen (1989) for changes in income after family events, Ott (1992a and 1992b) for the dependence of divorce rate and employment biography, Blossfeld & Jaenichen (1990), Frick & Steinhöfel (1991) and Witte (1990 and 1991) for marriage behaviour, and Wagner (1991) for exit of couples from the labour market.

Whereas right censoring is handled properly in the usual methods of event history analysis, dealing with left censoring is much more difficult (see Cox & Oakes 1984, p. 177). Often this problem is ignored in the literature, because it is assumed that the starting point of each history of interest is known. This is the case with process-produced data, such as medical case histories. In social science surveys, the origins of all actual episodes are often gathered retrospectively. However, this procedure yields another problem known as the 'waiting time paradox' or 'length-biased sampling' (see Cox & Isham 1980, pp. 7-8; Rao 1984).

7.5.1. Length-Biased Sampling

As shown in Figure 7.4, at the time of interview, episodes with longer duration are more likely to be in progress. Although, in each period, the number of short episodes starting is twice that of long episodes, at the time of interview, more long episodes are in progress. The problem can be illustrated by the duration of consensual unions (Table 7.3). During the first seven waves, the partners of 663 consensual unions participated in the SOEP. One-third of them (24 + 8.9 per cent) had already lived together before the start of the panel, i.e. before January 1, 1983. For most of them the starting time is unknown, because this information was only gathered from the third wave onward. For those with a known starting time, the average duration is 7.8 years.

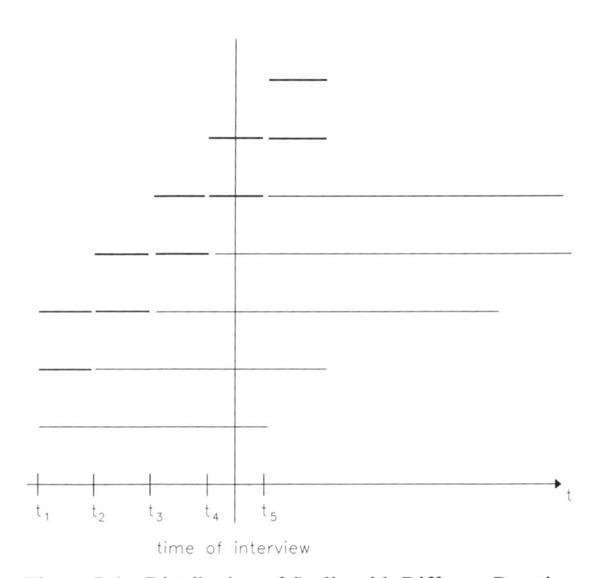

Figure 7.4. Distribution of Spells with Different Duration

Table 7.3. Start, End, and Duration of Consensual Unions

Origin		end			average	
		censored	separation	marriage	duration	total
before 1.1.1983						
	unknown	18	41	100		159
		11.3%	*25.8%*	*62.9%*		*24.0%*
	known	9	11	39	7.77	59
		15.3%	*18.6%*	*66.1%*		*8.9%*
after 1.1.1983		230	92	123	1.79	445
		51.7%	*20.7%*	*27.6%*		*67.1%*
total		257	144	262		663
		38.8%	*21.7%*	*39.5%*		*100.0%*

On the other hand, consensual unions which start during the observation period have an average duration of 1.8 years. But the proportion of censored data is very high here (51 per cent), i.e. where no end time is known until now. Therefore, the true average duration is certainly higher, but probably lies between the two values. Other consequences of that bias can be seen in the different distribution of the destination state. In all subgroups about the same proportion (circa 20 per cent) of separations is observed during the seven waves, but there are large differences in the proportion of marriages. This indicates different time dependencies for the two possible outcomes. Therefore the length bias cannot be ignored.

For adequately dealing with this problem, additional information is required on the distribution of the episodes during the time before the first interview (see Tuma & Hannan 1984, pp. 128-132; Cox & Oakes 1984, pp. 177-179). Otherwise only those episodes which are completely observed during the panel form a representative sample. In this case, long observation periods are necessary for reliable longitudinal analysis.

7.5.2. Selection by Increasing Non-Response

Another special problem in panel surveys is the increasing non-response rate in the course of time. Besides the initial non-response at the first wave, like in cross-sectional surveys, additional non-response occurs in all subsequent waves. Besides the reduction in sample size, there is usually a selection bias because this non-response is often dependent on processes of interest (Kalton *et al.* 1984, pp. 250-254; Löwenbein & Rendtel 1991). Attrition is much higher in split-off households and other households with changes in the composition

(Wagner *et al.* 1992). Particular problems arise for analyses of such phenomena that are only possible with a panel design like that of the SOEP.

This is illustrated by the follow-up of divorced or separated couples in Table 7.4. In each wave in the SOEP, there are about 35 to 40 divorces or separations, resulting in a split-off of one partner. In only about half of these cases can both partners (who now live in different households) be interviewed in the next wave. In addition, in about ten per cent of the cases, both partners are willing to participate again after a temporary attrition (i.e. a non-response in only one wave). In about 40 per cent of the cases, one partner refuses any further interviews, and in some single cases both partners refuse. Thus, out of the 225 cases observed in 7 waves, only 135 can be used for an analysis of the situation of both partners. Therefore, this type of analysis is not possible at present and many more waves are necessary.

Whereas some of the selectivity problems can be properly dealt with by using adequate weights (see Rendtel 1990), or by considering the selection process in estimating the model (Löwenbein & Rendtel 1991; Little & Rubin 1987), the reduction in the number of observed cases makes some types of analysis impossible at the moment. Nevertheless, a wide range of other questions can be investigated which gives some insights into household dynamics.

7.6. CHANGES IN HOUSEHOLD COMPOSITION AND FAMILY EVENTS

In order to give an impression of the extent of changes in the household structure, the frequencies of different types of changes are given in Table 7.5. In about 87 per cent of all households, no change in household composition

Table 7.4. Follow-up of Divorced/Separated Persons

	wave	wave	wave	wave	wave	wave	total
n	33	47	40	43	31	33	225
	percentage						
both partners interviewed	0.52	0.55	0.45	0.47	0.45	0.58	0.51
temporary attrition	0.09	0.09	0.08	0.09	0.10	0.00	0.08
one partner never interviewed	0.39	0.36	0.43	0.44	0.45	0.42	0.42
both partners never interviewed	0.00	0.00	0.05	0.00	0.00	0.00	0.00

occurs between two interviews. Additional members move into the household in about five per cent of the cases, and in about another five per cent, some household member splits off. Death occurs in one to two per cent, and less than one per cent of the households are dissolved. In about one per cent there is no information on household composition.

It can now be expected that most of the moves into and out of households result from an event in the family biography of one or more household members. Those events are reported by each household member for the period between two interviews. Table 7.5 shows the frequencies of family events reported by at least one household member for households with and without changes.

Children are born in nearly 40 per cent of the households with additional members, and in about 30 per cent a new partner has moved in. In about 20 per cent no family event is reported. In more than half of these cases a child has returned to the parental household. But in 18 per cent a new partner lives in the household with no report of this event. This cannot be explained solely by measurement and recall errors, but indicates differences in the subjective opinion of 'living together' to the usual household concepts.

In households with split-offs, the most frequent event is that a child has left home (about 40 per cent). A divorce or separation is mentioned in about 13 per cent. The number of other events is so small that it can be neglected. There is a strikingly high proportion of households where no event is reported. Many of these are non-response households.[5] Whereas about seven per cent of the households without changes or with additional members fall in this group, the non-response rate for households with split-offs is more than twice as high,[6] which yields the selectivity problems discussed in the previous section. Nevertheless, even without these cases, there is a large proportion of unexplained split-offs. This is surprising because nearly 90 per cent of all households are one-person households, couple households, or two-generation households. Therefore, in most cases, a partner or a child must have left without corresponding reports.

---

[5] Here partial non-response means that some but not all household members have refused an interview, which happens in only a few cases. If, in such a household, an event is reported by one of the members, the household is not put into this category but into that of the reported event.

[6] The non-response rate of the unweighted data is about ten per cent for households without change and about 23 per cent for split-off households.

Table 7.5. Changes in Household Composition and Family Events

		wave 2	wave 3	wave 4	wave 5	wave 6	wave 7
number of households in previous wave	n	5921	5619	5397	5266	5006	4863
no change in household composition	n	5069	4781	4546	4424	4256	4189
	%	*87.63*	*87.16*	*88.10*	*86.81*	*87.62*	*87.85*
events				percentages			
marriage/cohabitation		0.97	1.14	0.68	0.57	1.04	0.80
divorce/separation		0.63	0.46	0.60	0.29	0.69	0.73
children left home		0.27	0.50	0.33	0.26	0.23	0.41
birth		0.22	0.06	0.03	0.01	0.09	0.02
no event		87.13	87.64	92.47	90.86	90.26	90.29
(partial) non-response		10.77	10.23	5.99	8.05	7.77	7.72
households with additional members	n	319	336	340	322	340	322
	%	*4.56*	*4.98*	*5.49*	*5.20*	*5.30*	*5.47*
events				percentages			
marriage/cohabitation		24.08	34.87	34.57	37.52	29.98	40.83
divorce/separation		5.06	2.47	3.49	1.74	4.46	4.27
children left home		2.54	5.28	1.31	0.87	1.02	3.09
birth		34.83	39.06	34.22	44.09	38.84	34.67
no event		33.99	16.13	27.91	11.16	22.75	14.01
(partial) non-response		5.34	5.20	3.82	4.06	8.26	8.90
split off households	n	331	356	309	320	306	293
	%	*4.17*	*4.97*	*4.09*	*4.82*	*4.73*	*4.60*
events				percentages			
marriage/cohabitation		5.11	6.87	2.95	2.73	3.03	3.81
divorce/separation		13.38	12.85	14.51	13.63	16.64	11.03
children left home		43.20	39.33	39.01	40.41	38.11	42.32
birth		0.11	1.90	1.16	1.04	1.02	0.87
no event		25.99	27.38	28.65	27.54	27.44	29.87
(partial) non-response		14.60	16.24	17.07	16.20	15.81	16.71
households with deaths	n	74	94	85	95	87	76
	%	*1.64*	*2.33*	*2.03*	*2.83*	*2.41*	*2.37*
dissolved households	n	23	36	34	38	43	46
	%	*0.67*	*1.11*	*0.93*	*1.22*	*1.20*	*1.54*
no information on household composition	n	158	56	56	48	45	0
	%	*2.42*	*0.98*	*0.50*	*0.52*	*0.23*	*0.00*

Percentages are calculated based on weighted data.
Because more than one event can occur between two waves, the percentages can add up to more than 100.

7.7. FUZZINESS OF HOUSEHOLD MEMBERSHIP

This obvious imprecision in reported data indicates a 'fuzziness' of household membership. Using the data of the SOEP, the problem can be analyzed to a certain degree, because information on household composition is obtained from two different questions. In each wave, one household member is interviewed, usually the head. In addition, all household members report family events in the person questionnaire, including events like 'leaving home of children' or 'cohabitation with a partner'. Because the questions are in a different context and point to various aspects of household membership, the respondents may give seemingly inconsistent answers. This problem will be considered for the following two examples: the children leaving home and the cohabitation of non-married couples. But similar phenomena could arise in the households of the elderly if children or other persons who do not live in the same household take over some activities in the household.

7.7.1. Children Leaving Home

Table 7.6 shows two different approaches to this problem. During the period of seven waves, 1750 split-offs of children have been observed by the interviewer questioning the head. On the other hand, in 2215 households, one of the parents report the event 'a child has left home'. This is already a large difference. However, looking at conformity on the micro level, the differences are much larger. Overall, in only 46 per cent of the households where a child has left home has the event been reported by the parents. In the case of a temporary absence of the child, for instance, caused by an extended journey or military service, only seven per cent of the parents report an event. This indicates that parents consider 'leaving home' as the final event. But even when the child lives in a new household, only about two-thirds of the parents report this event.

On the other hand, a corresponding change in household composition is observed in only 32 per cent of all households where parents mention that a child has left home. In all other cases, all children who lived in the household in the previous wave are also household members in the present wave. There may be a short, temporary absence of the child between the two waves which is not discovered due to the design of the survey. But the high proportion of households with no child in the previous wave indicates that a child's split-off is often not regarded as 'leaving home' until much later.

Besides psychological factors, the reason for these differences is a certain ambiguity or fuzziness of 'household membership'. Usually 'leaving home' is not an event which happens at a certain date by a change into a new household, but a process over a longer period of time. During this process, some of the

Table 7.6. Child Split Off and Parental Responses

a) children left home (number of children)

child has left home	child has left home is mentioned	child has left home is not mentioned	total
total	799	951	1750
	46%	*54%*	*100%*
temporary absence	16	211	227
	7%	*93%*	*100%*
removal	783	740	1523
	51%	*49%*	*100%*
new household	662	373	1035
	64%	*36%*	*100%*
other / not answered	121	367	488
	24%	*76%*	*100%*

b) child has left home — response by the parents (number of households)

total	2215	*100%*
at least one child has left home	722	32%
no child has left home	639	*29%*
no child in household in previous wave	854	*39%*

household tasks are carried out in the new and others in the parental household, for example, washing the clothes of children studying away from home.

7.7.2. Consensual Unions

Similar problems in the identification of household membership are seen in the case of consensual unions. Here some household tasks are also done by persons who do not live together. Even living-apart-together couples have common areas besides their partnership, which usually count as common household tasks. Especially in these cases, the respondent's meaning of a 'household' is not clear.

These questions can also be analyzed with the SOEP data. All persons are asked whether they live with a partner in a consensual union, and, beginning with the third wave, how long they share a dwelling or whether they have separate dwellings. Nearly five per cent of all persons living in a consensual union[7] state that their partner does not live in the same dwelling, whereas in about ten per cent the interviewer does not find a partner in the same household (see Table 7.7). This indicates differences in the respondents' opinions on 'household membership', 'living together', and 'living in a common dwelling'.

Besides couples who do not want to reside together in the long run, in many cases these differences may result from the fact that forming a common household is also a process which takes a certain amount of time. During this process, some household tasks are carried out together and some in different households.

This also causes an additional problem. Because for a certain time interval the household is not yet clearly defined, the partners may have a divergent opinion of their household membership. Especially the beginning of a partnership may be seen differently. In 17 per cent of all consensual unions, the partner reported different dates for starting to live in a common dwelling.

These examples clearly illustrate the significance of fuzziness of household membership. This means that an exact dating of events is often impossible. In the case of consensual unions, the different dates of both partners may delimit the time interval in which cohabitation takes place. If this interval is short, the problem is similar to those with other imprecise dates, like in the case of measurement errors or grouped data. Here the usual methods of event history analysis can be used without causing any great problems (see Tuma & Hannan 1984, pp. 145-151; Heitjan 1989; Petersen & Koput 1992). If the interval for such a process is very long, like in the case of children leaving home, or if the

Table 7.7. Persons Living in Consensual Unions

	total	partner does not live in the household (reported by the interviewer)		same dwelling (reported by the person)	
wave 1	396	44	11,1%	-	-
wave 3	500	87	17,4%	7	1,4%
wave 4	499	68	13,6%	10	2,0%
wave 5	463	47	10,2%	22	4,8%
wave 6	519	54	10,4%	17	3,3%
wave 7	552	67	12,1%	21	3,8%

[7] This proportion has increased in the course of the panel, which may be traced back to a 'panel effect'. It could be supposed that, in the beginning, some respondents denied having separate apartments because of their opinions of the interviewer's expectations. Thus, these figures underestimate the true extent of 'living-apart-together' arrangements.

fuzzy household membership is a durable state like in the case of couples living-apart-together, fuzziness must be explicitly included in the model. For event history analysis, such models have been developed by, among other, Manton *et al.* (1991). But, until now, no general concept has been developed for regarding fuzziness in household modelling.

7.8. CONCLUSIONS

Using panel data in the analysis of household dynamics has some advantages with respect to methodological problems, as well as for substantive questions. In comparison to a cross-sectional survey with retrospective questions, a panel design provides an unbiased sample of biographies with complete life histories. A follow-up concept like that of the German Socio-Economic Panel allows for a wide range of longitudinal analyses of household and family processes. One can analyze changes in household composition, the progress of certain phases in the family life cycle, as well as the development of subgroups, even if they do not live in the same household over the total observation period. Nevertheless, some specific problems arise with panel data. Attrition leads to a particular selection bias and reduces the sample size. And especially for demographic processes, usually very long observation periods are necessary for a reliable longitudinal analysis.

In the analysis of household changes, panel data reveal effects which cannot be detected in cross-sectional data. Often changes in household composition between two waves do not match the reported events. This apparent inconsistency indicates a 'fuzziness' of household membership. Changes in household composition are not always an event at a fixed date, but a process which often needs a long period of time. During such processes not all household tasks are carried out by the same group, and some persons are only a 'household member' to a certain degree in different households. In this case, different household definitions may be useful for different problems.

REFERENCES

BAILAR, B.A. (1989), Information Needs, Surveys, and Measurement Errors. *In*: KASPRZYK, DUNCAN, KALTON & SINGH, pp. 1-24.

BERNTSEN, R. (1989), Einkommensveränderung aufgrund familialer Ereignisse. *In*: G. WAGNER, N. OTT & H.-J. HOFFMANN-NOWOTNY (eds.) *Familienbildung und Erwerbstätigkeit im demographischen Wandel,* Springer: Berlin, pp. 76-93.

BLOSSFELD, H.-P., & U. JAENICHEN (1990), Bildungsexpansion und Familienbildung — Wie wirkt sich die Höherqualifikation der Frauen auf ihre Neigung zu heiraten und Kinder bekommen aus? *Soziale Welt*, Vol. 41(4), pp. 454-476.

COX, D.R. &, V. ISHAM (1980), *Point Processes*, Chapman and Hall: London.

COX, D.R. &, D. OAKES (1984), *Analysis of Survival Data*, Chapman and Hall: London.

DAVID, M. (1989), Managing Panel Data for Scientific Analysis: The Role of Relational Data Base Management Systems. *In*: KASPRZYK, DUNCAN, KALTON & SINGH, pp. 226-241.

FRICK, J., & M. STEINHÖFEL (1991), Heiratsverhalten in der DDR und in der Bundesrepublik Deutschland — Der Zusammenhang von Heiratsalter und beruflichem Bildungsabschluß von Ehepartnern. *In*: PROJEKTGRUPPE "DAS SOZIO-ÖKONOMISCHE PANEL" (ed.), *Lebenslagen im Wandel — Basisdaten und -analysen zur Entwicklung in den Neuen Bundesländern,* Campus: Frankfurt, pp. 280-298.

GIESECKE, D. (1989), Die Auflösung von Familien und das Wohlfahrtsniveau geschiedener Frauen. *In*: W. PETERS (ed.) *Frauenerwerbstätigkeit — Berichte aus der laufenden Forschung,* SAMF Working Paper No. 1989-7, Paderborn, pp. 97-111.

HANEFELD, U. (1987), *Das Sozio-ökonomische Panel. Grundlagen und Konzeptionen*, Campus: Frankfurt.

HEITJAN, D.F. (1989), Inference from Grouped Continuous Data: A Review, *Statistical Science*, Vol. 4(2), pp. 164-183.

KALTON, G., D. KASPRZYK & D.B. MCMILLAN (1989), Nonsampling Errors in Panel Surveys. *In*: KASPRZYK, DUNCAN, KALTON & SINGH, pp. 249-270.

KASPRZYK, D. G. DUNCAN, J. KALTON & M.P. SINGH (eds.) (1989), *Panel Surveys*, Wiley: New York.

KEILMAN, N. (1988), Dynamic Household Models. *In*: N. KEILMAN, A. KUIJSTEN & A. VOSSEN (eds.) *Modelling Household Formation and Dissolution*, Clarendon Press: Oxford, pp. 123-138.

LITTLE, R., & D. RUBIN (1987), *Statistical Analysis with Missing Data*, Wiley: New York.

LÖWENBEIN, O., & U. RENDTEL (1991), Selektivität und Panelanalyse. *In*: U. RENDTEL & G. WAGNER (eds.), *Lebenslagen im Wandel: Zur Einkommensdynamik in Deutschland seit 1984*, Campus: Frankfurt, pp. 156-187.

MANTON, K.G., E. STALLARD & M.A. WOODBURY (1991), A Multivariate Event History Model Based Upon Fuzzy States: Estimation From Longitudinal Surveys With Informative Nonresponse, *Journal of Official Statistics*, Vol. 7(3), pp. 261-293.

OTT, N. (1990), Die Längsschnittanalyse von Haushalten und Familien im Sozio-ökonomischen Panel. *In*: BUNDESINSTITUT FÜR BEVÖLKERUNGSFORSCHUNG, *Methoden zur Auswertung demographischer Biographien*, Materialien zur Bevölkerungswissenschaft, Vol. 67, Wiesbaden, pp. 69-99.

OTT, N. (1992a), *Intrafamily Bargaining and Household decisions,* Springer: New York.

OTT, N. (1992b), Verlaufsanalysen zum Ehescheidungsrisiko. *In*: R. HUJER, H. SCHNEIDER & W. ZAPF (eds.) *Herausforderungen an den Wohlfahrtsstaat im strukturellen Wandel,* Campus: Frankfurt, pp. 227-253.

PETERSEN, T., & K.W. KOPUT (1992), Time-Aggregation Bias in Hazard-Rate Models With Covariates, *Sociological Methods and Research,* Vol. 21(1), pp. 25-51.

PISCHNER, R., & J. WITTE (1988), Strukturen von Erwerbs- und Familienbiographien. *In*: H.-J. KRUPP & J. SCHUPP (eds.) *Lebenslagen im Wandel — Daten 1987*, Campus: Frankfurt, pp. 138-171.

RAO, C. (1984), Weighted Distributions Arising Out of Methods of Ascertainment: What Population Does a Sample Represent? *In*: A. ATKINSON & S. FIENBERG (eds.) *A Celebration of Statistics*, Springer: New York, pp. 543-569.

RENDTEL, U. (1990), Weighting Factors and Sampling Variance in Household Panels, Working Paper of the European Science Network on Household Studies, No. 11, Essex.

RENDTEL, U. (1991), Die Behandlung des Selektivitätsproblems bei der Auswertung von Paneldaten. *In*: CH. HELBERGER, L. BELLMANN & D. BLASCHKE (eds.) *Erwerbstätigkeit und Arbeitslosigkeit. Analysen auf der Grundlage des Sozio-ökonimischen Panels*, Beiträge zur Arbeitsmarkt- und Berufsforschung, Vol. 144, Nürnberg, pp. 35-59.

SOLENBERGER, P., M. SERVAIS & G.J. DUNCAN (1989), Data Base Management Approaches to Household Panel Studies, *In*: KASPRZYK, DUNCAN, KALTON & SINGH, pp. 190-225.

TUMA, N., & M.T. HANNAN (1984), *Social Dynamics*, Academic Press: Orlando.

WAGNER, G. (1991), Der Rentenzugang von Ehepaaren — Anmerkungen zur Empirie und Regulierung. *In*: C. GATHER, U. GERHARD, K. PRINZ & M. VEIL (eds.), *Frauenalterssicherung*, Edition Sigma: Berlin, pp. 223-230.

WAGNER, G., J. SCHUPP & U. RENDTEL (1991), The Socio-Economic Panel (SOEP) for Germany — Methods of Production and Management of Longitudinal Data, DIW Discussion Paper No. 31a, Berlin.

WITTE, J. (1987), Haushalt und Familie. *In*: STATISTISCHES BUNDESAMT, *Datenreport 1987 — Zahlen und Fakten über die BRD,* Bundeszentrale für politische Bildung, Vol. 257, Bonn, pp. 368-376.

WITTE, J. (1988), Haushalt und Familie. *In*: H.-J. KRUPP & J. SCHUPP (eds.), *Lebenslagen im Wandel — Daten 1987,* Campus: Frankfurt, pp. 21-41.

WITTE, J. (1990), Entry into Marriage and the Transition to Adulthood Among Recent Birth Cohorts of Young Adults in the United States and the Federal Republic of Germany, DIW Discussion Paper No. 17, Berlin.

WITTE, J. (1991), Discrete Time Models of Entry into Marriage Based on Retrospective Marital Histories of Young Adults in the United States and the Federal Republic of Germany, DIW Discussion Paper No. 23, Berlin.

WITTE, J., & H. LAHMANN (1988a), Formation and Dissolution of One-Person Households in the United States and West Germany, *Sociology and Social Research*, Vol. 1, pp. 31-41.

WITTE, J., & H. LAHMANN (1988b): Residential Mobility of One-Person Households. *In*: BUREAU OF THE CENSUS, *Fourth Annual Research Conference — Proceedings*, Washington D.C., pp. 422-448.

8. EVENT HISTORY ANALYSIS OF HOUSEHOLD FORMATION AND DISSOLUTION

Daniel Courgeau

Institut National d'Etudes Démographiques
27, rue du Commandeur
F-75675 Paris Cédex 14
France

Abstract. This chapter shows how traditionally individual-oriented event history analysis may be generalized to model household history. These models raise an important distinction between different time scales. A first scale uses a common time origin for all episodes, while the second one introduces a waiting time for each change in household size. These two clocks lead to different results when applied to a retrospective life history survey undertaken in France. The discussion of these results leads to a preference for the waiting time model for household history analysis.

8.1. INTRODUCTION

Event history analysis, using individual data sets, has become widely accepted in demography during the last ten years. It has enabled the analysis of *interactions* between events and taking into account the *heterogeneity* of the studied population (Courgeau & Lelièvre, 1989 and 1992), by focussing on several different demographic events simultaneously which the researcher considers to be particularly important ones.

Some of these individual data sets, like the French "3B" survey, which reconstructed three biographies (family, profession, and migration), were collected to study *individual* life courses. However, these data give very detailed information on *household* changes (Klijzing 1988) and also permit a focus on household histories, even if some elements of these histories are missing.

We will take this household perspective to show how traditionally individual-oriented event history analysis may be generalized to analyse

Household Demography and Household Modeling
Edited by E. van Imhoff *et al.*, Plenum Press, New York, 1995

household histories. To support such a theoretical approach and to show its limits, we will use data from our French "3B" survey. This data set consists of retrospective life histories from a random sample of individuals between the ages of 45 and 69 inclusive, all of whom were living in France in 1981. From this survey, we obtained 4,602 completed questionnaires (2,050 for males and 2,552 for females), giving a response rate of 89 per cent. These individual histories contain information from the birth of the individual until the observation year 1981, and on a number of variables concerning parents' origin, childhood history, marriage and childbearing history, and detailed job and residencc history.

We will first present a general event history model for the analysis of household histories (Section 8.2). This model raises an important distinction between different time scales. We will distinguish two alternatives here. Firstly, we can introduce the time since the first establishment of the household as a common time horizon for the household history. Secondly, we can introduce a waiting time for each change in household size, with the clock reset to zero after each household change. These two clocks lead to different results. In Section 8.3 we try to apply such a model, from the more simple situation to more and more complex ones. This will give an idea of the model's usefulness in unravelling the interrelationships between household formation or dissolution and the various event histories of each household member.

8.2. A GENERAL MODEL FOR HOUSEHOLD FORMATION AND DISSOLUTION

We define a household as a primary group identified by the co-residence of its members. This group may consist of one or more individuals and may be both familial and non-familial (Kuijsten & Vossen 1988). However, such a definition is a cross-sectional one, and when one tries to give a proper longitudinal definition of a household many problems occur (McMillan & Herriot 1986; Duncan & Hill 1985). All possible longitudinal household definitions contain one or more decision rules which are to a large extent arbitrary, as well as different continuity rules which may produce different findings for the same situation (Keilman & Keyfitz 1988).

For these reasons, we prefer to follow an *individual-based* approach, which is also consistent with our "3B" survey. The time origin is the moment when the household is established with at least one member, the surveyed individual. Since the survey contains the occupation of each place of residence, we are able to detect a leave from the parental home. However, some of these departures may not be considered as the beginning of a household. For example, departures for military service or stays in student residences, in hospital (for long duration illness), etc., are not to be considered. In that case, we have to make a preliminary choice to be able to precisely define a departure from the parental home which leads to a new household. For our survey, we have considered the residential history and the professional one simultaneously, for which every period can be characterized by many possible states (studies, military service, war periods, etc.). The household may then follow a path of formation and dissolution, with successive increases or decreases in its size and changes in its composition.

Let us first focus on size changes. For a given household, we may define successive periods of time during which its size remains the same. Each of these periods has an end point, the time at which an event occurs, with the introduction of one or more newcomers or the departure of other members. From one given size, such a change may lead to any other size. For example, it may be the simultaneous arrival of a great number of individuals (a man marrying a widowed woman with many children all entering his household) or the departure of many individuals (a woman divorcing and leaving a household with all her children). Even in some cases, for example, when we are working in single-year discrete time, the resulting change in size may be zero as the death of a child may occur very close to the birth of another one.

The death of the household marker does not happen in our data set because the survey is retrospective. When using prospective data, such a death leads to a right censored observation of the household. As this censoring scheme is not independent of the household history, it creates problems that are difficult to solve. This issue will not be pursued in the present paper.

Let us now formalize such an approach. We are interested here in the following times at which changes in household size occur, considered as ordered random variables $0 = T_0 \leq T_1 \leq T_2 \ldots$, and the sizes of the household in successive episodes k between size changes characterized by a series of random variables $\{S_k;\ k=0,1,2\ldots\}$, $S_k \in \{0,1,\ldots,m\}$, where m is the maximum of the observed household sizes.

The corresponding stochastic process which describes the state of a household by a continuous time, discrete state process, can be decomposed into two related processes. First, a *duration process* which governs the time elapsed since the occurrence of some specific event. We will see later how it is possible to undertake different analyses by changing the definition of such a specific event. Second, a *transition process* which governs the moves between household sizes (Hamerle 1989). As stated above, such changes may lead to any other possible size.

Let us suppose that we have a vector $z_k(t)$ of relevant covariates measured at time t for every individual, including characteristics of the other household members, and episode k. Such a vector may contain dummy or metric variables, or both. These covariates may be time invariant or time dependent. In the latter case, the covariates may change during episode k. We also have to consider the previous history of the process until t_{k-1}. This history is collected in a vector

$U_{k-1}=\{s_0, z_0, t_1, s_1, z_1, \ldots, t_{k-1}, s_{k-1}, z_{k-1}\}$, where z_l is always a vector of covariates that may be time dependent in the interval (t_{l-1}, t_l); s_l is the size of the household; and t_l the time of change.

There are different ways to model the time dependence of event-specific hazard rates. We can first use a common process time for all episodes with a *common time origin*. As previously stated, the study of household formation and dissolution begins when the household is established with at least one member: this may be the time origin. In such a case, all subsequent starting and ending times of an individual's episodes are the durations between the time origin and the occurrence of household changes. Then, both the duration and the transition process can be simultaneously characterized by the following event-specific hazard rate:

$$h_k^{s_{k-1}\, s_k}(t;\, z_k(t),\, U_{k-1}) =$$

$$= \lim_{dt \to 0} \frac{P(T_k < t+dt,\ S_k = s_k \mid T_k \geq t \geq t_{k-1},\ S_{k-1} = s_{k-1},\ z_k(t),\ U_{k-1})}{dt} \quad (1)$$

for the kth move from household size s_{k-1} to s_k. Such a rate will be identically zero for $t < t_{k-1}$. There is a possibility that some household sizes are not attainable from certain sizes. This can be accommodated in equation (1) by restricting it to be identically zero for the appropriate (k, s_{k-1}, s_k) values. However, even with this possibility, the number of rates to estimate is so large that the observation of a very large sample is required.

This common time origin supposes that the time from the first establishment is the crucial variable affecting the future evolution of the household. In this case, we assume that the successive events occurring during the process of household formation and dissolution do not alter the rates for subsequent episodes: these are mainly directed by the duration since initial household establishment. Such an assumption does not permit a picture showing the dependence on time since the beginning of the kth episode, without introducing a multidimensional time scale. An alternative possibility is to consider that experiencing an event once alters the rates for subsequent episodes. In this case the 'common time horizon' loses importance, while we are taking a 'waiting time approach'. We will consider here that the clock is reset to zero after each event. The crucial variables are now the waiting times from successive household changes, each new position developing its own time scale.

Following this alternative approach, let us model the hazard rates in terms of *waiting times*, starting with time zero whenever a new episode begins. For the first episode, the two approaches lead to the same result as the starting time is identical. For the following ones, we will have to reformulate equation (1). For the kth change in household from size s_{k-1} to size s_k, the hazard rate may be written:

$$h_k^{s_{k-1}\, s_k}(t;\ z_k(t),\ U_{k-1}) =$$

$$= \lim_{dt \to 0} \frac{P(T_k - T_{k-1} < t+dt,\ S_k = s_k \mid T_k - T_{k-1} \geq t,\ S_{k-1} = s_{k-1},\ z_k(t),\ U_{k-1})}{dt} \quad (2)$$

The time t is now measured *from the beginning of each episode*, the other characteristics being the same as for the previous equation (1). Such a model would be suitable if the process of household formation and dissolution leads to hazard functions that can be more parsimoniously expressed in terms of gaps between failure times, rather than in terms of the total observation time (Kalbfleisch & Prentice 1980).

Let us see now in more detail how to include covariates and past history. As the household size changes, the covariates may depend on the various members which are present at each time. If the household has only one member, these covariates refer to him and his relatives: educational level, parents' occupation, past history, etc. If it is composed of more than one member, we can introduce the previous characteristics for each member of the household in the same way. It is also useful to introduce more complex characteristics that are related to the household as a whole: household composition, members becoming home-owners, etc.

It is often necessary to reduce the number of changes in households sizes. As we previously stated, it is not possible with survey data to consider every pair of sizes (s_{k-1}, s_k), as the number of such pairs is too large to permit a useful analysis of household changes. We will consider only two types of movements: to lower sizes and to sizes equal to or larger than the current one. As we are working in single-year discrete time, some changes in size may be zero. However, as such changes are very rare occurrences, we will not consider them separately and will include them arbitrarily with 'larger sizes'. In that case, we only have to estimate the two following kinds of rates:

$$h_k^1(t;\ z_k(t),\ U_{k-1}) = \sum_{s_k \geq s_{k-1}} h_k^{s_{k-1}\, s_k}(t;\ z_k(t),\ U_{k-1}) \quad (3)$$

and

$$h_k^2(t;\ z_k(t),\ U_{k-1}) = \sum_{s_k < s_{k-1}} h_k^{s_{k-1}\, s_k}(t;\ z_k(t),\ U_{k-1}) \tag{4}$$

It is easy to generalize these estimations to any other kind of more complex moves.

It is also necessary to formulate the hazard function with more precise specifications to be able to estimate its parameters. We will use a semi-parametric approach, which is a generalization of the Cox proportional hazard model. We assume that the hazard rates have a shape function that depends arbitrarily on the index number k of the event and that the various covariates act multiplicatively on these rates. We will introduce some of the characteristics of the previous history U_{k-1} that are considered as useful predictors of the hazard rate; they will be included in the $z_k(t)$ covariates.

If we follow a household from its first establishment, ignoring the waiting times between each step, we can write the following hazard rate, with $i=\{1,2\}$:

$$h_k^i(t;\ z_k(t),\ U_{k-1}) = h_{k0}^i(t) \times \exp\left(z_k^i(t)\beta_k^i\right) \tag{5}$$

where $h_{k0}^i(t)$ is an unspecified baseline hazard function and β_k^i is a column vector of regression parameters. To estimate these regression coefficients, a partial likelihood approach may be used (Cox 1972; Kalbfleisch & Prentice 1980). Such a partial likelihood[1] may be derived as:

$$PL = \prod_k \prod_i \prod_{l=1}^{d_k^i} \frac{\exp\left(z_k^{il}(t_k^{il})\beta_k^i\right)}{\sum_{r \in R(t_k^{il})} \exp\left(z_k^{ir}(t_k^{il})\beta_k^i\right)} \tag{6}$$

where t_k^{i1}, t_k^{i2}, ... are the d_k^i durations in which the kth episode ended by a transition to state i, and where $R(t_k^{il})$ is the set of households that are at risk of their k^{th} transition just before time t_k^{il}. For the common time horizon approach, such a risk set may increase as time goes on. For example, in a certain time interval, a large number of individuals can have their $(k\text{-}1)^{\text{th}}$ household change and then become at risk of their k^{th} change, while those who have their k^{th} change in the same interval may be less numerous. This may lead to an increase in the risk set.

---

[1] To make this estimation, we used the computer program TDA (Transition Data Analysis) written by Götz Rohwer (European University Institute).

In the second case, we follow the successive waiting times of an individual. The expression for the hazard rate is the same as (5), except that time t is defined in a different way. A similar partial likelihood like (6) may be written to estimate the regression parameters. But in this case the risk set is always decreasing as time goes on, as at each episode everybody begins at time zero. The two partial likelihoods are not directly comparable, and cannot be used to tell which approach is the better one. To do that, we will have to use some other methods.

Until now, we only dealt with *changes in the size* of the household. We will now have to introduce *changes in composition*. According to the household position of the entering or departing members, we may observe very different changes in composition for a given size change. A divorce and a child's death may lead to the same decrease in household size, but have quite different consequences for subsequent household development. As we consider household changes, we may introduce the occurrence of one or more different events: marriage, a child's birth, a divorce, etc. A split of the destination state according to the events leading to it permits such an introduction. This leads to a generalization of previous hazard rates without introducing new difficulties. For example, rate h_k^1 may be split in h_k^{1m}, if the change is due to a marriage, h_k^{1b}, if the change is due to a child birth, etc. We can then introduce comparative rates on household composition changes, according to the destination state.

8.3. AN EXAMPLE OF ANALYSIS OF HOUSEHOLD FORMATION AND DISSOLUTION

As previously stated, the "3B" survey gives detailed information on some changes in household composition. But as it was undertaken with an individual perspective, some elements may be missing. Let us see first how to approach a household history.

8.3.1. The "3B" Survey in a Household Perspective

Since the survey contains the residential history and the professional one, we are able to define very precisely a departure from the parental home in the way we gave in Section 8.1 of this chapter.

When leaving his parental home, an individual may already be married, with children, and the initial size of his household may be greater than one. Also, once in a new household, an individual may come back to his parental home, and we have to consider the possibility of several successive households for the same individual. In every case, we have to follow the increase or the decrease of this household size, according to marriage, birth of children, divorce, departure of children from the parental home, etc. The major part of these events are to be found in the part of the questionnaire on the matrimonial history and on children of the surveyed person.

However, when taking a household perspective, some events are missing from our survey. Cohabitation was not taken into account, but since it was a rare situation for the surveyed generations during their youth, this will not introduce an important bias. It is evident that if we study more recent generations, cohabitation will have to be considered. Also, only the children or the adoptive children of the surveyed individual have been considered. We may miss some children from previous marriages of the spouse that are living in the household. We also lose some ancestors, siblings, cousins, etc., or other non-kin members of the household.

If we want to have a complete event history of households, we must undertake a survey with a household perspective, as the individual perspective misses some events. However, we will consider the household histories obtained from our "3B" survey, even though this is incomplete.

Before undertaking any analysis, it may be useful to observe the evolution through time of different kinds of households, in order to illustrate our individual-based household concept. Figure 8.1 gives four household histories from our sample.

The first household has a very simple history: from its beginning (a male leaving his parents' home at age 22), it is characterized by regular increases due to marriage and birth of children, to reach its maximum size of 8 members after a duration of 20 years. After that, a very quick decrease occurs due to the departure of children and the wife's death, to end with the male and one child at survey time.

The second household has a somewhat more complex history. First, a household with a single male member from 19 to 32 years old is formed, followed by a return to the parental home for one year. He again leaves his parental home in order to settle independently with a wife, has two children followed by the death of the first one, quickly followed by the birth of two more children. Then the usual decrease follows due to departure of the children.

The third household is started by a marriage, quickly followed by the birth of a child. But eleven years later a divorce occurs, the child remaining with his father. This is followed by a remarriage seven years later, with the birth of a new child, etc. The fourth household is even more complex: twice there is a return to the parental home after the first household formation. Next, a first marriage followed by divorce and departure of the first wife with her child, a remarriage, etc.

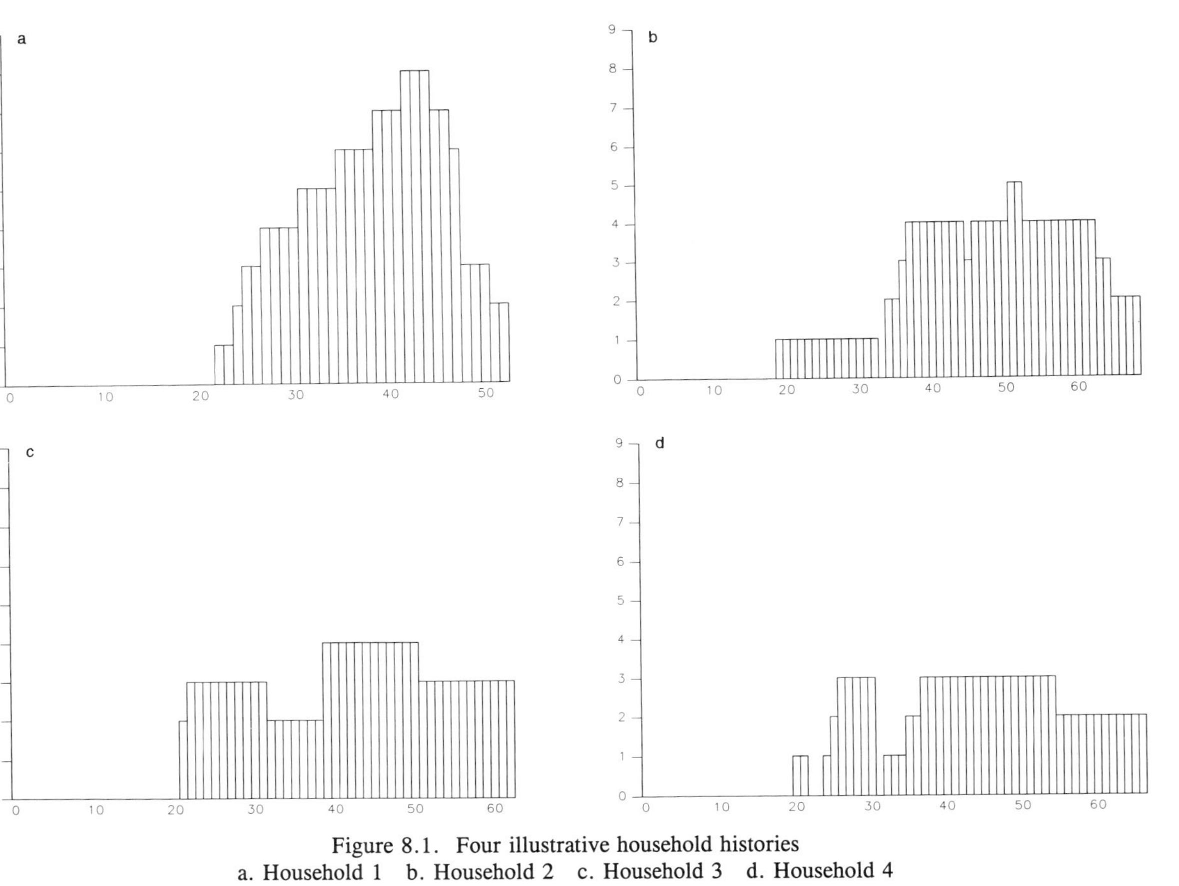

Figure 8.1. Four illustrative household histories
a. Household 1 b. Household 2 c. Household 3 d. Household 4

8.3.2. A Nonparametric Approach to Household History

The maximum number of spells for one household history is 24 for the male sample and 27 for the female sample, leading to a total number of spells of 10,211 for males and 13,596 for females. However, the major part of these histories lies in a more restricted limit: less than 7.5% of the spells are of serial number greater than eight for males (10.0% for females).

With this limit size of eight, we will now consider what are the main transitions. Table 8.1 gives the destination in per cent for each size of origin. As the household instigator may return to the parental home (size zero) and leave it again afterwards, we also have a line and a column with this number. We also have to consider those in a household of a given size who remain in this situation until censoring occurs.

We first observe that the major part of household changes consists of the addition or the subtraction of one member. For households with an origin size of less than four members, the addition of a new member is much more important than subtraction, for both sexes. Afterwards the reverse is true. Differences between sexes appear for origin sizes equal to zero and one. First the females returning to their parents' home are proportionately less numerous than males, and afterwards the probability of attaining size 1 is smaller than the probability of attaining any of the sizes 2, 3, and 4 (0.198, 0.273, 0.233 or 0.201, respectively), which cannot be observed for males. Secondly, the proportion of censored women in household size equal to one is much larger than that for men (46.4% against 23.7%). This is related to the more important number of women observed alone in our survey (mainly widowed ones). These observations lead us to conclude that it is necessary to observe the surveyed individuals according to their sex.

Let us now observe the cumulative hazard rates to higher (or similar) and to lower household size, for males and females separately. If we first consider *waiting times*, we can use formulae (3) and (4) without rank distinction to give an overall view of changes in household size. Figure 8.2 gives these results.

Again, differences between males and females can be observed. When we observe the rates to higher size, the male plot is above the female one, the reverse being true for rates to lower size. However, the main difference between these plots lies in their shape. For movements to higher household size, the rate regularly decreases with duration, for both sexes. It remains near zero from 25 years onward. For movements to lower household size, the plot is more complex. It begins with a low but nearly constant rate until a duration of 15 years. Afterwards this rate increases from 15 to 25 years, with a new decrease coming afterwards. These first results clearly show that we have to consider these two kinds of movements separately.

We can also consider the *duration from first establishment of the household*. Figure 8.3 gives these results.

Table 8.1. Destination sizes of uncensored households for each size of origin (percentages)

Origin size	Destination size 0	1	2	3	4	5	6	7	8	Number censored	Total number
					Males						9841
0	0.0	35.4	25.2	20.9	11.7	3.9	2.4	0.5	0.0	0	206
1	12.9	0.4	75.2	11.2	0.3	0.0	0.0	0.0	0.0	321	1353
2	1.3	12.8	2.8	81.6	1.5	0.0	0.0	0.0	0.0	964	2373
3	0.9	1.4	42.1	1.5	53.3	0.8	0.0	0.0	0.0	378	2416
4	0.5	0.7	3.1	51.4	1.8	42.0	0.5	0.0	0.0	200	1726
5	0.1	0.5	0.5	6.3	53.2	0.9	37.9	0.6	0.0	80	948
6	0.2	0.4	0.9	1.5	6.5	50.4	2.6	37.0	0.5	31	491
7	0.0	0.0	0.4	0.4	1.7	9.4	48.3	3.0	36.8	14	248
8	1.4	0.0	0.0	1.4	0.0	0.0	12.1	79.7	5.4	6	80
					Females						13522
0	0.0	19.8	27.3	23.3	20.1	7.9	0.8	0.8	0.0	0	253
1	13.2	0.7	74.0	12.1	0.0	0.0	0.0	0.0	0.0	755	1626
2	2.3	27.4	2.6	66.5	1.2	0.0	0.0	0.0	0.0	1204	3459
3	1.4	1.2	47.3	1.8	47.7	0.6	0.0	0.0	0.0	356	3342
4	0.9	0.1	4.4	55.0	1.6	37.3	0.7	0.0	0.0	108	2357
5	0.7	0.0	0.4	6.5	54.8	2.0	35.1	0.5	0.0	50	1319
6	0.0	0.1	0.0	0.3	6.3	55.0	2.4	35.6	0.3	18	684
7	0.0	0.0	0.0	0.8	1.1	9.9	51.6	1.1	35.5	5	360
8	0.0	0.0	0.0	0.0	0.0	0.8	14.3	79.8	5.1	3	122

Source: French "3B" survey.

The patterns in Figure 8.2 and 8.3 are quite different. First, the differences between men and women only begin to appear after a duration of 15 years. But afterwards they are similar to those found for the waiting-time approach. Secondly, the movements to lower household sizes do not exhibit a decrease in hazard rates after 25 years, as for waiting times. The rates remain quite constant from 20 to 40 years after the first establishment, and afterwards only show a slight decrease.

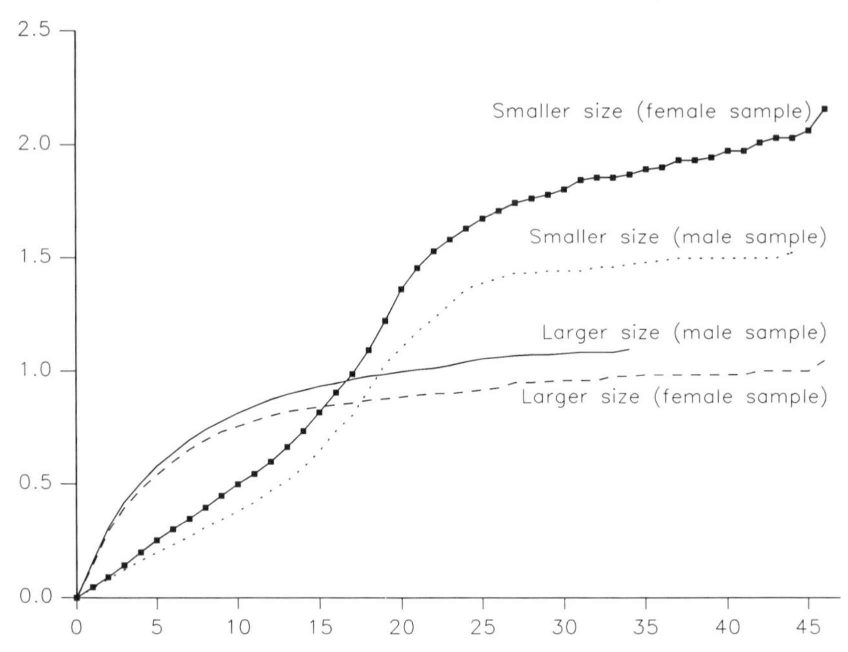

Figure 8.2. Cumulative transition rates to larger (or similar), smaller household size: Waiting time approach

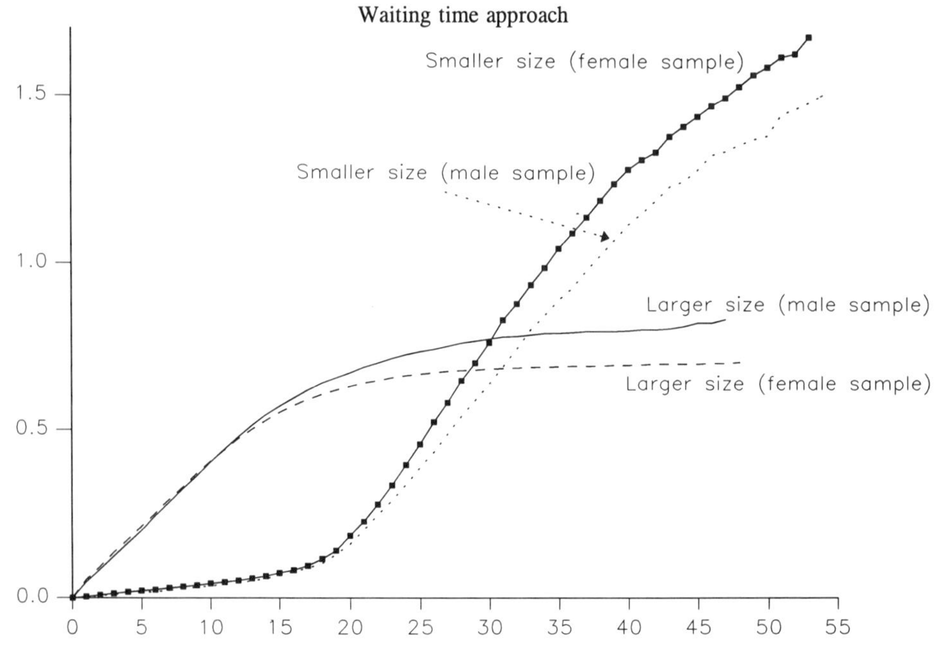

Figure 8.3. Cumulative transition rates to larger (or similar), smaller household size: Time since first establishment approach

The two approaches give us different perspectives on the same phenomena, but for the moment we have no reason to prefer one to the other.

Let us go further and observe, for each successive change the cumulative rates to larger (or similar) and to smaller household size, for males only, considering the different waiting times. Figures 8.4 and 8.5 give these rates.

We see that these hazard rates are highly dependent on the order of the studied transition and that they do not show any typical shape for all groups. For example, the cumulative hazard rates to smaller household size show a linear trend for the first change, an S-shaped trend for the second change, a J-shaped trend for the third and fourth change, and again a linear trend with a steep slope for the higher order changes.

8.3.3. A Semiparametric Approach to Household History

Such a dependence on the order of the studied transition does not allow us to use a simple parametric model for each order. It is the reason why we prefer to use a semiparametric model, which allows an arbitrary baseline hazard for each event order.

To give a simple but clear example of such a semiparametric analysis, we only investigate the effect of two covariates. The first one is time invariant: it is a binary variable of value 1 when the surveyed individual has no diploma, 0 if they do. The second one is time dependent and is a characteristic of the household as a whole: it is also a binary variable of value 1 when the household is home owner.

The β parameter estimates are given in Figure 8.6, for each order of the transition, k, up to eight, for the two destination states ($i=1$, if movements are to lower size households, $i=2$, if they are to sizes equal or higher) and for the two considered models (waiting times or times from the first establishment of the household). We have used square markers where the results were significantly different from zero (at the 5% level).

We can see from these plots that the model which uses waiting times gives much more clear and significant results than the model which uses time from first establishment. This result is particularly clear when we look at waiting times to larger household size.

Being a home owner reduces the probability of increasing ones household size already from the second step, and this reduction increases with the order of the step, when we look at waiting times. The other approach shows significant results only for step 3 and 4, with a less important effect. To have no diplomas increases the probability of moving to a larger household size also from the second step. In that case, the differences between the two models are less clear, with only a non-significant result at step 7 for the model with time from first establishment.

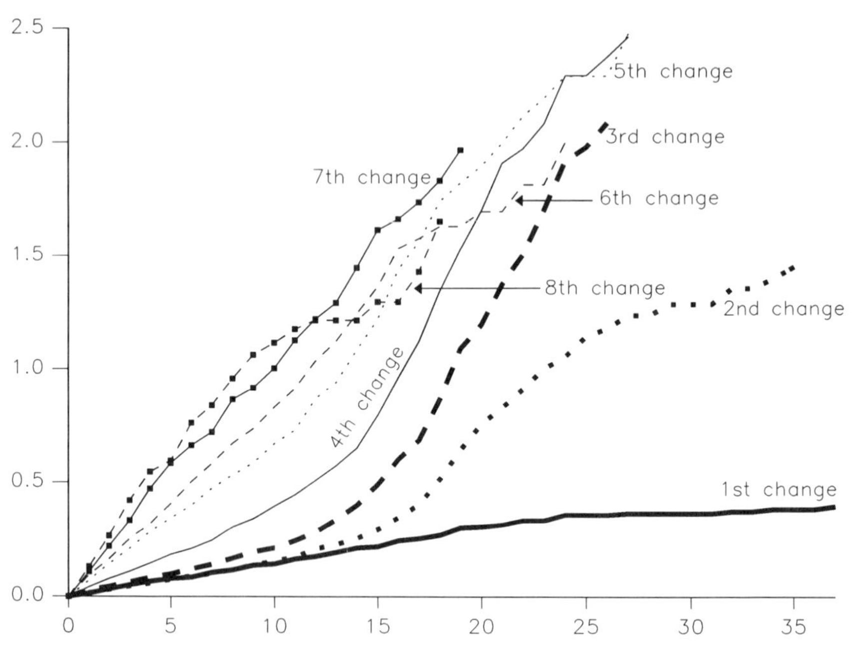

Figure 8.4. Cumulative transition rates to smaller household size: Waiting time approach

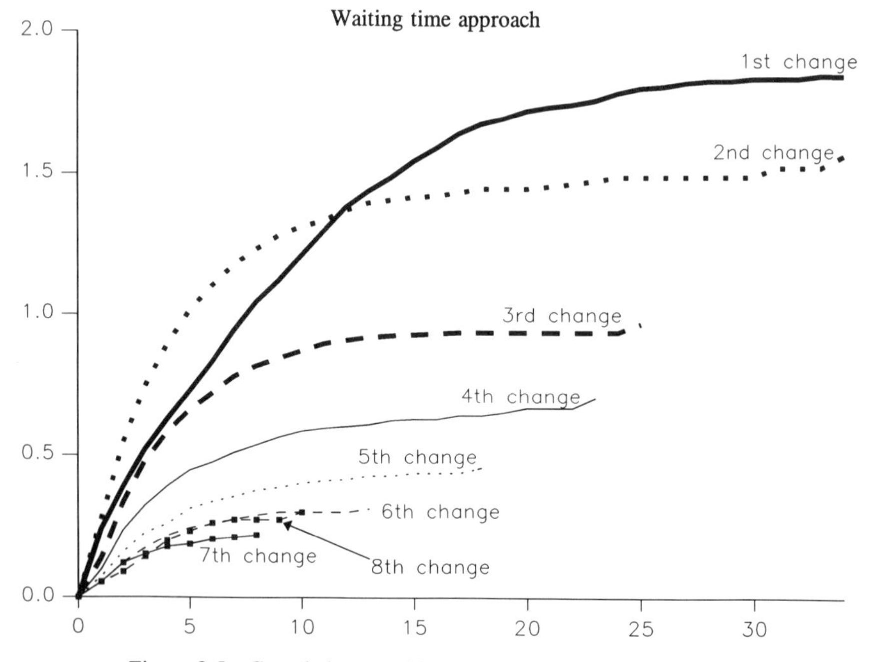

Figure 8.5. Cumulative transition rates to larger household size: Waiting time approach

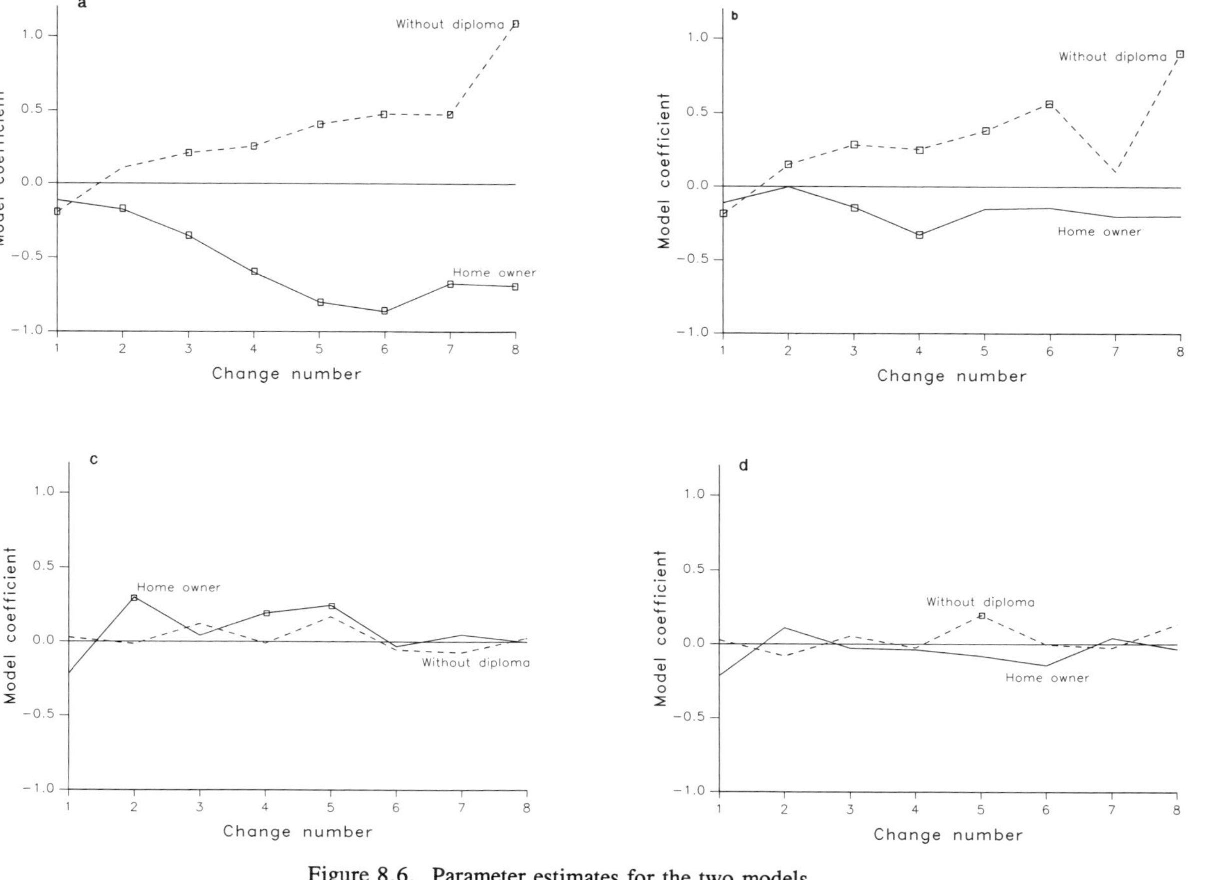

Figure 8.6. Parameter estimates for the two models
a. Waiting time to larger household size b. Time from first establishment to larger household size
c. Waiting time to smaller household size d. Time from first establishment to smaller household size

For movements to smaller household sizes, the waiting time model always gives better results, even if these are less significant than for movements to larger household sizes.

Let us conclude this section with a more detailed discussion of the reasons for preferring the waiting-time model to the common time-origin model for the study of household formation and dissolution. When using a common clock for all episodes, some difficulties of interpretation appear. How can we interpret, for example, the hazard rates for households leaving their i^{th} state after a duration, t, from its formation? Such rates are estimated on the whole population entering the ith state, but only on a part of these households in their i^{th} state at duration t. This risk set has a complex definition, and the interpretation of the corresponding rates is not clear. This difficulty is increased as each household at risk has a different duration of stay in the ith state. Thus, it is not possible to have a clear picture showing the dependence on time of these rates from the beginning of the ith episode. Such an approach, expressed in terms of total observation time, seems to be too narrow to understand such a complex phenomenon as household formation and dissolution.

A more flexible solution is to use a waiting time approach. First, it replaces the common clock by a number of waiting times for each of these episodes. We include all the households entering the ith state before the survey, and we follow them from the beginning of each interval. Such hazard rates have a clear meaning, as the duration variable is conditional on the time of entry in the ith state. It is also easy to introduce a dependence of these rates on the past history, such as the duration since household formation, the occurrence of some related events, etc. Finally, this last approach gives statistical results that are more easy to interpret than the previous ones, and also more significant with the variables used in this paper.

8.4. CONCLUSIONS

In order to study household formation and dissolution, we have developed an event history approach, with an application to our "3B" survey. In this chapter, we tried more precisely to follow a complex entity over time, which may change its size and its composition. We have not considered changes in localization, even if we consider this topic an important one. A further paper will have to take this issue into account.

Such an approach has led us to explore multiple-spell models for duration data. This chapter provides a general formulation of hazard rates. However, for a complete specification, we must define a time scale which measures the time dependence of these rates. Two different clocks have been used which lead to different underlying stochastic processes.

If time is measured as duration of stay since the establishment of the household formation, such a time scale leads to nonhomogeneous Markov processes. A Markov process assumes that the state at some future time depends on its state at the present time, but not on its past states nor on the duration spent in the current state. However, as the process is nonhomogeneous, it may depend on other relevant aspects of its history. We have given a general formalization of the hazard rate in this case (formula (1)). If time is measured as waiting time in the current state, the clock is reset to zero after each transition. Such a time scale leads to a different underlying stochastic process which is semi-Markov. A semi-Markov process also assumes that the state at some future time depends on its state at the present time, but not on its past states. However, it also allows dependence on the duration spent in the current state. Again we have given a general formalization of the hazard rate in this case (formula(2)).

The application of these two approaches to our data set showed that the second model is more suitable to show the dependence of rates on covariates, particularly time-dependent variables. Obviously this is only a first step. We will have to study, in greater detail, the means to test the most suitable model using more precise criteria.

It seems to us that we will have to introduce models with multiple time scales: a first time scale from the beginning of household formation with secondary time scales from the beginning of each transition. Further research will tell us if such a complex model may be estimated from our data sets.

REFERENCES

COURGEAU, D., & E. LELIÈVRE (1989), *Analyse Démographique des Biographies*, INED: Paris.

COURGEAU, D., & E. LELIÈVRE (1992), *Event History Analysis in Demography*, Oxford University Press: Oxford.

COX, D.R. (1972), Regression Models and Life Tables, *Journal of the Royal Statistical Society B*, Vol. 34, pp. 187-220.

DUNCAN, C.J., & M.S. HILL (1985), Conceptions of Longitudinal Households: Fertile or Futile?, *Journal of Economic and Social Measurement*, Vol. 13, pp. 361-375.

HAMERLE, A. (1989), Multiple-Spell Regression Models for Duration Data, *Applied Statistics*, Vol. 38, pp. 127-138.

KALBFLEISCH, J.D., & R.L. PRENTICE (1980), *The Statistical Analysis of Failure Time Data*, John Wiley and Sons: New York.

KEILMAN, N., & N. KEYFITZ (1988), Recurrent Issues in Dynamic Household Modelling. *In*: KEILMAN, KUIJSTEN & VOSSEN.

KEILMAN, N., A. KUIJSTEN & A. VOSSEN (eds.) (1988), *Modelling Household Formation and Dissolution*, Oxford University Press: Oxford.

KLIJZING, E. (1988), Household Data from Surveys Containing Information for Individuals. *In*: KEILMAN, KUIJSTEN & VOSSEN.
KUIJSTEN, A., & A. VOSSEN (1988), Introduction. *In*: KEILMAN, KUIJSTEN & VOSSEN.
MCMILLAN, D.B., & R. HERRIOT (1985), Toward a Longitudinal Definition of Households, *Journal of Economic and Social Measurement*, Vol. 13, pp. 349-360.

9. COMPETING RISKS AND UNOBSERVED HETEROGENEITY, WITH SPECIAL REFERENCE TO DYNAMIC MICROSIMULATION MODELS

Heinz P. Galler

Department of Econometrics and Statistics
Martin-Luther University of Halle-Wittenberg
Grosse Steinstrasse 73
D-06099 Halle
Germany

Abstract. The idea of competing risks provides a useful conceptual basis for multistate models in general, and microsimulation models as a special case. However, some methodological problems arise when the different risks are not independent. Besides causal relations, such dependencies may be caused by unobserved heterogeneity of the units. In this case, not all relevant explanatory variables are included in the model, which may cause stochastic dependencies between different partial processes. Approaches for dealing with this problem are discussed both in a continuous time and a discrete time framework.

9.1. THE NOTION OF COMPETING RISKS

In dynamic microsimulation models, an attempt is made to simulate socio-demographic processes at the level of elementary micro units. Based on a stochastic model of the relevant processes, trajectories are generated for the state of each individual unit in a representative sample of the population considered, using Monte Carlo techniques. The structure of the population and its dynamics are then inferred from the structure of the sample generated by the simulation model. Since comparatively rich information is available on the state of each individual unit in the sample, rather complex model structures can be used and detailed analyses of the processes under consideration become feasible.

On the micro level, many of the characteristics of the units are of a qualitative nature and are represented by qualitative variables. Such variables do not change smoothly over time, but jump at some point in time from one

Household Demography and Household Modeling
Edited by E. van Imhoff *et al.*, Plenum Press, New York, 1995

state to another. Thus, in order to simulate the dynamics of the *state* of a micro unit, the *events* that lead to such changes must be simulated. If demographic models are considered, these are events like the birth of a person, a marriage or the formation of a consensual union, a separation from a partner, transitions between different households, changes in labour force participation, or the death of an individual. Simulating the time path of the state of the unit then amounts to simulating the event history with regard to the relevant events. For such models, the concept of competing risks has proven to be very useful.

The basic notion of *competing risks* refers to event histories in which not only the timing of a single event is of interest, but also which event out of a set of *mutually exclusive outcomes* may be observed at a point in time. In this case, the different outcomes may be regarded as different events 'competing' with each other for becoming observed. A simple example is the problem of occupational choice, in which an individual may choose between different types of occupations, but only one transition can be observed at a given point in time. In this case the different transitions to other occupations that are possible from the original state may be modelled as different mutually exclusive events. These events 'compete' in the sense that only one event can be observed. Thus, in competing risk models, not only the *timing* of the event is of interest, but also the *type* of event being observed.

The competing risks approach can also be used to account for *dependencies* between different processes that are not mutually exclusive. In many cases, dependencies exist of the form that if one event is observed there is a high (or low) conditional probability that also the other event will be observed. An example is the relation between fertility and marriage: There is no strict dependency between age at first marriage and age at the birth of the first child, but a strong positive correlation can be observed.

The decisions to marry and to have a child are not independent, but in many cases can be regarded as a joint decision. Such dependencies between different events cannot easily be taken into account in formal models. Tuma and Hannan (1984, pp. 266ff) have proposed to model such events as compound events and to model these events in a competing risks framework. In the example, each combination of the marriage decision and of fertility would be regarded as a single compound event. In this case the different combined outcomes are mutually exclusive since they are defined as the Cartesian product of the sets of possible elementary events and can be treated as competing risks.

This approach can be generalized further to model *multivariate processes with state dependence*. In this case the partial processes corresponding to the different events depend on the state of the unit. If the state changes the probability to observe a particular event also changes. Again, the dependency of fertility on marital status is an example. If marital status changes, fertility will change as well. At the same time, the propensity to marry may depend on the number of (minor) children beside other variables. Thus the event of a marriage will affect the probability of childbirth and a childbirth will alter the probability to observe a marriage. A simple way to model such dependencies is to split the time path of the unit into time intervals in which the values of the state variables remain constant, and to consider all processes that will alter a state variable as competing risks. If one of the events has occurred, the state of the unit changes and a new interval starts with the new specific probabilities to observe the different events.

Thus, the concept of competing risks is not restricted to modelling events with mutually exclusive outcomes, but it provides a conceptual basis for a broad range of multistate models. In such models transitions of units between different discrete states are considered. Since by definition, at a given point in time, only one transition can take place, the different transitions that are possible constitute competing risks, even if the basic events are non-exclusive in the sense that the observation of one event does not exclude that also another event may be observed. But this is not possible at the same time. As a consequence, the different events may be modelled as competing risks with regard to the transition in the state of the unit.

Models of competing risks are based on the idea of an underlying stochastic process that generates the events considered. Modelling competing risks then requires the specification of an appropriate stochastic process and the estimation of its parameters. From such a model, specific probabilities can be derived to observe a given event. These probabilities can be used in a microsimulation model to generate event histories for the units considered by Monte Carlo methods. Numerically, a sequence of events is simulated for a unit by performing the corresponding random experiments.

In contemporary microsimulation models, *continuous time* and *discrete time* approaches are used to simulate competing risks. An example of continuous time models is the SOCSIM model (Hammel 1990) that has been developed at Berkeley[1]. The discrete time modelling concept has been used in the DYNASIM model (Orcutt *et al.* 1976) and the models derived from it, the FRANKFURT model (Galler 1988), the DARMSTADT model developed by Heike *et al.* (1988a, 1988b; also Hellwig 1988) and the model of the Hungarian Statistical Office derived from it (Csicsman & Pappne 1987), and in the Dutch NEDYMAS model (Nelissen 1989). In the group of discrete time models, a

---

[1] Actually, the SOCSIM model is not truly time continuous but a month is used as the basic time unit, i.e. durations are measured in units of months and no fractions are considered (Hammel 1990, p. 6). However, since mean durations are comparatively large for the demographic processes being simulated, this can be regarded as a close approximation to a continuous time model.

further distinction can be made according to whether events are simulated in a fixed sequence or a randomization approach is used.

A common problem to all models that has not yet been solved satisfactorily, even on a conceptual basis, is the problem of modelling *unobserved heterogeneity* of the micro units. Since it is not possible to exactly model socio-demographic processes in a quantitative model, some unobserved heterogeneity will always remain in the form of differences in the individual behaviour of the micro units that are not explained in the model. Due to the comparatively low explanatory power of models, unobserved heterogeneity will be substantial in many cases, and should therefore be taken into account when modelling and simulating socio-demographic processes.

In the following, some methodological problems of competing risks models are discussed in more detail. Special attention is given to the consequences of unobserved heterogeneity for the model structure. First, the specification of continuous time competing risk models is considered. Then, discrete time models based on a multivariate probit structure are discussed. Such models provide an alternative approach to multistate modelling that allows for rather flexible model structures that can be used to improve the explanatory power of the models.

9.2. COMPETING RISKS IN CONTINUOUS TIME MODELS

9.2.1. The Basic Modelling Approach

Simulation of competing risks in contemporary time continuous micro-simulation models like the SOCSIM model is based on the concept of conditionally independent stochastic processes for different risks. In this approach the different possible outcomes are treated as different partial risks. Formally, for each partial process, a random variable T_i is assumed giving the time the unit remains in the original state till the partial event i takes place (cf. Lawless 1982, pp. 478ff). However, only the first partial event can be observed and all other durations are censored at that time. As a consequence, the durations corresponding to the different partial events are latent variables that cannot be observed directly except for the event occurring first. In this way a multistate process can be modelled by a multivariate stochastic process resulting from the combination of a set of partial processes[2].

---

[2] Using this approach, general problems of censoring can also be modelled as competing risks with the partial risks consisting of the event considered and the event of censoring respectively. If the censoring process is not stochastically independent from the process of interest, the censoring process must be modelled explicitly in the model in order to obtain

Usually *conditional stochastic independence* is assumed for the partial processes, given the values of the explanatory variables. This can be justified by a non-identifiability property of competing risks models. For any competing risk problem with dependent partial processes, there always exists a model specification with conditionally independent partial risks that is *observationally equivalent* to the dependent specification (Tsiatis 1975). Since there is a well-developed statistical theory for models with independent competing risks and models for independent risks also have a comparatively simple structure, usually conditional independence is assumed in empirical analyses of competing risk problems. This is also true for microsimulation models. However, the independence assumption has some implications that should be discussed in more detail. But first, the basic theory of conditionally-independent risks will be presented.

If there are m possible outcomes of a process in a competing risks problem, m partial durations T_i, $i=1,\ldots,m$ can be defined corresponding to the different outcomes. The duration till the event being observed equals the *minimum of the partial durations,* $T=\min(T_i, i=1,\ldots,m)$. If for simplicity the explanatory variables are assumed to be given constants, the conditional joint distribution of the partial durations T_i given the vector x of explanatory covariates is given by the conditional multiple survival function

$$S(t_1,\ldots,t_m \mid x) = \text{Prob}\{t_1<T_1,\ldots,t_m<T_m \mid x\}.$$

It defines the probability that no event is observed before the durations $t_1,\ldots,t_m$. However, since in the case of competing risks the observation of the other partial processes is censored at the time of the partial event occurring first, only durations $t=t_1=\ldots=t_m$ can be observed. Correspondingly, the joint probability density of the latent durations is not observable for all values of $t_1,\ldots,t_m$, but only subdensities for $t_1=\ldots=t_m$ can be obtained from empirical data.

In analogy to univariate problems, for competing risks, a *(partial) hazard rate* $r_i(t \mid x)$ can be defined for each partial risk. It has the interpretation of a conditional instantaneous probability density to observe the partial event i at time t given that no event has been observed before time t. It is defined as the negative partial derivative of the logarithm of the multiple survival function.

$$r_i(t|x) = \lim_{dt\to 0}\frac{1}{dt}\text{Prob}\{t<T_i<t+dt|x,T_j>t,j=1,\ldots,m\} =$$

unbiased estimates (e.g. Lawless 1982, pp. 38ff).

$$= \left. \frac{-d\log S(t_1,...,t_m|x)}{dt_i} \right]_{t_1=...=t_m=t} \tag{1}$$

Since in a continuous time framework the probability to observe more than one event at a point of time is zero, the partial events are mutually exclusive and *the compound hazard rate* to observe any of the events at time t equals the sum of the partial hazards,

$$r(t|x) = \sum_{i=1}^{m} r_i(t|x).$$

Thus the value of the joint survival function for $t=t_1=\ldots=t_m$, i.e. the probability that no event is observed before time t, can also be defined by the following expression as the product of partial survival functions $S_i(t \mid x)$ corresponding to the partial events:

$$S(t,...,t|x) = \exp\left\{-\int_0^t r(\tau|x)d\tau\right\} = \exp\left\{-\int_0^t \sum_{i=1}^{m} r_i(\tau|x)d\tau\right\} =$$

$$= \prod_{i=1}^{m} \exp\left\{-\int_0^t r_i(\tau|x)d\tau\right\} = \prod_{i=1}^{m} S_i(t|x) \tag{2}$$

The partial survival functions $S_i(t \mid x)$ give the probability that the i-th event is not observed before time t if only this single event is considered. The product form of the expression for the joint survival function implies stochastic independence of the partial durations T_i conditional on the values of the covariates x. Thus, it is always possible to specify a model with conditionally independent risks that for $t=t_1=\ldots=t_m$ is equivalent to a given joint survival function. Both models imply the same conditional probabilities to observe an event for $t=t_1=\ldots=t_m$. In general, this is not true for combinations of different partial durations, $t_i \neq t_j$. But such combinations cannot be observed empirically. Therefore, it is not possible to distinguish between both models on the basis of empirical data only. Additional prior restrictions are required in order to identify a model with dependent competing risks. Since in most cases no precise theoretical models of the processes considered are available, a specification based on the assumption of conditionally independent risks may appear to be reasonable.

Therefore, it is common practice, not only in microsimulation models but also in general models of competing risks, to use a model specification based on conditional independence of the risks. This assumption allows the hazard rate of one partial event to be modelled independent of the specification adopted for other processes. Given longitudinal data, the parameters of the model can be estimated by standard methods of event history analysis.[3] Then, the joint survival function is defined as the product of the partial survival functions of the partial risks.

An example of this approach is the SOCSIM microsimulation model. In this model, a constant partial hazard rate is assumed for each process, conditional on a subset of the state variables and on the age group of the individual. This implies a piecewise exponential partial survival function. Then, given a starting point in time, for each process i, the time span T_i till the next event is simulated by Monte Carlo techniques: A pseudo random number uniformly distributed in the interval [0,1] is generated, and using the inverse of the partial survival function that duration is computed for which the value of the partial survival function equals the value of the random variable (Hammel 1990, pp. 14ff). After all partial durations have been computed, that event with the smallest partial duration is selected and the corresponding event is executed in the simulation model, i.e. the state of the individual is altered correspondingly. Then in the next simulation step again the time of the next event is simulated for all partial risks, conditional on the new state at that point in time. Again the next event is determined by selecting the smallest duration. This procedure is repeated until the death of the individual has been simulated.[4] This simulation procedure is a straightforward implementation of the continuous time competing risks model.

From a substantive point of view, a drawback of a specification of conditionally independent partial processes is that the probability densities of two events change proportionally in response to some exogenous factor that does not enter directly into the hazard rate functions of the two events considered. If x_i are the covariates in the hazard rate function for the i-th partial risk, the following expression is obtained for the probability density for an event by taking the partial derivative of the joint survival function:

---

3 However, if the partial hazard rate functions have some covariates in common, the parameters for the partial hazard functions must be estimated simultaneously because censoring is not independent for the different risks.

4 Since in SOCSIM II mortality is assumed to be independent of the current state of the individual, the total life span of the individual can be determined at the time of birth of the individual. However, for state-dependent processes like fertility, the sequential procedure of simulation is applied.

$$f_i(t|x) = -\frac{\partial}{\partial t_i} S(t_1,...,t_m|x)\Bigg]_{t_1=...=t_m=t} =$$

$$= -\frac{\partial}{\partial t_i}\prod_{j=1}^{m} S_j(t_j|x_j)\Bigg]_{t_1=...=t_m=t} = r_i(t|x_i)\prod_{j=1}^{m} S_j(t|x_j) \tag{3}$$

The ratio of two probability densities depends only on the rate functions of the two events and is independent of covariates that do not directly influence these two rates. This implies that a change in a covariate that only affects a third hazard rate will result in proportional changes in the two probability densities considered.[5]

$$\frac{f_i(t|x)}{f_j(t|x)} = \frac{r_i(t|x_i)\prod_{k=1}^{m} S_k(t|x_k)}{r_j(t|x_j)\prod_{k=1}^{m} S_k(t|x_k)} = \frac{r_i(t|x_i)}{r_j(t|x_j)} \tag{4}$$

This property is similar to the independence from irrelevant alternatives property of the independent multinomial logit and probit models for qualitative data. Thus, if for instance in a demographic simulation mortality changes, the conditional probabilities for all other events will change proportionally. This may not be very plausible in all cases, since for instance one might assume that individuals who survive longer due to improved medical services will differ in their behaviour from individuals with comparatively better health. However, this cannot be modelled using conditional independence, but a model which allows for dependencies is required.

9.2.2. The Problem of Unobserved Heterogeneity

The continuous time approach to competing risks loses much of its simplicity if there is unobserved heterogeneity in the individual units. This is a common problem in empirical studies since the models used in practice do not generally include all relevant explanatory variables. Usually, some variables cannot be controlled for or are not even known to be of importance for the problem. Such factors can be modelled as latent random variables that affect the hazard rates, in addition to the impact of the explanatory variables

---

[5] The same is true for the conditional probability densities of the events, given that one of the events occurs at time t.

considered explicitly in the model. For each individual, unobserved heterogeneity results in deviations of the actual hazard rates for the unit from the values derived from the systematic part of the model. However, since the latent factors are not observed, they cannot be controlled for when the model is estimated.

In the most simple case, there is only one latent variable u that influences the different hazard rates, but that is not included explicitly in the model structure. In this case the true joint survival function depends on both the vector x of explanatory variables of the model and on the latent variable u. However, since the variable u is not controlled for, the model defines the survival function that is marginal on u. Given the distribution $M(u)$, this marginal survival function is given by the following expression, with E_u being the expected value with regard to u:

$$S(t_1,...,t_m|x) = \int S(t_1,...,t_m|x,u)dM(u) = E_u\{S(t_1,...,t_m|x,u)\} \tag{5}$$

Given the observed data, only the hazard rates corresponding to this marginal survival function can be estimated, since the survival function of the basic model depends on unobserved factors. The hazard rates corresponding to the marginal survival function in general depend on all covariates in the model in a complex way, even if in the basic model the partial risks are conditionally independent given x and u, i.e. the joint survival function can be factored into the product of partial survival functions. This can be easily seen if, for the basic model, independent partial survival functions $S_i(t \mid x_i,u)$, $i=1,\ldots,m$ are assumed, conditional on the covariates x_i and the heterogeneity factor u. In this case the following relation holds for the hazard rates of the marginal distribution:

$$r_i(t|x) = -\frac{d}{dt_i}\log E_u\left\{\prod_{k=1}^{m} S_k(t_k|x_k,u)\right\}\Bigg]_{t_1=...=t_m=t} =$$

$$= \frac{E_u\left\{r_i(t|x_i,u)\prod_{k=1}^{m} S_k(t|x_k,u)\right\}}{E_u\left\{\prod_{k=1}^{m} S_k(t|x_k,u)\right\}} =$$

$$= \frac{E_u\left\{ r_i(t|x_i,u) \prod_{k=1}^{m} \exp\left[-\int_0^t r_k(\tau|x_k,u)d\tau \right] \right\}}{E_u\left\{ \prod_{k=1}^{m} \exp\left[-\int_0^t r_k(\tau|x_k,u)d\tau \right] \right\}} \tag{6}$$

The expression for the hazard rates of the marginal model is rather complex. It does not equal the expected value of the hazard rate of the basic model, but the marginal hazard rate in general depends on all covariates of all risks in the model.[6] If unobserved heterogeneity is present, the marginal hazard rates that can be estimated from the observed data have no simple functional form. Also, the coefficients of the covariates have no direct causal interpretation, since covariates that affect only one hazard rate in the basic model will also influence all other hazard rates in the marginal model. It is not possible to distinguish between direct causal effects of the covariates on the hazard rate and indirect effects that are caused by unobserved heterogeneity. In this respect, models of competing risks with unobserved heterogeneity are comparable to the reduced form of an unidentified system of simultaneous equations in which also all explanatory variables of the model enter into each equation and the coefficients cannot be given a causal interpretation.

According to the non-identifiability property, the hazard rates of the marginal model can be used to define a model with conditional independence that is observationally equivalent to the original model. But, if unobserved heterogeneity is present, this model generally has a rather complex form with no causal interpretation. Thus, the use of the independence assumption becomes questionable. From a substantive point of view, a structural model that allows for unobserved heterogeneity is preferable over the marginal model since it allows one to infer about the causal effects of the explanatory variables, while this is not possible on the basis of a marginal model if no restrictive assumptions are made like the absence of unobserved heterogeneity.

In principle, models with explicit unobserved heterogeneity can be derived as mixtures of the basic model using an appropriate mixing distribution to

---

6 Only if there are different, stochastically independent heterogeneity factors, each affecting only one partial risk, the marginal survival function can be factored into the product of marginal functions for the different risks, and the hazard rates depend only on the covariates affecting that risk in the basic model. Given a structural model for the hazard rates, the effects of covariates can be used to identify the distribution of the heterogeneity factor (cf. Heckman & Honoré 1989).

describe unobserved heterogeneity. However, such mixture models generally have a very complex structure. Even if only simple bivariate problems are considered, only some rather restrictive specifications can be handled even in theory (e.g. Oakes 1989, Hougaard *et al.* 1992). Given the statistical methods available, more general models are infeasible because of their complexity. At present, unobserved heterogeneity of the micro units leads to severe problems for continuous time models of competing risks. But it should be stressed that this is primarily a general problem of quantitative modelling and not a special problem of microsimulation, since it is not difficult to simulate dependent processes if an appropriate model is available.

9.3. DISCRETE TIME MODELS OF COMPETING RISKS

9.3.1. The Basic Statistical Structure

If a discrete time framework is used for modelling competing risks, some differences to the continuous time approach must be taken into account. First of all, in such models, the state of a unit is only modelled for discrete points in time, with no information on the state within the corresponding time intervals. The individual biography is described by a finite sequence of state vectors z_t corresponding to the periods $t=0,1,\ldots$ with the state variables giving the state at the start of the period, at the end of the period, or some average for the period. In general, this allows one to simplify the model substantially since the timing of events within a period is disregarded. The data requirements of discrete time models are also less demanding. However, a discrete time approach also implies that only information on the state at discrete points in time can be used to model the behaviour of the unit. Especially changes of the state variables that occur within the time interval cannot easily be taken into account in a discrete time model.

A second property of discrete time models is that, within a given finite time interval $[\tau_1,\tau_2]$, *more than one event* may occur with a positive probability. Even if only one event can take place at a given point in time, a sequence of more than one event may be observed over a time interval. Repeated observations of the same event are also possible within the same interval. Thus, in a discrete time framework, events are not competing in a strict sense; the possibility of multiple events must be taken into account. In addition, dependencies between different events may occur if the partial processes considered depend on the state of the unit. In this case an event taking place in a time interval will alter the state of the unit, and thereby induce a change in the probability to observe another event in the same time interval. This creates stochastic dependencies between the different possible events. An example is the dependency between marriage and fertility: If a marriage takes place in a given time interval, the probability to also observe a childbirth in the same period will be higher than in the absence of marriage.

In theory, the structure of a discrete time model can be derived from a corresponding continuous time model by integrating the model for the time intervals considered. However, in general, this results in rather complex relations that are not manageable even in theory. All possible sequences of events and the corresponding changes in the state of the unit must be considered in order to derive the joint probability distribution of the events that are possible in a finite time interval. At present, this is not possible if not very restrictive assumptions are introduced with regard to the properties of the continuous time processes (cf. Allison 1982). Therefore, discrete time models have been developed almost independently from continuous time models.

Basically, modelling the dynamics of a unit in a discrete time framework amounts to modelling changes in the state of the unit between given points in time. The state at time t is given by the vector z_t of the values of the state variables. Given the state vector z_t at time t, the transition of the unit to a state z_{t+1} at time $t+1$ must be explained. Since, due to our limited knowledge, such transitions cannot be fully explained, a stochastic approach is used with the probability to observe a given transition depending conditionally on the original state of the unit and eventually on other exogenous variables. Thus, the main problem in a dynamic discrete time model is to specify the probability distribution of the state vector z_{t+1}, conditional on the state at time t and eventually also conditional on additional explanatory variables x_t that are not elements of the state vector. These additional variables may, for instance, describe environmental factors that are not part of the state of the unit, but affect the transitions between states.[7] If in addition a semi-Markov process is assumed, the dynamics of the process are given by conditional probabilities of the form $f(z_{t+1} \mid z_t, x_t) = \text{Prob}\{Z_{t+1}=z_{t+1} \mid z_t, x_t\}$.

In Markov type transition models, these transition probabilities are modelled directly. But since each combination of elementary events defines a new state, and since the number of possible transitions that must be considered in the model is defined as the product of the number of states at two points in time, the size of such a model increases rapidly with the number of different events considered. This is the problem of combinatorial explosion of transition models that makes models with a large number of different events infeasible in practice. In addition, usually insufficient empirical information is available to estimate probabilities for all different transitions separately. Thus, more

[7] For example, the probability to observe a change of occupation may depend on the number of job vacancies in different occupations.

restrictive model structures must be used that are less demanding in terms of both the size of the model and empirical data requirements.

In practice, some simplifications are obtained by imposing additional restrictions on the conditional probabilities, especially by introducing some independence assumptions. For this purpose, it is useful to consider the joint probability distribution of the elementary events and not the transitions probabilities for the state vector. If $Y_{i,t}$, $i=1,\ldots,m$ are random variables describing the outcome of the m partial processes being considered for the time interval $[t,t+1]$, and (z_t,x_t) is the vector of explanatory variables at the start of the interval, the joint outcome of all processes can be described by the conditional joint probability function

$$f(y_{1,t},\ldots,y_{m,t} \mid z_t,x_t) = \text{Prob}\{Y_{1,t}=y_{1,t},\ldots,Y_{m,t}=y_{m,t} \mid z_t,x_t\}$$

giving the probability that a vector of outcomes $y_t=(y_{1,t},\ldots,y_{m,t})$ is observed given the explanatory variables (z_t,x_t).

In general, the joint conditional probability function of the elementary events will have no simple structure that can be handled easily in a simulation model. Especially stochastic dependencies may occur between different events. Therefore, usually a factorization of the joint probability function is used to simplify the model. By definition, the joint probability function can always be factored *recursively* into a product of the marginal probability of one event and the univariate probabilities of the other events conditional on the vector x_t, and on the events that have been modelled before in the sequence:

$$f(y_{1,t},\ldots,y_{m,t}|z_t,x_t) = f_1(y_{1,t}|z_t,x_t)\times\prod_{i=2}^{m} f_i(y_{i,t}|z_t,x_t,y_{1,t},\ldots,y_{i-1,t}) \tag{7}$$

Such a factorization simplifies the model structure to some extent, since it allows one to account for dependencies recursively. This is the approach that has been used in discrete time microsimulation models like the DYNASIM model, the FRANKFURT microsimulation model, or the NEDYMAS model. In each period the events are simulated in a recursive order, conditional on the state at the start of the period and on the events simulated before in the simulation sequence. Technically, the simulation is performed for each event by first calculating the conditional probability to observe the event, given the state of the unit and the other events simulated before. Then a pseudo random number is generated and, depending on the outcome, the event is assumed to take place or not. This procedure is repeated recursively for all elementary events that are considered in the model.[8] The new state vector is derived by evaluating the consequences of the different events being simulated.

This approach is fairly simple to implement in a simulation model. However, the specification and estimation of the model poses some methodological problems. First of all, the recursive factorization of the joint probability in general has no causal interpretation. The dependencies between the events are of a correlative nature and depend on the sequence of events being adopted. But, in principle this sequence can be chosen arbitrarily. An attempt is made in the simulation models to specify the sequence of events in such a way that it can be given a plausible causal interpretation. However, if the true model structure is not recursive, i.e. if there are dependencies between the different events, the recursive decomposition of the joint probability distribution is a semi-reduced form of a simultaneous system. In such a system, the coefficients of the explanatory variables have no direct causal interpretation, since they represent a mixture of direct and indirect effects of the corresponding variable.

9.3.2. Fixed Versus Random Order of Events

An additional problem results from the fact that the recursive structure of the model implies that the probability of a given event is modelled conditional on the events that are modelled before in the recursive sequence and marginal with regard to all events following later on.[9] But in many cases the empirical data available do not allow such probabilities to be estimated without additional restrictive assumptions. In general, a micro data set is required for such estimates that contains longitudinal information on all processes that are considered in the model. Only then is it possible to estimate probabilities that are conditional with regard to some of the events, and marginal with regard to the others without further assumptions. But most data bases available to research do not contain information on all events considered in the model, but are restricted to a subset of events or even to only one event. Typically, information is available on the state of the unit before an event, but little or even nothing is known on other events that have occurred in the same period.[10] This is for

---

[8] For some events the procedure can be simplified if a given outcome will be observed with a probability equal to one. In this case no random experiment is required, but the outcome can be determined directly. An example is the event of divorce for an unmarried person which has probability zero.

[9] This is similar to the recursive structure of the canonical solution of a simultaneous equation system in which, in each equation, all endogenous variables explained before in the sequence may occur explicitly as explanatory variables, but variables being explained later on in the sequence have been eliminated from the equation.

[10] Not only information on other events that occurred before the event considered is required, but also events that take place later on in the same period. But data on events like marriages

instance true for register data on marriages or on fertility that do not contain information, for example, on occupation. This problem becomes even more severe if demographic processes are considered in the context of families and households with dependencies between events for different family members.

As a consequence, in practice, probabilities can be estimated conditional on the state of the unit but not conditional on other events. For example, there are many data sets containing information on labour supply given marital status, but there are only a few from which the probability of a change in labour supply can be estimated conditional on a change in marital status in the same period. More comprehensive information on events is available, to some extent, from survey data, and especially from panel surveys in which all or at least most of the events are recorded. From such data, in principle, the conditional probabilities required for the model can be estimated. However, a drawback of these data sets is the comparatively small number of observations and, at least in some cases, the selectivity of the sample that makes it difficult to derive precise and unbiased estimates.

If no information on other events is available in the data, conditional independence of the events given the state of the unit at the start of the time interval must be assumed for the model. Given this assumption, estimates conditional on the state are sufficient to model the process, since the probability of an event does not change depending on another event. However, this is a rather restrictive assumption in demographic models. A counter example is the death of the individual that obviously will affect the probability to observe other events in the same period. To a smaller extent such dependencies also exist for other events, like the formation of marriages and partnerships, labour supply, and fertility. Thus, the assumption of independent events is not suitable for such models and other solutions should be developed.

An improvement can be achieved if the approach of a continuous time model is considered and the assumption of conditional independence of the events is maintained but the effects of changes in the state variables within the time interval are taken into account. In such a specification, an event has no direct effect on the probability to observe other events. But these probabilities may change conditional on the change in the state variables induced by the event. Given this assumption, the probability to observe an event given other events can be computed as the integral of the corresponding probability density of a continuous time model for that interval, conditional on the sequence of states of the unit in that interval. In general, this results in very complex expressions. But if some simplifying assumptions are introduced, at least an approximate solution can be obtained.

---

or births usually only contain information on the past and not on the future, since the information is collected at the time of the event.

If the conditional probability density for an event given the state vector is assumed to be constant, the integral of this conditional probability over the time interval equals the sum of the conditional probability densities for the different states weighted by the duration in these states. Thus, the probability to observe an event conditional on another event can be computed as a mixture of the probabilities conditional on the state before and after the conditioning event. The idea behind this is that the individual is exposed to the risk to experience the event considered for some time in two (or more) different states. If there is no proper effect of the event, the conditional probability given a change of the state can be computed as the sum of the probabilities conditional on the state weighted by the time spent in the different states.

This approach is widely used in demography to calculate rates taking into account the risk of death in a discrete time framework. It is also applied in microsimulation models like the FRANKFURT model to account for the effects of the death of a unit on other processes. If an individual is simulated to die, the probability to experience any other event in the same period is reduced in the FRANKFURT model to one half, based on the assumption that individuals dying live half the period on average. This assumption corresponds to the practice of calculating crude rates by dividing the number of events observed by the average number of individuals at risk in a given period. This procedure could easily be generalized by first simulating the exact time of death within the period and then changing the probabilities for other events accordingly.

However, except for death, this procedure is not applied to other events since it would be necessary to consider all possible sequences of events and to compute the corresponding marginal probabilities. This would be very complex and make the approach infeasible in practice. Therefore, with regard to other events, a more simplified approach has been used: State-specific probabilities are used, conditional on the state of the individual, given the events that have been simulated earlier in the sequence. In this way, the effects of events simulated earlier in the sequence can be taken into account for those processes simulated later on. However, no feedback to events simulated earlier in the model is possible in the same period. Such dependencies can only be taken into account in the simulation of the next period.

This approach implies the assumption that the events take place at the very beginning of the period and that the resulting state vector is valid for the whole time interval. Obviously this is not fully consistent since except for mortality this assumption is applied to all events simulated in the recursive order. But changes in the state variables that are caused by events that are simulated later on in the sequence are not taken into account for those events simulated before. Therefore, estimates derived from the model are biased to some extent, since the conditional probabilities being applied are not fully correct. However, this bias appears to be small as compared to other errors in the models that result, for instance, from an erroneous model specification or from poor estimates. This must be compared with the advantages of a fixed recursive order that allows a rather simple structure of the simulation.

A different approach has been adopted in the DARMSTADT model and in the model of the Hungarian Statistical Office. In these models, probabilities are also applied that are marginal with regard to other events, but depend on the state of the individual at the given simulation step. For technical simplicity, the events are also simulated in a recursive order. But, in order to balance the effect of the recursive simulation approach, a randomization of the simulation sequence is applied. For each individual in the simulation, the sequence in which the events are simulated recursively is determined by a random procedure. Then state-specific probabilities are applied to the different processes, conditional on the state resulting from the events having been simulated before in the sequence. This implies that, for an individual unit, biased probabilities are applied, since the timing of the events and the different states within the time interval are not considered. But this bias will tend to average out over the sample of microunits due to the randomization of the simulation sequence (Heike *et al.* 1988a).

The randomization approach makes it unnecessary to calculate the probabilities of the events conditional on a given sequence of events. Rather, simple state-specific probabilities are used. This simplifies the estimation of probabilities. However, the approach also rests on the assumption that the probabilities are independent of other events and are determined only by the state of the individual. Thus it provides no final solution to the primary problem of dependencies between different events in a discrete time framework.

9.3.3. Unobserved Heterogeneity in Discrete Time Models

Like in the continuous time framework, dependencies between events may be caused by unobserved heterogeneity of the individuals. It can be taken into account if a latent variable approach is used to model the probabilities of the elementary events. In such a model, a latent variable $y_{i,t}^*$ is associated with each event i at time t. The event is observed if the latent variable exceeds a threshold value. This latent variable is usually modelled to be a linear function of the explanatory variables and a random error term. The probability to observe an event is then given by the probability that the latent variable associated with the event exceeds the threshold value.

In a general model specification, unobserved heterogeneity, state dependence, or autocorrelated errors can be taken into account, in addition to the effects of exogenous explanatory variables. If a dummy variable y_{it} is used to indicate event i, z_t is the state of the unit at the start of the interval, x_t is a vector of exogenous explanatory variables, u is a latent variable representing the effect of unobserved heterogeneity of the unit, and e_{it} is a random error term, the following specification with parameters α, β and γ results for the i-th event for a given unit:[11]

$$
\begin{aligned}
y_{i,t}^{*} &= \alpha_i' z_t + \beta_i' x_t + \gamma_i u + e_{i,t} \\
y_{i,t} &= 1 \quad \text{if } y_{i,t}^{*} > 0 \\
& 0 \quad \text{otherwise} \\
\text{Prob}\{y_{i,t}=1 \,|\, z_t, x_t, u\} &= \text{Prob}\{e_{i,t} > -(\alpha_i' z_t + \beta_i' x_t + \gamma_i u)\}
\end{aligned}
\tag{8}
$$

Since the latent variable is not observable, the threshold can be defined arbitrarily, for instance to equal zero. Also, the expected value and the variance of the error term cannot be identified. Usually identically distributed error terms e_{it} are assumed, with $E\{e_{it}\}=0$ and $\text{Var}\{e_{it}\}=1$, that are mutually uncorrelated and independent of the other variables in the model.[12] Similarly, an expected value of $E\{u\}=0$ must be assumed for the heterogeneity factor in order to identify a constant term in the equation. For simplicity, the heterogeneity factor u is also assumed to be independent of the variables in x_t of all periods.[13] In order to ensure identifiability of γ, the variance of the heterogeneity terms is restricted to σ_u^2. If the random variables are distributed according to a multivariate normal distribution, a random effects panel data probit model is obtained. In principle, such a model can be estimated from panel observations (cf. Hsiao 1986, pp. 167ff).

The basic theory for panel data probit models has been developed for single equation models. However, the approach can be generalized to problems with more than one event. In this case, a system of probit equations must be considered with correlated random terms. If the same basic structure with a common random factor u representing unobserved heterogeneity is assumed for the equations corresponding to the different events, the full model can be given in matrix notation. The variables are collected in the vectors y_t^*, z_t, x_t and e_t while the matrices A and B and the vector γ contain the coefficients of the model, and Φ is the standard normal distribution function. The error terms e_{it}

---

[11] Assuming only a first order dependency on the state z_t at the start of the interval is not restrictive, since earlier states before t can be included by introducing additional state variables giving lagged values of state variables. Also note that since z_t represents the state at the start of the interval, and therefore does not depend on the events in the interval, the consistency of the model is maintained.

[12] A more general error structure with autocorrelated errors e_{it} can also be considered (cf. Hsiao 1986, pp. 167f).

[13] In a more general specification, u can be made a linear function of all x variables plus a random error. Cf. Chamberlain (1984, p. 1271f); Hsiao (1986, p. 165f).

are assumed to be independent random variates, uncorrelated over equations, periods, and individuals:

$$\begin{aligned} y_t^* &= A' z_t + B' x_t + \gamma u + e_t \\ y_{i,t} &= 1 \quad \text{if } y_{i,t}^* > 0 \\ & 0 \quad \text{otherwise} \\ \text{Prob}\{Y_{it}=y_{i,t} \mid z_t, x_t, u\} &= \Phi\big((2y_{it}-1)(\alpha_i' z_t + \beta_i' x_t + \gamma_i u)\big) \end{aligned} \tag{9}$$

Since the heterogeneity factor u is not observable, it is useful for the simulation to consider the specification of the marginal model with regard to u. In this model the error terms are defined by $v_t = \gamma u + e_t$ with the following properties:

$$\begin{aligned} E\{v_t\} &= 0 \\ E\{v_t v'_s\} &= \gamma\gamma' + I \quad \text{for } t=s \\ & \gamma\gamma' \quad \text{for } t \neq s \end{aligned} \tag{10}$$

Since, given the assumption of multivariate normality, the marginal model for y_{it} also defines a probit model, the coefficients in A, B and γ can be estimated using the single equation probit model for panel data for each equation individually (Chamberlain 1984, pp. 1296ff). Due to the restrictions in the model specification, the correlation structure of the error terms of the marginal model is fully defined if the autocorrelation γ_i^2 of the error term is known for all equations. Since this quantity can be estimated from a single equation if panel data are available, a single equation approach is sufficient.

Given the specification (10) of the error structure, the marginal model can be estimated even from cross-sectional data, if information on z_t and x_t is available. If the model is reparametrized so that the variance of the error term of each equation equals one, $E\{v_t v'_t\} = \Sigma$ with $\sigma_{ii} = 1$, the coefficients in A and B can be estimated by the single equation probit method. Given these estimates, the correlation between the error terms of two equations can be estimated by a procedure proposed by Muthén (1983, 1984). If a pair of events is considered, the joint probability to observe the events results from a bivariate probit model with correlation coefficient $E\{v_{it} v_{jt}\}$. Thus, this correlation can be estimated from such a model restricting α_i, α_j, β_i, and β_j to the values estimated before. This approach has been implemented in the LISCOMP program that can be used for estimation. It can also be used to estimate the model from panel data (cf. Muthén 1983).

Thus, in a discrete time framework, it is possible to model dependencies between events that result from unobserved heterogeneity. The model specification used here is rather restrictive. It can be generalized further to some extent by allowing for a less restrictive error structure. Also, the approach can be generalized to model both quantitative and qualitative dependent variables in a simultaneous equation framework (cf. Maddala 1983, pp. 117ff). In such a hybrid model, dependencies between qualitative events and other quantitative socio-economic variables, like labour supply or income, can be considered explicitly.

When estimating a dynamic model of this kind, the initial state of the units poses serious problems, since the processes considered in the model are not usually observed from their very beginning. Basically, the initial state z_0 is an outcome of the processes considered, and is therefore a random variable that is correlated with the heterogeneity term u. Monte Carlo experiments by Heckman (1981) suggest that, in this case, treating the initial state as exogeneously fixed may result in severely biased estimates, especially if only a small number of panel observations is available for each unit in the sample. In theory, unbiased maximum likelihood estimates can be obtained from left censored sample observations, using the conditional probability density of the initial state given the heterogeneity term (cf. Hsiao 1986, pp. 169ff). In practice, however, this approach is infeasible in most cases, since the conditional distribution of the initial state cannot be derived.

As an approximate solution to this problem, for a single equation model, Heckman (1981) has suggested to add an additional probit equation for the initial state of the unit, depending on all exogeneous variables and without any restrictions on the random term. In this way, the initial state is treated as an endogeneous variable. In principle, this approach could also be applied to a multivariate problem. Since in this case the state of a unit is represented by a vector of state variables, a system of reduced form equations should be specified for the initial state of the unit. Then the whole model including these auxiliary equations is estimated. However, the properties of such an approach are still to be investigated.

9.4. CONCLUSIONS

Summing up, the model structures used in contemporary microsimulation models for competing risks are not fully satisfactory. Especially unobserved heterogeneity is not taken into account in the simulation models. Considering the deficiencies of the models, this appears to be a major problem. Simulation results are probably biased to some extent since dependencies between different events resulting from unobserved heterogeneity are not properly taken into account. However, this is not only a problem of microsimulation models, but it is also a shortcoming of the theoretical models used for the model specification.

If the continuous time and the discrete time approach are compared, the latter has some advantages with regard to modelling such dependencies. While in a continuous time framework mixture models that allow for dependencies have a rather complex structure and in general are infeasible, the latent variable approach to discrete time models, in combination with two-step estimation procedures, accommodates at least some types of dependencies in a feasible manner. Since the approximation of the joint probability density by recursive factorization can be discarded if a model specification allowing stochastic dependencies between the events is used, the recursive structure of contemporary models can also be overcome in this way. Such a model structure could be implemented in a microsimulation model without difficulties.

An additional advantage of the discrete time approach is that it allows multivariate models to include not only qualitative dependent variables, but also continuous and truncated dependent variables in a hybrid model structure. In this way, very flexible model structures can be obtained, that explain demographic events jointly with other quantitative or qualitative socio-economic variables. An example is the use of income as an explanatory variable in a model of child birth. This may improve the explanatory power of such models. Corresponding mixed model structures are presently not available for the continuous time approach.

REFERENCES

ALLISON, P.D. (1982), Discrete-Time Methods for the Analysis of Event Histories. *In*: S. LEINHARDT (ed.) *Sociological Methodology 1982*, Jossey-Bass: San Francisco.

CHAMBERLAIN, G. (1984), Panel Data. *In*: Z. GRILICHES & M. INTRILIGATOR (eds.) *Handbook of Econometrics*, Vol. 2, North Holland: Amsterdam, pp. 1247-1318.

CSICSMAN, N., & PAPPNE, N. (1987), The Software Developed for the Hungarian Microsimulation System. *In*: *Proceedings of the IIASA Conference on Demographical Microsimulation*, Budapest.

GALLER, H.P. (1988), Microsimulation of Household Formation and Dissolution. *In*: N. KEILMAN, A. KUIJSTEN & A. VOSSEN (eds.) *Modelling Household Formation and Dissolution*, Clarendon Press: Oxford, pp. 139-159.

HAMMEL, E.A. (1990), *SOCSIM II*, Graduate Group in Demography, University of California, Working paper No. 29, Berkeley.

HECKMAN, J. (1981), The Incidental Parameter Problem and the Problem of Initial Conditions in Estimating a Discrete Time-Discrete Data Stochastic Process. *In*: CH. MANSKI & D. MCFADDEN (eds.) *Structural Analysis of Discrete Data with Econometric Applications*, MIT Press: Cambridge, pp. 179-195.

HECKMAN, J., & HONORÉ, B. (1989), The Identifiability of the Competing Risk Model. *Biometrika*, Vol. 76, pp. 325-330.

HEIKE, H.D. *et al.* (1988a), Das Darmstädter Mikrosimulationsmodell - Überblick und erste Ergebnisse. *Allgemeines Statistisches Archiv*, Vol. 72, pp. 109-129.

HEIKE, H.D. *et al.* (1988b), Der Darmstädter Pseudo-Mikrosimulator, Modellansatz und Realisierung, *Angewandte Informatik*, Vol. 1/1988, pp. 9-17.

HELLWIG, O. (1988), *Probleme und Lösungsansätze bei der Datenbeschaffung und Parameterschätzung für Mikrosimulationsmodelle der privaten Haushalte*, PhD thesis, Technical University of Darmstadt.

HOUGAARD, P. *et al.* (1992), Measuring the Similarities between the Lifetime of Adult Danish Twins Born between 1881-1930, *Journal of the American Statistical Association*, Vol. 87, pp. 17-24.

HSIAO, C. (1986), *Analysis of Panel Data*, Cambridge University Press: Cambridge.

LAWLESS, J. (1982), *Statistical Models and Methods for Lifetime Data*, John Wiley: New York.

MADDALA, G. (1983), *Limited Dependent and Qualitative Variables in Econometrics*, Cambridge University Press: Cambridge.

MUTHÉN, B. (1983), Latent Variable Structural Equation Modelling with Categorical Data, *Journal of Econometrics*, Vol. 22, pp. 43-65

MUTHÉN, B. (1984), A General Structural Equation Model with Dichotomous Ordered Categorical, and Continuous Latent Variable Indicators, *Psychometrika*, Vol. 49, pp. 115-132.

NELISSEN, J. (1989), *NEDYMAS: Microsimulation of Household and Education Dynamics*, Growth Dynamics University Institute Discussion Paper No. 89/3, Erasmus University: Rotterdam.

OAKES, D. (1989), Bivariate Survival Models Induced by Frailties, *Journal of the American Statistical Association*, Vol. 84, pp. 487-493.

ORCUTT, G. *et al.* (1976), *Policy Exploration through Microanalytic Simulation*, The Urban Institute: Washington D.C.

TSIATIS, A. (1975), A Nonidentifiability Aspect of the Problem of Competing Risks, *Proceedings of the National Academy of Sciences* (USA), Vol. 32, pp. 20-22.

TUMA, N., & M. HANNAN (1984), *Social Dynamics, Models and Methods*, Academic Press: Orlando.

Part III

Models

10. ALTERNATIVE OPTIONS FOR LIVING ARRANGEMENT MODELS: A SENSITIVITY ANALYSIS

Christopher Prinz,[1] Åke Nilsson,[2] and Hakan Sellerfors[2]

[1]European Centre for Social Welfare Policy and Research
Berggasse 17
A-1090 Vienna
Austria
[2]Population Unit
Statistics Sweden
S-70189 Örebro
Sweden

Abstract. Projections of the living arrangements structure of the Swedish population are carried out using actual marital status (in contrast to legal marital status) as the main criterion. The analyses and projections are based on an excellent Swedish data set which gives unique information on consensual unions as compared to marriages. By comparing demographic differentials in fertility, mortality, and couple formation and dissolution, both by actual and legal marital status, it is concluded that de facto marital status is by far more discriminatory than de jure status. Fertility among cohabiting women is found to be higher than usually recognized. Several alternative status selections as input to the multi-state projection model are suggested and tested, ranging from the 'minimum model' which only considers actual marital status to the 'complete model' which considers all actual and legal marital status combinations. It is found that, for the projection of living arrangements, the inclusion of legal marital status information is superfluous. Based on conservative assumptions, the proportion of consensual unions among all unions is projected to increase from about 23 in 1985 to 34 per cent by 2020 among the working age population, and from three per cent in 1985 to 12 per cent by 2020 among the elderly population (aged 65 years and over). While the number of consensual unions will increase remarkably, even by 2020 the majority of Swedes would still live in marital unions. Projected developments in the legal marital status distribution in no way reflect expected changes in actual marital status, i.e. in living arrangements.

Household Demography and Household Modeling
Edited by E. van Imhoff *et al.*, Plenum Press, New York, 1995

10.1. INTRODUCTION

This chapter is concerned with the projection of the living arrangements structure of the Swedish population, both looking at methodological issues and at implications of results. Estimating current and future actual living arrangements of a population was simple as long as marital status was largely identical with living arrangements. Data on nuptiality, divorce, and widowhood were easily available and reasonably complete. Due to the upswing in new, non-traditional living arrangements, like for example consensual unions, flat-sharing communities, or relationships where both partners maintain their own homes (so-called 'living apart together' relationships), it is increasingly difficult to estimate the living arrangements structure of the population. If it is already problematic to get information on the structure, this is even more strongly the case for information on dynamics, a fact which makes living arrangement projections even more difficult. Sophisticated multistate projection models require detailed information both on stocks and flows, for which data are not easily available. The use of marital status projection models becomes less and less justifiable, although they are still useful for certain applications, as for example for comparative studies or for studies that specifically analyse legal marital status, e.g. for social security authorities.

The idea of this chapter is to start from a conventional marital status model and to extend this model, making it as simple as possible, but at the same time as sophisticated as necessary to cover the most essential types of living arrangements. A main difference is whether people live alone or in company. Even nowadays in most cases to live in company means to live as a couple. Among couples, the main distinction is that between marital and consensual unions. Thus, three major living arrangement groups, or groups of 'actual marital status' can be identified: single (i.e. living alone), married, and cohabiting. Each of those groups may be subdivided further. Singles and cohabitees may be classified by legal marital status, couples in general may be classified by order of their union, be it marriages or cohabitations.[1]

A critical decision when modelling living arrangements using a multistate projection model[2] is the determination of the states to be used. Essentially, the states selected should both make sense from a substantive point of view and be discriminatory from a statistical point of view. Depending on the specific data

---

[1] One might also classify cohabiting and married couples — and even singles — by their number of children. In this paper, such a disaggregation is, however, not investigated further.

[2] For a description of the multistate projection model see for example Rogers (1975), Willekens & Drewe (1984) or Van Imhoff (1990).

set, there are also limits to the possible maximum number of states. In the current paper, the states single, married, and cohabiting are regarded as a minimum requirement, while additional selections arise from subdividing those three groups further. The question of heterogeneity heavily influences the choice of the state space. Observed heterogeneity is usually dealt with by breaking down aggregates into homogeneous subgroups, both for purposes of analysis and of projections. Unobserved heterogeneity is intrinsic and not to be reduced by gathering more data.[3] A good strategy to deal with heterogeneity seems to take an aggregation approach, i.e. simplify the state space as much as possible to still obtain reasonable results. The following analysis gives an example of different selections of states from a given data set and tries to argue why one selection may be better than another. If possible, the optimum selection should be determined.

In the following section, the unique Swedish data set used for the study is discussed in the context of general data requirements. Section 10.3 is concerned with the choice of the state space for the multistate projection model. It carefully analyses demographic differentials between various legal and actual marital status groups to finally suggest several alternative model selections. Next, in section 10.4, results of the alternative living arrangements projections are presented. Eventually, conclusions with regard to the optimum choice of a model are drawn.

10.2. THE DATA

Data on the population, both by legal marital status and by actual marital status, are available from the periodical censuses in Sweden. However, a census can account only for the prevalence of a phenomenon like consensual unions, and even a sequence of quinquennial censuses (as carried out in Sweden) offers only limited information on the dynamics of the process by which such a feature spreads in a population. In this respect, data from surveys using retrospective interviews offer a much better opportunity for analysis. On the other hand, these retrospective surveys presuppose that timing and type of events may be recalled accurately. Moreover, they can give information over the period covered by the questionnaire for survivors only; mortality and emigration cannot be captured by such a data collection system. Surveys of some thousand respondents can give interesting results, but due to the low precision of the estimates and possibly sample bias they have limited suitability for the total population. A third option, which is using events from vital statistics, leads to

---

[3] For a more comprehensive discussion on observed and unobserved heterogeneity see, for example, chapter 14 in Keyfitz (1985).

another problem when calculating transition rates: flow data for the numerator stem from vital statistics and give information on legal marital status only, whereas stock data for the denominator stem from censuses and give information on actual marital status as well. Hence, numerator and denominator do not fit, e.g. a dissolution of a consensual union, clearly a change in living arrangements, is not reported in vital statistics. A mathematical solution to this problem is not trivial. The ideal data situation would be a continuous registration system on both the type of each event (including the formation and dissolution of consensual unions) and its time of occurrence.[4]

An unique Swedish data set, prepared by Åke Nilsson and Hakan Sellerfors from Statistics Sweden, gives a solution to the described problems. The data stem from four sources: the register from the population census 1985 in which people are classified according to legal and actual marital status; the register of marriages; the register of internal migration; and the register of deaths. Linking the registers has been possible on account of the Swedish system of personal identity numbers. It is this system which makes a combination of census data and statistics on couple formation and dissolution possible. To study couple formation, the census register was updated with marriages and migration during the year before the census, and to study couple dissolution, the census register was updated with migration and deaths during the year after the census. Thus, the data set gives unique information on formation and dissolution flows of consensual unions as compared to marriages and detailed data on mortality by living arrangements. It is not a continuous registration system itself, but by linking the respective registers it has all the features of such a registration system: it gives information for the total population and thus avoids any sample bias, there are no numerator/denominator inconsistencies, no memory biases, and emigration and mortality are accurately captured. Stock and flow information was obtained for the following states:

- single persons (not living as a couple) by legal marital status and marriage order,
- cohabiting persons (living in consensual unions) by the same characteristics, and
- married persons (living together or separated) by marriage order.

In addition, for the first time, age-specific fertility rates were estimated by actual living arrangements, that is, separately for single, married, and cohabiting people.

---

[4] A more comprehensive discussion on various types of possible data sets is found in Keilman & Prinz (1995).

The quality of the estimated stock data is very good. Differences with other sources are mainly due to differences in the definition of cohabitation. In this study, a cohabiting person was defined as a person who declared on the census form that he or she lives in a consensual union with someone of the opposite sex.[5] Two additional conditions have to be fulfilled: both man and woman must be registered in the same household, and both must declare that they live in a consensual union. Thus, the definition of cohabitation is strict, i.e. these data include only 'established' cohabitations. The quality of estimates on flows is more difficult to assess. Possible comparisons, like the number of divorces in vital statistics and the number of separations among married in this study, show a reasonable agreement.

10.3. THE CHOICE OF A MODEL

The choice of a model is first of all influenced by the intended purpose of the projection results. If one is interested in divorce, the respective information should be available from the model. From a theoretical point of view, the specification of the model, i.e. of the states to be used in the model, is mostly influenced by statistical criteria concerning the possible states. The starting point to determine a selection is the comparison of status-specific fertility, mortality, and couple formation and dissolution rates. Similar to cluster analysis, which could actually be used to determine the groupings statistically, states with highly different fertility, mortality, and couple formation and dissolution levels should be separated, while states with comparable levels may be grouped together.

10.3.1. Demographic Differentials

To compare the respective rates for each of the states available from the Swedish data set, standardized ratios have been computed. This method is widely used for mortality comparisons, but can be applied to other demographic variables as well. A Standardized Mortality Ratio (SMR) gives the ratio of the number of observed deaths within a given status and the number of expected deaths in this status if age-specific mortality rates of the reference group (e.g. the total population) were applied.[6] The SMR is usually multiplied by 100. With the same procedure, Standardized Fertility Ratios (SFR), Standardized Marriage Formation Ratios (SMFR), Standardized Marriage Dissolution Ratios

---

[5] Homosexual unions are not included.

[6] For this chapter, direct standardization was chosen to adjust for differences in age patterns, while for some purposes indirect standardization seems more appropriate.

(SMDR), Standardized Cohabitation Formation Ratios (SCFR), and Standardized Cohabitation Dissolution Ratios (SCDR) can be calculated, based on the number of observed and expected births, and couple formations and dissolutions.

Standardized ratios for the minimum living arrangements model, which differentiates between the states single, married, and cohabiting, are compared first, using the total population of the respective gender as reference group.[7] Table 10.1 gives standardized ratios for fertility and mortality, together with status-specific total fertility rates (TFR) and life expectancies at birth (E_0).

Surprisingly, the standardized fertility ratio is eight per cent higher for cohabiting than for married women, but it is only 34 for single women. While this result suggests higher birth rates among cohabiting women, this conclusion is clearly disproved by looking at total fertility rates calculated over each state separately. The TFR is 40 per cent higher for married than for cohabiting women, and it is again around one-third of the total TFR for single women. The strong discrepancy between the two comparative fertility indicators becomes clear when looking at Figure 10.1, which gives age and status-specific fertility rates.

Age-specific fertility rates differ strongly between married and cohabiting women. At young ages, marital fertility by far exceeds cohabitational fertility; after age 30, however, the situation is reversed. While the TFR uses equal weights for each of the age-specific rates, the SFR gives higher weights to the age groups with a high occurrence of births. The TFR is especially vulnerable

Table 10.1. Comparative fertility and mortality indicators

	Actual marital status			
Indicator	single	married	cohabiting	Total
Stand. fertility ratio	34	134	145	100
Stand. mortality ratio, women	104	89	107	100
Stand. mortality ratio, men	122	87	101	100
Total fertility rate	0.64	3.86	2.72	1.78
Female life expectancy	78.6	81.3	79.2	80.2
Male life expectancy	69.7	76.4	73.7	74.2

Note: Life expectancies are conditional on not experiencing a marital status change.

[7] The choice of a reasonable standard population is essential as different standards might lead to different conclusions. In general, one should take a population similar to the populations analysed as standard population.

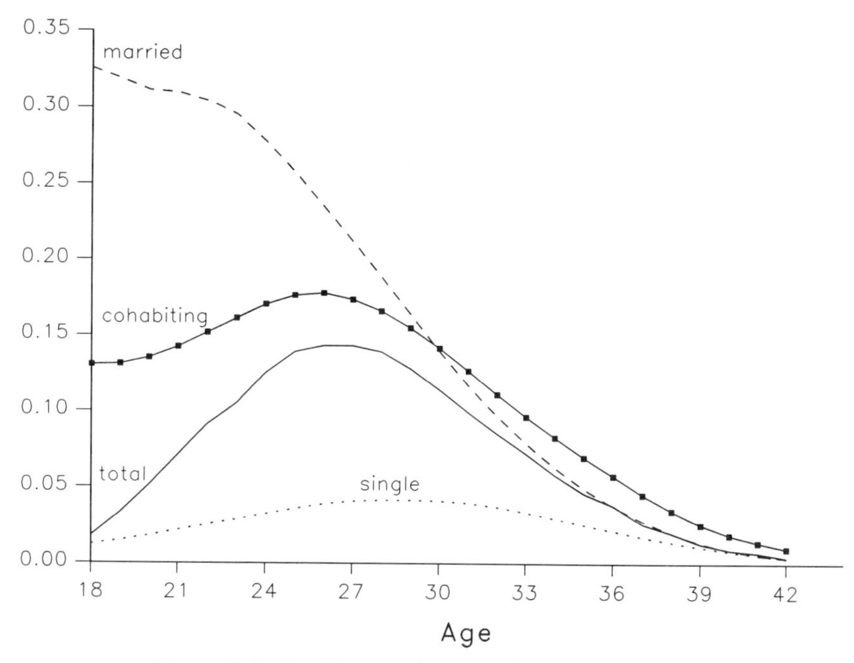

Figure 10.1. Fertility rates by age and actual marital status.

to high fertility rates among young women, and thus the SFR seems to be a better overall measure of differential fertility. We can conclude that fertility among cohabiting women is roughly as high as among married women, although with a very different age pattern, and that the fertility level of single women is in no way comparable to either of the two groups.

The standardized mortality ratio suggests different conclusions for men and women. For both sexes, SMRs for married persons are below 90, while for single and cohabiting persons SMRs are above 100 (see Table 10.1). For men, cohabitation clearly reduces the risk of dying when compared to persons living alone, for women the opposite conclusion seems to hold. Figures 10.2a and 10.2b shed more light on these mortality differentials by presenting age-specific SMRs, i.e. calculating the ratio of observed and expected numbers of death for each age separately. Between age 20 and 60, when death rates are low, mortality among cohabiting persons is far below that of singles and clearly closer to that of married persons. Beyond that age, however, differentials in mortality become generally smaller and mortality rates of cohabitees are increasingly closer to those of singles. Among women, mortality rates among cohabitees are even the highest among all groups. Since the great majority of deaths occur at ages above 65, overall SMRs of cohabitees are closer to and, for women, even higher than

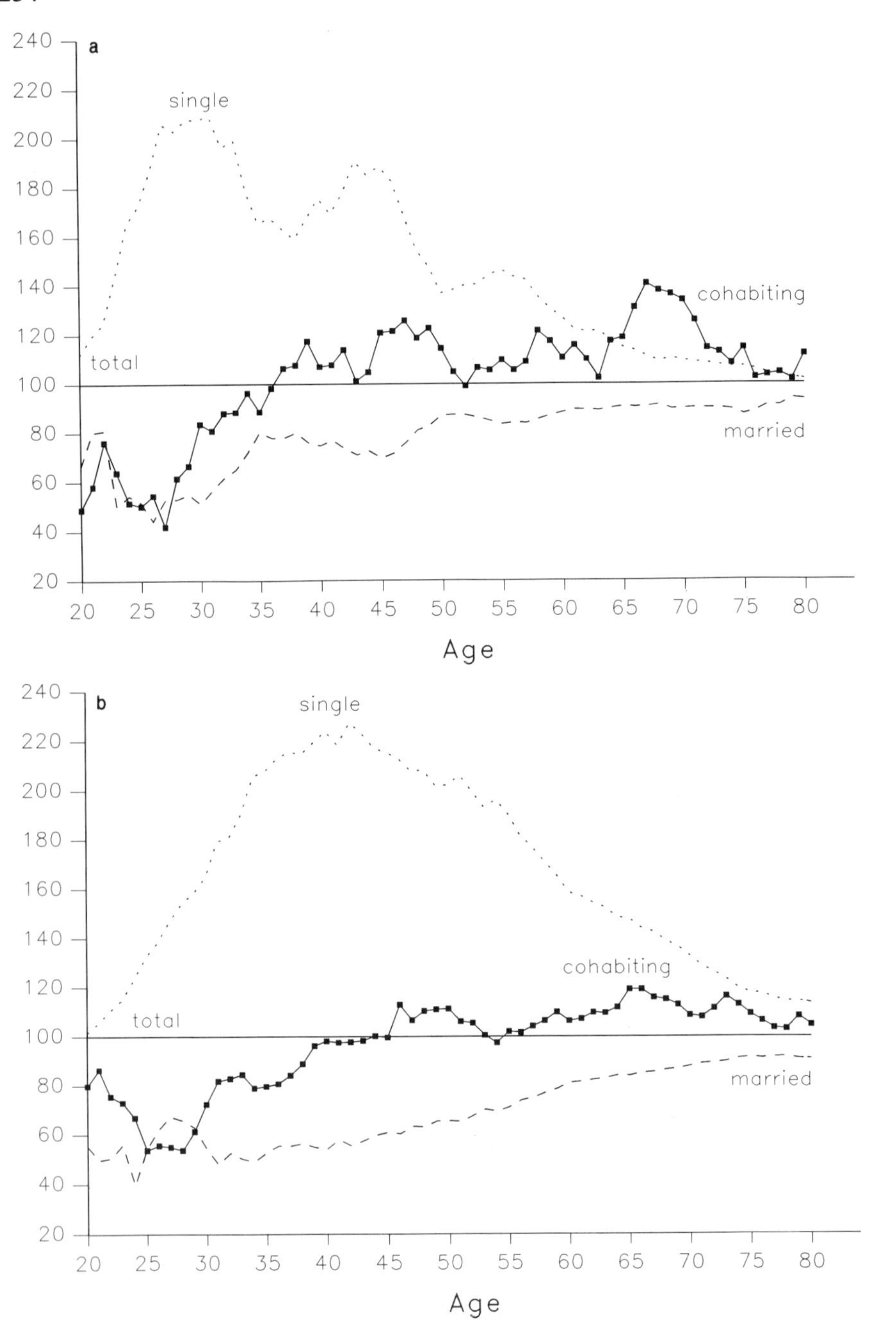

Figure 10.2. Standardized mortality ratios by age and actual marital status
a. Males b. Females

SMRs of singles. This result has to be interpreted with some caution as today still a very small group of the population aged 65 and over is living in a consensual union — 2.5 per cent among men and only 1.5 per cent among women. This group may be highly selective and not representative for future generations with possibly increasing numbers of elderly living in such unions.

Table 10.1 also gives status-specific life expectancies at birth calculated as if an individual would live in one of the three states throughout his/her whole life (beyond age 18). These calculations show strong mortality differentials among men. Life expectancy of cohabiting men is close to the average and somewhat closer to that of married men. Among women, differentials are much less significant and cohabitees are only slightly better off than singles.

Differentials in couple formation and dissolution are shown in Table 10.2. In the case of couple formation, the total marriage rate is taken as reference, i.e. the sum of the single and cohabiting population is the respective exposure; in the case of couple dissolution, the total dissolution rate is taken as reference, i.e. the respective exposure is the sum of the married and cohabiting population. For both sexes, marriage formation rates of already cohabiting persons are much higher than of singles. For the former group, SMFRs are close to 200, for the latter, less than 50. Differentials — in terms of age-specific SMFRs which are not presented — are strongest among the very young (below age 25 for men, and below age 22 for women), and even more so among those aged 50 years and over. However, the risk for singles to form new consensual unions is again significantly higher, with SCFRs of around 300 for both sexes. While direct marriage without cohabitation is unusual, marriage is more likely to happen once

Table 10.2. Standardized formation and dissolution ratios

		Actual marital status		
Indicator	Gender	single	married	cohabiting
Stand. marriage formation ratio	women	46	-	173
	men	45	-	190
Stand. cohabitation formation ratio	women	301	-	-
	men	305	-	-
Stand. marriage resp. cohabitation dissolution ratio	women	-	72	143
	men	-	69	141

a person cohabits. On the other hand, the rate of couple dissolution is also much higher among cohabitees when compared to married couples, which is shown by SCDRs around 140 as opposed to SMDRs around 70, for both men and women.

One can conclude that cohabitation in Swedish society in 1985 is seen as a true alternative and not just a substitute to marriage. Cohabitation in terms of demographic behaviour is in no way comparable to being single or being married. Single individuals seeking partnership choose between two options: marriage or cohabitation. Those who have chosen cohabitation obviously maintain a higher degree of independence associated with higher risks of mortality and union dissolution, they opt for a different timing but comparable level of childbearing, and they also keep the option to marry eventually.

Differentials by legal marital status and by order of union within these groups of actual marital status are far less pronounced, although we have not been able, due to data limitations, to check this for the case of fertility. As concerns mortality, the only significant characteristic is that — both among singles and among cohabitees — legally widowed men and women have lower mortality rates, i.e. seem to keep their previous life style to a larger extent than divorcees do. Mortality differentials give little indication on which states other than the three actual marital status categories should be included in the model. Legal marital status differentials in transition rates within the three groups single, married, and cohabiting are also less pronounced than between the three groups, but are discussed in more detail below. Table 10.3 gives standardized marriage and cohabitation formation ratios for singles by legal marital status.[8]

Both SMFRs (122-125) and SCFRs (142-151) are high among previously married men, confirming the fact that men who have already lived in a union are exposed to a higher chance to form another union, in particular as concerns cohabitation. Among women, this conclusion holds only for divorcees considering cohabitation. Marriage is highest among never-married women, and — partly due to their age — particularly uncommon among widowed women.

Table 10.4 gives standardized marriage formation and cohabitation dissolution ratios for cohabitees, again by legal marital status. Both the risk to end a consensual union by marriage and to end it by dissolution of the union is significantly higher among legally divorced cohabitees. This again shows that those who have already dissolved a union are more likely to form and also dissolve additional unions. Tables 10.3 and 10.4 show that despite the age of widowed women, their propensity to form and also dissolve a consensual union

---

[8] Due to the respective choices of standard populations, cross-comparisons between marriage and cohabitation formation of singles, between marriage of singles and marriage of cohabitees, and between men and women are not possible by looking at the ratios in Tables 10.3 and 10.4.

is not much lower than for never-married women, while this is not true as far as marriage is concerned. Except for widows and widowers, legal marital status differentials among cohabitees within actual marital status groups are very similar for men and women.

Table 10.3. Standardized formation ratios for single persons

		Legal marital status		
Indicator	Gender	never-married	divorced	widowed
Stand. marriage formation ratio	women	105	95	24
	men	93	122	125
Stand. cohabitation formation ratio	women	95	122	89
	men	92	151	142

Note: For each sex separately, the total single population was taken as reference group, thus both SMFR and SCFR for all single men resp. women equal 100.

Table 10.4. Standardized formation and dissolution ratios for cohabiting persons

		Legal marital status		
Indicator	Gender	never-married	divorced	widowed
Stand. marriage formation ratio	women	96	131	32
	men	96	128	113
Stand. cohabitation dissolution ratio	women	93	146	79
	men	93	147	100

Note: For each sex separately, the total cohabiting population was taken as reference group, thus both SMFR and SCDR for all cohabiting men resp. women equal 100.

Although not shown in the tables, standardized marriage dissolution ratios differ strongly by marriage order. For both sexes, SMDRs are around 92 for first marriages and above 200 for marriages of higher order, which is just in conformity with the conclusions drawn above concerning the higher risk of union dissolution for those who have been married previously.

The following conclusions can be drawn from the statistical analysis of fertility, mortality, and couple formation and dissolution:

- The three actual marital status groups single, married, and cohabiting are enormously different in terms of demographic variables and behaviour but at the same time they are sufficiently homogeneous to be regarded as minimum choice for a living arrangements model.
- Differentials within the three actual marital status groups are smaller, but still significant as far as couple formation and dissolution is concerned. Breaking actual statuses down further seems promising though not absolutely necessary.

An interesting measure of the importance of each actual marital status and of each legal marital status within those actual states is the life table proportion of time spent in each state on the basis of 1985 mortality and couple formation and dissolution rates. Table 10.5 gives the number of years and the resulting proportion of lifetime spent in each of the states, for men and women separately.

Table 10.5. Lifetime spent in different marital statuses

	Years spent in each status		Proportion of lifetime (%)	
Status	Women	Men	Women	Men
Single	44.3	40.6	55.4	55.2
never-married	33.8	35.6	42.3	48.4
divorced	5.5	3.6	6.9	4.9
widowed	5.0	1.4	6.3	1.9
Married	24.2	22.1	30.3	30.0
order 1	22.3	19.9	27.9	27.0
order 2+	1.9	2.2	2.4	3.0
Cohabiting	11.5	10.9	14.4	14.8
never-married	9.9	9.3	12.4	12.6
divorced	1.5	1.5	1.9	2.1
widowed	0.1	0.1	0.1	0.1
Total	80.0	73.6	100.0	100.0

On the basis of 1985 rates, both men and women would spend around 55 per cent of their life living without a partner, 30 per cent in a marital and 15 per cent in a consensual union. Disregarding their childhood, they would still spend some 40 per cent of their life without a partner, another 40 per cent married, and some 20 per cent cohabiting. Due to differences in mortality and remarriage, single adult life is subdivided quite differently by sex. Single women would spend one-fifth as divorcees, one-fifth as widows, and the remaining three-fifths never-married. Single men would spend one-sixth as divorcees, only six per cent as widowers, and eventually four-fifths never-married. As the proportion of lifetime spent as a legally widowed cohabitee is negligible for both sexes, considering this status in a projection model seems irrelevant.

10.3.2. Alternative Model Selections

Since information on the order of consensual unions was not available from the data set, the actual marital status 'cohabiting' — in accordance with the state 'single' — can only be subdivided by legal marital status, while in the case of the state 'married', including marriage order may be fruitful. The analysis suggests several selections of states, from the most aggregated to the most detailed alternative. Table 10.6 summarizes all the states and the resulting direct transitions for each of the alternative selections.

Selection 1 is the simplest, the so-called 'minimum model'. It provides actual marital status information disregarding legal marital status. Selection 2, the 'widowhood model', separates voluntary singles/cohabitees (never-married or divorced) from widowed singles/cohabitees. The motivation for model 2 is twofold: first, widows behave differently than others, and second, widowhood is a valuable characteristic in its own right (e.g. for social security reasons).

Selection 3a separates never-married from previously married singles/ cohabitees; this is the model suggested by the statistical analysis since differentials within actual marital status groups are strongest between the groups never-married and previously married. It is therefore called the 'default model'.

Selection 3b, the 'marriage order model', is an extension of model 3a by dividing the group married further into those previously never-married (order 1) and those previously married (order 2+). It is a consistent legal-actual marital status model using the criterion never versus previously married. Finally, selection 4 uses all the available legal and actual marital status information; it is thus called the 'complete model'.[9]

---

[9] For two reasons, cohabitees were not treated in a consistent way: firstly, the group 'legally widowed cohabitees' seemed to be too small to be considered as a separate state. Secondly, the computer program used for projecting the population allows for only six states when using single year age groups.

Table 10.6. States and direct transitions for five alternative selections

no.	States	Direct transitions
1	a. single	a. single → married
	b. married	b. single → cohabiting
	c. cohabiting	c. married → single
		d. cohabiting → single
		e. cohabiting → married
2	a. single, not widowed	a. single, not widowed → married
	b. single, widowed	b. single, not widowed → cohabiting, not widowed
	c. married	c. single, widowed → married
	d. cohabiting, not widowed	d. single, widowed → cohabiting, widowed
	e. cohabiting, widowed	e. married → single, not widowed
		f. married → single, widowed
		g. cohabiting, not widowed → single, not widowed
		h. cohabiting, not widowed → married
		i. cohabiting, widowed → single, widowed
		j. cohabiting, widowed → married
3a	a. single, never-married	a. single, never-married → married
	b. single, previously married	b. single, never-married → cohabiting, never-married
	c. married	c. single, previously married → married
	d. cohabiting, never-married	d. single, previously married → cohabiting, prev. married
	e. cohabiting, prev. married	e. married → single, previously married
		f. cohabiting, never-married → single, never married
		g. cohabiting, never-married → married
		h. cohabiting, previously married → single, prev. married
		i. cohabiting, previously married → married
3b	a. single, never-married	a. single, never-married → married, order 1
	b. single, previously married	b. single, never-married → cohabiting, never-married
	c. married, order 1	c. single, previously married → married, order 2+
	d. married, order 2+	d. single, previously married → cohabiting, prev. married
	e. cohabiting, never-married	e. married, order 1 → single, previously married
	f. cohabiting., prev. married	f. married, order 2+ → single, previously married
		g. cohabiting, never-married → single, never-married
		h. cohabiting, never-married → married, order 1
		i. cohabiting, previously married → single, prev. married
		j. cohabiting, previously married → married, order 2+
4	a. single, never-married	a. single, never-married → married
	b. single, divorced	b. single, never-married → cohabiting, never-married
	c. single, widowed	c. single, divorced → married
	d. married	d. single, divorced → cohabiting, previously married
	e. cohabiting, never-married	e. single, widowed → married
	f. cohabiting, prev.married	f. single, widowed → cohabiting, previously married
		g. married → single, divorced
		h. married → single, widowed
		i. cohabiting, never-married → single, never-married
		j. cohabiting, never-married → married
		k. cohabiting, previously married → single, divorced
		l. cohabiting, previously married → single, widowed
		m. cohabiting, previously married → married

Living arrangement projections of the population were prepared using the DIALOG personal computer software developed at IIASA, a flexible multistate projection model.[10] Although the model allows for complex scenario setting, in the following, only constant rates assumptions are tested. This decision was necessary to make the different selections of states comparable and to avoid masking effects of different selections by effects of different scenario assumptions.

One drawback of the DIALOG model is that it does not easily allow one to impose consistency relationships.[11] Without going into detail, one should keep in mind that any legal and/or actual marital status model requires consistency in the sense that, for example, the number of males marrying should equal the number of females marrying, the number of males entering a consensual union should equal the number of females entering such a union, and so on. While, in theory, missing consistency constraints may cause problems, particularly in the long run, applying the model has shown surprisingly robust results. Even in the long run the projected numbers of events are almost consistent.

10.4. PROJECTION RESULTS

Surprisingly, notwithstanding differentials in couple formation and dissolution rates by legal marital status, projections of the actual marital status structure of the Swedish population are independent of the choice of the model. Given the significant differences in behaviour, e.g. between never and previously married individuals, one would have expected different results with different model selections.

For two reasons this is not the case, as can be seen from Figure 10.3 which gives the projected number of cohabiting women of all ages: firstly, higher formation and higher dissolution rates among those previously married partly offset each other, and secondly, changes in the legal marital status distribution within actual marital status groups are small relative to changes in the actual marital status distribution. From 390 thousand cohabitees in 1985, all models project an increase to about 560-600 thousand. Models which distinguish never-married from previously married individuals result in larger numbers of cohabitees, as could be expected from the comparison of differentials in couple formation and dissolution rates.

---

[10] For more details on the projection model see Scherbov & Grechucha (1988).

[11] A multistate projection model with flexible consistency device was developed by Van Imhoff & Keilman (1991).

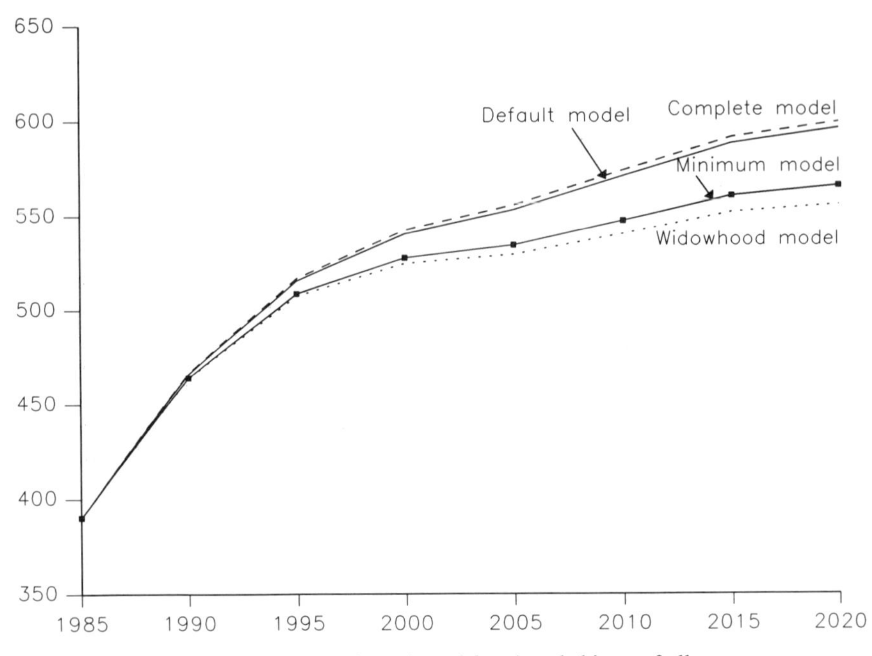

Figure 10.3. Projected number of female cohabitees of all ages.

From that perspective, the choice of the model, i.e. the choice between those five selections tested, no longer depends on statistical analysis or the robustness of certain results, it only depends on the legal marital status information one is interested in.

10.4.1. The Minimum Model

If one is only interested in living arrangements, or equivalently in the future actual marital status distribution of the population, the minimum model provides the required information. Tables 10.7a (for females) and 10.7b (for males) give the actual marital status structure both for the population aged 15-64 years, the so-called working age population, and the population aged 65 years and over, the elderly group, in absolute figures and in percentages for the years 1985, 2000 and 2020. In 1985, among both sexes, 14 per cent of the working age population were living in a consensual union, 50 per cent (among women) and 45 per cent (among men) were married, and the remaining 36 per cent (among women) and 41 per cent (among men) lived as — mostly never-married — singles. Among the elderly, the actual marital status distribution of the population was quite different. An almost negligible part lived in consensual

Table 10.7a. Actual marital status structure of the female population. Absolute figures (in 1000) and percentages

	Single		Married		Cohabiting		Total
Year	1000	%	1000	%	1000	%	1000
Population aged 15 to 64 years:							
1985	962	36	1324	50	378	14	2664
2000	954	36	1198	45	507	19	2659
2020	989	39	1034	41	521	20	2544
Population aged 65 years and over:							
1985	506	61	319	38	12	1	837
2000	498	60	314	38	21	2	833
2020	549	58	352	37	45	5	946

Table 10.7b. Actual marital status structure of the male population. Absolute figures (in 1000) and percentages

	Single		Married		Cohabiting		Total
Year	1000	%	1000	%	1000	%	1000
Population aged 15 to 64 years:							
1985	1121	41	1227	45	383	14	2730
2000	1108	41	1114	41	514	18	2735
2020	1136	43	963	37	515	20	2614
Population aged 65 years and over:							
1985	197	31	418	66	15	3	630
2000	198	33	373	62	28	5	598
2020	252	35	405	57	57	8	714

unions. Among men, two-thirds were married and almost one-third was single, while among women 60 per cent were single, mainly widowed, and less than 40 per cent married.

For the working age population, the main change during the next three decades will be a significant decline in the number and proportion of married couples, together with a corresponding increase in the number and proportion of cohabiting couples. While the total population in the age group 15 to 64 will — on the basis of our assumptions — decline some five per cent until the year 2020, the population married will decrease by 22 per cent and the population cohabiting will increase by 36 per cent. As can be seen from Table 10.8, this change will result in an increase in the share of consensual unions among both types of unions from around 23 per cent in 1985 to more than one-third by the year 2020. The proportion of singles will also increase by some two to three percentage points, which is much less than one would expect from the high rates of couple dissolution among married and in particular cohabiting couples.

For the elderly population, expected changes are partly different for men and women. For both sexes, the total population will increase by some 13 per cent until the year 2020 (see Tables 10.7a and 10.7b), while the cohabiting population will even quadruple during that period. This huge relative increase in the number of cohabiting elderly is reflected in the increase of consensual unions among all unions in the age group 65 years and over (see Table 10.8): from 3.5 per cent in 1985 to around 12 per cent by 2020. Since the rate of consensual union formation is low at that age, the increase in cohabitation among the elderly is mainly due to the increase in long-lasting cohabitations.

Among women, the huge increase in the number of elderly cohabitees is accompanied by an increase in the number of single (by eight per cent) and married (by ten per cent) women, although the proportions of the two groups will decline slightly. Among men, however, the single elderly population will increase both absolutely (almost by 30 per cent) and relatively, while the proportion of married men will decline by almost ten percentage points (see Table 10.7b).

These sex differentials in changes in the elderly population will become clearer when we look at the results of those selections that distinguish several sub-groups of the three actual marital status groups considered.

A few more conclusions resulting from all models can be drawn. The annual number of births will decline from 115,000 in 1985 to around 100,000 by the turn of the century. Since this decline is exclusively due to a decline in

Table 10.8. Proportion of consensual unions among all unions

	Population 15-64		Population 65+	
Year	females	males	females	males
1985	22.2	23.8	3.6	3.5
2000	29.7	31.6	6.3	7.0
2020	33.5	34.8	11.3	12.3

births among married couples, already by 1990 the majority of children will be born in a consensual union. The number of annual marriages will increase from around 39,000 in 1985 to more than 45,000 by 1995, and it will remain at that level after the turn of the century. This increase is entirely due to an increase in the number of marriages among cohabitees, while the number of direct marriages remains at the 1985 level (11,000-12,000 marriages). The number of newly-formed cohabitations is very stable throughout the projection period, fluctuating around 78,500 annually. Conclusions concerning the annual number of couple dissolutions can best be drawn from models that distinguish legal marital status groups.

10.4.2. Alternative Models

The proportion married is projected to decline among both sexes for the working age groups 15 to 64, and only among men for the age groups 65 and over. Results of the marriage order model show some interesting trends concerning marriage order. In 1985, only seven and three per cent of the total population married among the working age and the elderly population, respectively, belonged to the group 'married order 2 and over'. Among the 15-64 years group, the projected decline in the proportion married will entirely be due to a decline in the population married order 1, while the population married order 2+ remains stable and thus increases its share. Among the elderly population, a remarkable decline for the group married order 1 is projected, in particular among men (from 64 per cent of the total elderly population in 1985 to only 45 per cent by 2020), together with a significant increase for the group married order 2+. For women the two changes offset each other, while for men the decline in married order 1 dominates the results. By 2020, almost one-fifth of those men married beyond age 65 will already have experienced a second marriage.

The number and proportion of cohabitees is projected to increase significantly for both sexes and all age groups. Today, and even more so in the future, the great majority of cohabitees at age 15-64 belongs to the group never-married, while the two groups never-married and divorced each contribute equally to the group cohabiting among the elderly. The number of widowed cohabitees is and will be very small, even among the elderly. Figure 10.4 gives the proportion of cohabitees among unmarried women according to their legal marital status. For women of all ages together, the proportion cohabiting increases among the never-married population, while it decreases among divorcees and widows. In contrast to 1985, by 2020, the proportion cohabiting will be largest among never-married women. This conclusion also holds for the elderly population, although the proportion cohabiting increases both among

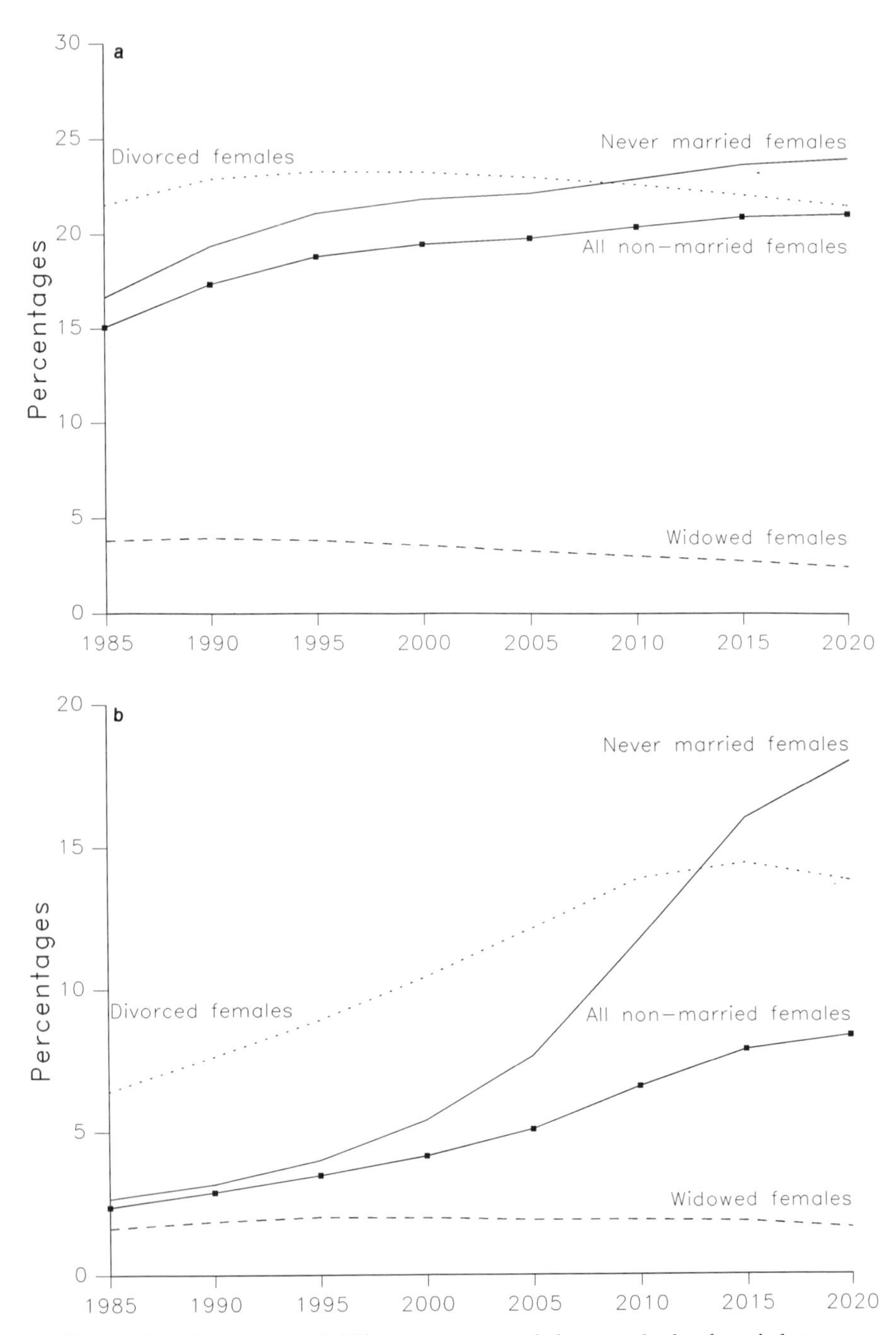

Figure 10.4. Proportion cohabiting among unmarried women by legal marital status
a. Ages 15 and over b. Ages 65 and over

divorcees and, more particularly, among never-married women. It is almost constant among elderly widows.

The single population aged 15-64 years is projected to increase among both sexes. This is due to an increase in the population of never-married singles, which constitutes already the great majority of the single population in 1985 and which increases its share to 80 per cent and almost 90 per cent of the single working age population for women and men, respectively, by 2020. Due to the gradual improvement of the (currently rather imbalanced) sex ratio, the number and proportion of widowed singles declines at all ages, in particular among the elderly. Among women aged 65 and over, this decline is even strong enough to more than offset the trebling of the number of divorced singles, resulting in a decline in the overall proportion of elderly singles. Among men, the overall proportion of elderly singles increases due to the fact that the increase in the number of divorced singles is stronger than the decline in the number of widowed singles. Interestingly, in 2020, around 80,000 elderly men belong each to the groups single never-married, single divorced, and single widowed and again some 70,000 each to the groups married order 2+ and cohabiting, altogether comprising 55 per cent of the male elderly population.

From those models considering legal marital status sub-groups, we can derive the relevant information concerning couple dissolutions. The annual number of divorces will decline from around 20,000 in 1985 to around 14,000 by 2015, mainly as a consequence of the lower proportion of married couples. At the same time, the number of dissolutions of consensual unions will increase from around 31,000 annually in 1985 to around 37,000 by 2020. While today 40 per cent of all union dissolutions are captured by vital statistics, in 2020 only some 27 per cent will be. The annual number of newly-widowed men and women remains at around 10,000 and 25,000, respectively, keeping the ratio of 2.5 widows per widower constant.

10.5. CONCLUSION

The current and, more so, the future population distribution by actual marital status, i.e. by living arrangements as defined for this study, suggests that legal marital status projections have ceased to be a suitable surrogate for living arrangement projections, at least in Sweden. Consensual unions as a relatively new form of living together are increasing at such a speed that they will form one-third of all unions of people between age 15 and age 64 already in the medium term even under conservative constant rates assumptions. Also among the elderly, this form of partnership will constitute a significant share in the long run.

Without considering consensual unions, one would project largely irrelevant and on top of that simply wrong information, which is proven by the following comparison. Until the year 2020, the model used in the present study projects a decline in the number of married men aged 15 years and over by around 15 per cent, together with an increase in the number of cohabiting men by around 42 per cent. Taking the two opposite trends together, this means that the proportion of singles would only increase by some six per cent. A legal marital status projection carried out on the basis of the Swedish 1985 population distribution and the same assumption of constant rates, however, projected a decline in the number of married men of the same age of around 25 per cent (Gonnot 1995). This decline suggested an increase in the proportion of single men of around 24 per cent, while on the basis of an actual marital status model it turns out that the proportion of men living in couples will decline only marginally. Not only can the number of couples not be projected with a legal marital status model, but even the number of married couples is significantly underestimated.[12] This underestimation is due to the fact that marriage rates are significantly higher among cohabitees as compared to singles (around three times higher), which plays an important role when the share of cohabitees in the total unmarried population increases as it does in Sweden.[13]

Another answer that the analysis in this paper promised to give concerns the level of aggregation and the optimum model selection for an actual marital status projection model. Both from the comparison of legal marital status differentials within actual marital status groups and from the robustness of results concerning the actual marital status distribution, it seems obvious that the minimum model is at the same time the optimum model. If one is interested in the future actual marital status distribution, that is, if one is interested in the future size of the population living alone, living in a consensual union and living in a marital union, incorporating legal marital status information does not improve the results; it only complicates the model. Following the rule to aggregate as much as possible to still obtain reasonable results, choosing only the three actual marital states gives all the required information about the future actual marital status distribution. Such a model will give most of the demographic information one would need, for example, to estimate housing demand.

An additional advantage of this three-state actual marital status model is that it is as simple as the legal marital status model. Legal marital status information can easily be incorporated, as was shown by the alternative model

---

[12] A more comprehensive comparison of actual and legal marital status projections, also considering different scenarios, can be found in Prinz (forthcoming).

[13] This problem is related to the discussion on heterogeneity in this paper (see introduction). Such heterogeneity can also be observed in countries where cohabitation is still on a lower or even much lower level than in Sweden today.

selections. The major difficulty for such a model is to find adequate data. While this unique Swedish data set allows actual marital status projections for Sweden, for the moment such data are not available for any other country in Europe. The results of the projections for Sweden suggest that more emphasis should be put on indirect data estimation from scarce data. This problem is very much related to the problem of demographic estimation of data for statistically underdeveloped countries. In other demographic areas, such as mortality, related problems have long been solved by model life tables and model stable populations (Coale & Demeny 1983), or by several other procedures based on data from censuses and surveys, or from sample registration areas (Shryock *et al.* 1976). Very recently, methods have been developed to estimate events in family demography (Preston 1987). These new expressions are based upon a generalization of stable population relations, and their basic idea is to substitute a second cross-sectional observation on the age and status distribution of a population for certain (missing) flow data. While those procedures are relatively complex, a straightforward indirect estimation would be to substitute some of the existing Swedish flow estimates for missing flows for another country. This procedure seems plausible for countries which are similar to Sweden with regard to trends in cohabitation, like Denmark or Norway, while those estimates would have to be adjusted for other countries. Such indirect estimation approaches are tested in another study (Prinz, forthcoming).

If one is also interested in developments concerning the legal marital status distribution, other model selections should be preferred. Depending on the specific interests, each of the alternative models seems reasonable. The widowhood model lays emphasis on the future number of widow(er)s, the default model on never-married people, the marriage order model on married people by marriage order, and the complete model on all legal marital statuses. The complete model also has the advantage that one can eventually rearrange the data, i.e. the projection results, in two different ways: according to legal marital status or according to actual marital status. Such further disaggregated models give the demographic information to answer social, legal, and social security questions which require additional information on legal marital status. However, it is to be expected that also in these areas of policy concern, actual marital status will more and more replace the traditional de jure marital status criterion.

REFERENCES

COALE, A., & P. DEMENY (1983), *Regional Model Life Tables and Stable Populations*, Second Edition, Academic Press: New York.

GONNOT, J.P. (1995), Demographic Changes and the Pension Problem: Evidence from Twelve Countries. *In*: GONNOT, KEILMAN & PRINZ.

GONNOT, J.P., N. KEILMAN & CH. PRINZ (eds.) (1995) *Social Security, Household, and Family Dynamics in Ageing Societies*, Kluwer Academic Publishers: Dordrecht.

KEILMAN, N., & CH. PRINZ (1995), Modelling the Dynamics of Living Arrangements. *In*: GONNOT, KEILMAN & PRINZ.

KEYFITZ, N. (1985), *Applied Mathematical Demography*, Second Edition, Springer-Verlag: New York.

PRESTON, S. (1987), Estimation of Certain Measures in Family Demography Based upon Generalized Stable Population Relations. *In*: J. BONGAARTS, TH. BURCH & K. WACHTER (eds.) *Family Demography: Methods and their Applications*, Clarendon Press: Oxford.

PRINZ, CH. (forthcoming), Modelling Legal and Actual Marital Status: A Theoretical Note with Some Examples. *In*: CH. HÖHN (ed.) *Demographic Implications of Marital Status*, Schriftenreihe des Bundesinstituts fur Bevölkerungsforschung, Wiesbaden.

PRINZ, CH. (forthcoming), *Cohabiting, Married, or Single? Portraying, Analysing and Modeling New Living Arrangements in the Changing Societies of Europe*, Avebury: Aldershot.

ROGERS, A. (1975), *Introduction to Multiregional Mathematical Demography*. John Wiley: New York.

SCHERBOV, S., & V. GRECHUCHA (1988), *DIAL — A System for Modelling Multidimensional Demographic Processes*, Working Paper WP-88-36, IIASA: Laxenburg.

SHRYOCK, H., J. SIEGEL & E. STOCKWELL (1976), *The Methods and Materials of Demography*, Academic Press: New York.

VAN IMHOFF, E. (1990), The Exponential Multidimensional Demographic Projection Model, *Mathematical Population Studies*, Vol. 2(3), pp. 171-182.

VAN IMHOFF, E., & N. KEILMAN (1991), *LIPRO 2.0: An Application of a Dynamic Demographic Projection Model to Household Structure in the Netherlands*. NIDI CBGS Publications 23, Swets & Zeitlinger: Amsterdam/Lisse.

WILLEKENS, F., & P. DREWE (1984), A Multiregional Model for Regional Demographic Projection. *In*: H. TER HEIDE & F. WILLEKENS (eds.) *Demographic Research and Spatial Policy: the Dutch Experience*, Academic Press: London.

11. NATIONAL HOUSEHOLD FORECASTS FOR THE NETHERLANDS

Joop de Beer

Statistics Netherlands
Department of Population
P.O. Box 4000
2270 JM Voorburg
The Netherlands

Abstract. The Population Forecast of Statistics Netherlands (NCBS), which is revised annually, projects the population by age, sex, and marital status. As marital status is increasingly a less accurate indication of life style, the NCBS has decided to publish Household Forecasts as well. The first of these was issued in June 1992. Six household positions are distinguished: living in an institution, living in the parental home, living alone, living with a partner, lone parents, and other members of private households. A semi-dynamic model is used. The forecasts of the population by age, sex, and marital status are based on a dynamic model. The distribution of these population categories by household position is based on a static model. The number of households is forecast on the basis of the distribution of the population by household position. The main developments in the forecast period 1990-2010 are the increase in one-person households and in non-marital cohabitation. While one out of four households consisted of one person in 1990, this will be one out of three in 2010. As a result, the average will decline from 2.45 to 2.27 persons per private household.

11.1. INTRODUCTION

Statistics Netherlands (NCBS) has regularly published national population forecasts since 1950. Since 1984 these forecasts have been revised annually. In the population forecasts, the population is not only distinguished by age and sex but also by marital status. Four categories are distinguished: never-married, married, divorced, and widowed. In the past, marital status was a reasonable indicator of the composition of households. Most households consisted of married couples with or without children. Lone-parent families formed a small minority, mainly young widows or widowers with their children. Most elderly

Household Demography and Household Modeling
Edited by E. van Imhoff *et al.*, Plenum Press, New York, 1995

unmarried, divorced, and widowed persons lived alone. Nowadays marital status has become a less accurate indicator of household composition. Differentiation in household formation has increased. Whereas the majority of unmarried young people used to live with their parents until they married, at present, after leaving the parental home, many young people live alone for some time before they decide to live together with a partner.

Both for the government and the market sector, information on future developments in the number of households is often just as important as information on the future number of people. Therefore, Statistics Netherlands has decided to publish national household forecasts, in addition to the population forecasts. The first household forecasts were published in June 1992.

The 1992-based National Household Forecasts are consistent with the 1991-based National Population Forecasts: the population by age, sex, and marital status in both forecasts is equal. The forecast of households is based on forecasts of household positions. Assumptions are made on the division of the population by age, sex, and marital status into six categories: living in the parental home, living alone, living with a partner, lone parent, living in an institution, and other.

Section 11.2 describes the household positions. Section 11.3 discusses the forecast model. The subsequent section makes a comparison with two other Dutch household projection models. Section 11.5 sets out how the age profiles for the target year are assessed. The uncertainty variants are dealt with in section 11.6 and section 11.7 outlines the extrapolation procedure. Section 11.8 describes the main assumptions underlying the forecasts in very broad lines. Section 11.9 presents the main results. The final section points to how the forecasts may be improved in the future.

11.2. HOUSEHOLD POSITIONS

The first NCBS National Household Forecasts start out from the results of the 1991-based National Population Forecasts: the population by age, sex, and marital status in each of the years 1991-2010. In the Population Forecasts for each forecast year, 706 population categories are forecast (100 ages $\times$ 2 sexes $\times$ 4 marital states - 94 categories that are zero by definition). The distribution by household position is determined in the Household Forecasts for each category.

One criterion for the choice of household positions is that it should be appropriate for deriving the number of households. Another is that the number of categories resulting from the combination of marital status and household position should not be too large.

On the basis of these criteria, it was decided to distinguish six household positions. First those people are distinguished who are not part of a private household, but who have lived in an institutional household for over a year (e.g. a home for the elderly, a social home, or a prison). Secondly, those who form a one-person household are distinguished. Everyone else belongs to a multi-person household. Within this category, children living with their parents are distinguished: all never-married own or step children who do not have children themselves. People who form a household together with a partner — whether or not they are married — and possibly other people are labelled 'living together'. Lone parents who have a partner belong to this category. The other lone parents make up a separate category. People without a partner who live in a multi-person household form the category 'other'.

In summary, the six household positions are:

- living in an institution;
- living alone;
- child living at home;
- living with a partner;
- lone parent;
- other member of private household.

In combination with the four marital states, these six household positions yield 21 categories (by definition, people living in the parental home are never-married).

The starting point for the distribution of the population by household position is the situation on 1 January 1990, according to the Housing Demand Survey 1989/1990.

On the basis of the forecasts of the numbers of persons in the various household positions, the number of households can be forecast under some simplifying assumptions. Someone living together counts for half a household, someone living alone for one household, children living at home and people living in institutions do not form a separate household, and lone parents count for one household. Other members of private households count for either no or one household. Heads of a household (who do not live alone or together or are a lone parent) count for one household and the others for zero. It is assumed that the percentage of other members of households who are head of a household in the forecast period, equals that in the last observation year. When translating numbers of persons into numbers of households, multi-family households are not taken into account in view of their small number (in 1990, 0.2% of all households).

11.3. FORECAST MODEL

The 1992 National Household Forecasts are based on a semi-dynamic model. That is to say, the forecasts are partly based on transition probabilities (dynamic model) and partly on the distribution of the population over various categories at a specific moment (static model). On the one hand, the forecast of the population by marital status is based on a dynamic nuptiality model. This model is part of the national population forecast model. In this model, assumptions are made on the probability of first marriage, the probability of divorce, the probability of becoming a widow or widower, and the probability of remarriage for divorced or widowed persons. These probabilities are specified by age. For a description of this model see NCBS (1984). On the other hand, assumptions are made on the distribution of the population in each marital status by household position in each forecast year. This strongly resembles a headship-rate model. No assumptions are made on the transitions of persons between the various household positions. By contrast, in a dynamic model, the number of people in a given household position is the result of assumptions on transitions to and from that household position.

One problem with using a static model is that it does not explicitly take into account how the distribution of the population by household position has come about. The forecast is based on an assumption on changes in the structural distribution at different points in time which are the result of various demographic events with possibly opposite effects. In a dynamic model, the assumptions on the size of the various transitions are explicitly stated. Hence one advantage of a dynamic model is that the parameters in such a model (usually the parameters of a function describing transition probabilities by age) tend to be more suitable for analysing past developments than the parameters of a static model (describing a distribution at a given moment).

One problem with applying a dynamic model is the lack of reliable data for a number of transitions. The number of transitions is considerably larger than the number of household positions. Consequently, for various transitions, the numbers of people experiencing that event are relatively small. This is particularly problematic as data on transitions are usually available from sample surveys only. Even though the Housing Demand Survey is a rather large sample (54 thousand respondents), some transitions are observed for very few or no respondents. Therefore, when analysing data on transitions, it is very difficult to discern 'signal' from 'noise'. In particular, it is a problem to assess whether changes over time are systematic or random. Moreover, in surveys such as the Housing Demand Survey, usually not all transitions are measured via retrospective questions, so that a number of transition probabilities have to be estimated on the basis of structural distributions. Furthermore, time series of a reasonable length are usually lacking. For each cohort there are only a few measurements. This makes it difficult to separate age, cohort, and period effects. These problems hamper the analysis of changes in transition probabilities. In practice, when using dynamic models, the values of the transition probabilities are often kept constant in the forecast period.

One advantage of a static model is that it is much more robust than a dynamic one. A small change in the parameters of a dynamic model tends to yield widely different outcomes, whereas in a static model small changes in the assumptions yield only small changes in the results. This is particularly important in view of the data problems.

The label 'static' does not imply that in the National Household Forecasts it is assumed that there will be no changes in the future. On the contrary, for the future distribution of the population by household position, assumptions are made on how the distribution will change in the future. These assumptions are not based on comparing structural distributions in different years only. Social backgrounds of the observed changes are also taken into account. Moreover, as noted above, the Population Forecasts underlying the Household Forecasts are based on a dynamic model.

In a static model, the age patterns describe a transversal picture. When analysing changes in the age patterns in the past and extrapolating them to the future, it should be taken into account that the form of the age pattern in a calendar year may differ from that of cohorts. For the forecast based on a static model, this is not necessarily a problem, if the process of postponement at young ages and catching up at older ages has already started. If it is not assumed that the postponement behaviour has come to an end, the observed changes at young and old ages can be projected into the forecast period. If it is expected that the postponement will end soon, the forecast should take into account that the decrease at young ages will stop before the increase at older ages.

Unlike a dynamic model, only few consistency conditions have to be satisfied in a static model. The first obvious consistency condition is that in every year, for each age-sex group, the proportions in the household positions add up to 100. Secondly, the number of married men and women living together should be about equal. In the Household Forecasts, these numbers are assumed to be exactly equal. However, there is a difference between the total numbers of married men and women: sometimes one partner lives abroad. Thirdly, it is assumed that the number of unmarried men living together is equal to the number of unmarried women living together. From the Housing Demand Survey, it turns out that there are more cohabiting widows than widowers, whereas there are more cohabiting divorced men than divorced women. For the forecast period, it is assumed that the surplus of cohabiting widows about equals the surplus of cohabiting male divorcees. The number of never-married cohabiting men is assumed to be about equal to the number of never-married cohabiting women. It is assumed that the number of persons living together with someone of the same sex is about equal for men and women.

11.4. OTHER HOUSEHOLD FORECAST MODELS

In the Netherlands, two household projections have been published recently based on dynamic models, viz. the PRIMOS model (Heida 1991) and the LIPRO model (Van Imhoff & Keilman 1991).

The PRIMOS model distinguishes 14 household positions:

- child living in the parental home;
- never-married person living in a one-person household;
- partner in a two-person household;
- partner in a three-person household;
- partner in a four-person household;
- partner in a household with five people or more;
- divorced living in a one-person household;
- divorced without partner living in a two-person household;
- divorced without partner living in household with three or more people;
- widowed person living in a one-person household;
- widowed person without partner living in a two-person household;
- widowed person without partner living in household with three or more people;
- living in an institution;
- other member of a private household.

Unlike the NCBS Household Forecasts, the PRIMOS forecasts make no distinction between married couples and consensual unions. Hence the categories 'divorced' and 'widowed' do not refer to the previous marital status. If one partner of an unmarried cohabiting couple dies, the surviving partner is labelled 'widowed'. Cohabiting couples that split up are labelled as divorcees. Another difference with the NCBS forecast is that the size of the household is taken into account.

The 14 household positions in the PRIMOS model imply 129 different transitions (including birth, death, and migration). However, in the PRIMOS model, probabilities are not different for all those transitions. A number of probabilities are assumed to equal each other. The distribution of the population by age and sex in the PRIMOS forecast equals that in the NCBS forecast. In section 11.9, the results of the PRIMOS forecast will be compared with those of the NCBS forecast.

The LIPRO model distinguishes 11 household positions. These are discussed by Van Imhoff in chapter 12 of this volume. The positions in the LIPRO model imply 112 events (including birth, death, and migration). One difference with both the NCBS and the PRIMOS model is that different categories of children are distinguished. Furthermore, couples with and without children are distinguished. This does not affect the number of households, but it allows the average size of different types of households to be assessed.

Van Imhoff and Keilman (1991) do not employ the LIPRO model for forecasts but for scenarios used to estimate the effects of various demographic developments on social security. Even though one of the scenarios, labelled the 'realistic scenario', aims to describe a demographic future corresponding closely to that of the NCBS Population Forecast, there are considerable differences (see chapter 12 of this volume).

11.5. AGE PROFILES

For each sex and marital status, age profiles are assessed for the various household positions. The age profiles are described by so-called cubic splines. One advantage of cubic splines is their great flexibility: they can describe all kinds of curves.

Parametric functions are usually used in dynamic models. For example, in the PRIMOS model, the probability of leaving the parental home is described by a lognormal distribution and the probability of divorce by a negative exponential function. One advantage of such distributions is that the values of the parameters can be interpreted. For a static model this is not true, as the age patterns are the net result of several underlying processes. As a consequence, the interpretation of the patterns is less straightforward.

A cubic spline is a succession of third-degree polynomials. The curve described by the spline is smooth, because at the so-called knots the levels and slopes of the adjacent curves are equal. In choosing the number of knots, there is a trade-off between smoothness and goodness of fit. If only a few knots are chosen, the curve is smooth, but the deviations of the observed values from the curve may be large.

The coefficients of the spline can be determined in two ways. One possibility is to choose those values that describe the observations as well as possible, given the number of knots, e.g. by means of minimizing the sum of squared residuals. One problem with this is that it limits the usefulness of the age profile for forecasting purposes, as it is a 'black box'. The coefficients are not interpretable. Therefore, the model is not a good instrument to determine the shape of the curve for future years. Of course, it is possible to determine the curve for a future year by means of extrapolation of changes between curves for different years in the past. But then it is not possible to explicitly take into account the changes in the shape of the curve which are the result of cohort effects.

The other possibility is to impose some conditions, e.g. by a priori fixing some points of the curve. This allows the broad shape of the curve to be determined beforehand. It is suitable for forecasting purposes, as the forecaster only needs to assess the values of a limited number of — interpretable — parameters. In the National Household Forecasts the second method is applied, using a Hermite interpolation. This implies that for a small number of ages the percentage of the population in a given household position and the size of the change between successive ages are fixed. See the Appendix for details.

In the Dutch Population Forecasts, assumptions for fertility, mortality, migration, and nuptiality are specified for so-called target years. Until the target years (2000 for migration and 2010 for the other components), the assumptions aim to describe the most plausible development. After the target year the age-specific rates (numbers for migration) are kept constant. This does not imply that no changes are expected after 2010, but that it is considered impossible to indicate the most probable development beyond that year.

For the Household Forecasts the target year is 2010. For that year the age profile of the household positions for each marital status is assessed. In section 11.8, the main assumptions underlying these profiles are discussed.

For each year in the period 1991-2009, the percentage distribution of the household positions is determined by means of a cubic interpolation. For this purpose, a Hermite interpolation is applied (for a description, see NCBS 1984, Appendix 3). The interpolations are determined by four conditions: starting value, starting slope, end value, and end slope. For the starting values, the percentages observed in the Housing Demand Survey 1989/1990 are used, after smoothing the age patterns by means of cubic splines. The starting slopes are assumed equal to the average annual changes between 1 January 1986 and 1990, based on the Housing Demand Surveys of 1985/1986 and 1989/1990, again after smoothing the age curves. The end values are determined by the age profiles for the target year 2010, which are based on an extrapolation of observed changes (the extrapolation procedure will be discussed in section 11.7). The end slopes are assumed to be zero. This implies that it is assumed that the changes will gradually become smaller during the forecast period. One reason for this assumption is that obviously the extent to which percentage distributions can change in the same direction is limited. Another reason is that the size of most changes has decreased during the last decade.

In contrast to the Population Forecasts, the target year of the Household Forecasts is the end year of the forecast. As already noted in section 11.3, the age profiles are transversal cross-sections of cohorts. Changes between cohorts result in a change in age patterns. If recent cohorts have shown different behaviour at young ages than older cohorts at the same ages, it can be assumed that the distribution by household position of the recent cohorts at older ages, i.e. after the target year, will also be different. As a result, the age curves will change after the target year. Hence an extrapolation after the target year by assuming the age patterns to be constant would result in systematic forecast errors.

11.6. VARIANTS

In the National Population Forecasts, uncertainty about future developments is taken into account by means of specifying Low and High variants in addition to the most probable Medium variant. In the High variant fertility, migration and marriage are high and mortality is low. In the Low variant the opposite applies. The Low and High variants are 'extreme' variants: of all possible combinations of variants of the components, the Low and High variants represent the smallest and largest population respectively.

It is not useful to specify Low and High variants for the Household Forecasts that correspond to those for the Population Forecasts in the same way as the Medium variant. The reason for this is that the Low and High variants of the Population Forecasts do not result in 'extreme' variants of the numbers of households. In the High variant of the Population Forecasts, the size of the population is large (leading to many households), but as the marriage rate is high, the percentage of unmarried persons is low (leading to relatively few households, as most of the unmarried adults form one-person households).

For this reason, the High variant of the Household Forecast is based on a different variant of the Population Forecasts, namely a variant with high population growth (high fertility and migration and low mortality), and a low marriage rate and high divorce rate. For the Low variant the opposite applies. Consequently, in the High and Low variants of the Household Forecast, the population by age and sex is consistent with the High and Low variants of the Population Forecasts, but the population by marital status is different.

Given the distribution of the population by marital status, the development of the number of households is mainly determined by the development of the number of one-person households and the number of consensual unions; therefore it was decided to specify variants for those two types of households only. In the High variant, the number of one-person households is assumed to be high and the number of couples to be low, whereas in the Low variant the opposite applies.

The width of the interval between the variants is determined differently than for the Population Forecast. In the latter, for each component, the interval is aimed to describe a two-thirds confidence interval, i.e. the odds are assumed to be two to one that the interval will cover the true future value. This assumption is based on an analysis of previous forecast errors. However, as the NCBS did not publish household forecasts previously, there is no information on the size of forecast errors. So it was decided to determine the width of the interval differently.

In the High variant, it is assumed that the percentage of persons living in consensual union in each marital status remains constant at the percentages observed in 1990. The differences with the percentages in the Medium variant are added to the number of one-person households. Since the percentage of cohabiting couples is assumed to increase in the Medium variant (as will be argued in the next section), the number of households in the High variant will be larger than in the Medium variant. For the Low variant, it is assumed that the percentage of non-married cohabitees is higher than in the Medium variant to the same extent that it is lower in the High variant.

11.7. EXTRAPOLATION PROCEDURE

As already noted, assumptions on the proportions of household positions in the population broken down by sex, age, and marital status are stated for the year 2010. These assumptions are based on time-series analyses of the development of household positions in the past on the one hand, and on socio-demographic analyses providing explanations of those changes on the other.

The main source of information for the time-series analyses are the Housing Demand Surveys of 1981, 1985/1986, and 1989/1990. For the socio-demographic analyses, data from the Netherlands Fertility Surveys of 1982 and 1988 are used (NCBS 1991).

Broadly speaking, a comparison of the three Housing Demand Surveys shows that changes in the second half of the 1980s were smaller than in the first half. For the forecast period, it is assumed that the changes will continue in the same direction, but that the speed of the changes will decline in the coming years.

The procedure for specifying the quantitative assumptions consists of seven — partly iterative — steps.

1. For each marital status, age profiles of household positions are determined by smoothing the Housing Demand Survey data using cubic splines.
2. The proportions of household positions for 2010 are obtained by assuming that the changes between 1990 and 2010 equal the changes between the (smoothed) age profiles of 1986 and 1990.

3. For the years 1991-2009, the proportions are obtained by means of cubic interpolation, assuming the slope in 2010 to be zero. Steps 2 and 3 imply a decreasing change. As already noted, the reason for this choice is that, first, the 1980s show decreasing changes and, secondly, changes in percentage distributions cannot continue in the same direction unrestrictedly.
4. For each forecast year, the numbers are adjusted in such a way that the consistency requirements are satisfied.
5. It is examined whether aggregation of the distributions of the household positions over the marital states yields a plausible pattern. If not, the age profiles are adjusted and step 4 is repeated.
6. The age profiles are compared with the broad patterns that are expected on the basis of the socio-demographic analyses. If there are differences, the shape of the curves is adjusted, after which steps 4 and 5 are repeated. Section 11.8 discusses the main conclusions from the socio-demographic analyses.
7. The numbers of people in various household positions are compared with those according to the PRIMOS forecast, taking into account differences in definitions. If there are no explicit arguments for the differences, the age profiles are adjusted and steps 4 and 5 are repeated. The reason for this procedure is that the Advisory Committee for Population Forecasts, which coordinates demographic projections produced by Dutch government agencies, asked Statistics Netherlands to make sure that the results of the National Household Forecasts would differ from the PRIMOS forecasts (which are produced under the authority of the Ministry of Housing) only if there were explicit reasons to do so. In section 11.9, the differences between the NCBS and PRIMOS forecasts will be discussed.

11.8. MAIN ASSUMPTIONS

After a decrease in the preceding decades, the first half of the 1980s witnessed an increase in the home-leaving age, which stabilised in the second half of the decade. One explanation for this development is the postponement of living with a partner due to the lengthening education period and the increasing tendency of women to invest in a career. Moreover, the relationship between parents and children has changed: the generation gap has narrowed. In addition, the smaller average family size means more available room for children in the parental home. For the forecast period, it is assumed that the home-leaving age will remain constant.

As a consequence of the increase in educational attainment and labour force participation, the economic independency of women has increased. A rising number of women are financially capable of maintaining their own household without being dependent on the income of a male partner. Consequently, women are increasingly choosing to live alone after leaving the parental home. This does not imply that they do not have a partner or that they do not want to live together in the long run, but that maintaining a household together with a male partner is postponed, e.g. preceded by a period of LAT relationship. Although non-marital cohabitation has risen strongly, the increase has been smaller than the decrease in the propensity to marry. Apart from the delay in relationship formation, the phenomenon of serial monogamy plays a role. Between successive relationships, the separated partners live alone for some time before they decide to live together with a new partner. As it is expected that the economic independence of women will continue to increase, the household forecast assumes that the percentage of unmarried persons forming a one-person household will increase. As the data from the Housing Demand Surveys did not show a clear-cut rise in the percentages of never-married persons living alone at various ages during the 1980s, the age schedules used in the household forecast differ from the pattern that would result from a straightforward extrapolation of the observed age curves observed. This is the main adjustment of the age schedules in step 6 of the procedure discussed in section 11.7.

Furthermore, an increase in the percentage of never-married persons forming a consensual union is assumed, but the size of the increase is assumed to be smaller than the decrease in the propensity to marry. The percentage of lone parents among divorced and widowed persons has decreased, caused by the postponement of the first birth and the increase in permanent childlessness. A slight further drop in the percentage of lone parents is expected.

During the 1980s, the government policy concerning housing for the elderly changed. Now government policy is aimed at letting elderly people live in their own house as long as possible. Both emancipatory and financial considerations play a role in this. Consequently, in spite of the ageing of the population, the number of persons living in institutional households has not risen in recent years. For the forecast period, the numbers of elderly people living in institutions are assumed to be constant. In view of the ageing of the population, this implies decreasing percentages.

As already noted, the future population by marital status is determined by the 1991 Population Forecasts. In these forecasts, the probability of first marriage is assumed to decline from 90% for women born in the 1950s, via 80% for women born in the 1960s, to 75% for women born in the 1970s. For men the percentages are about 5% lower.

The probability of divorce of married persons born in the 1950s is about 30%. It is assumed that this percentage will be the same for younger cohorts.

The probability of remarriage of divorcees is about 60% for men and about 50% for women. These percentages are assumed not to change. For widowed persons the probability of remarriage is well under 10%. These percentages are not assumed to change either.

In the second half of the 1980s, changes in the distribution of the population by household position were smaller than in the first half of the 1980s. In summary, it is assumed that in the forecast years trends will continue (i.e. increase in the proportion of one-person households, increase in non-marital cohabitation, stabilisation of home-leaving age, decrease in the percentage of lone parents, decrease in the proportion of persons living in institutions), but that the size of the changes will diminish.

As the household forecast model is based on period age schedules, the possibility should be taken into account that the form of the age pattern may change due to cohort effects. In the NCBS Household Forecasts, the age schedules of the percentages of never-married, divorced, and widowed persons living together in the forecast period differ from the age patterns that would result from an extrapolation of observed changes. The main difference is that the percentages of unmarried persons living together at the middle ages are assumed to be higher than the percentages that would result from an extrapolation of the percentages observed in the Housing Demand Surveys. The reason for this is that it is assumed that the increase of unmarried cohabitation at young ages observed in the 1980s can at least partly be attributed to a cohort effect and that this will result in a future increase at older ages.

11.9. MAIN RESULTS

The forecasts of changes in the distribution of the population by household position are the result of changes in the distribution of the population by age, sex, and marital status on the one hand, and changes in proportions of household positions in each population category on the other.

Changes in the distribution of men and women by marital status according to the 1991 Population Forecasts are shown in Figure 11.1. Clearly the proportion of never-married persons will rise. The proportion of divorced persons will also rise. As a consequence, the proportion of one-person households will rise for most ages (Figure 11.2). For the youngest ages, the main cause is the postponement of marriage. For those in their thirties and forties, the decline of the probability of first marriage. Moreover, the percentage of one-person households among never-married persons is assumed to increase. For persons in their fifties and sixties, the rise in the percentage of divorced persons plays a role. In addition, the percentage of divorced persons living alone

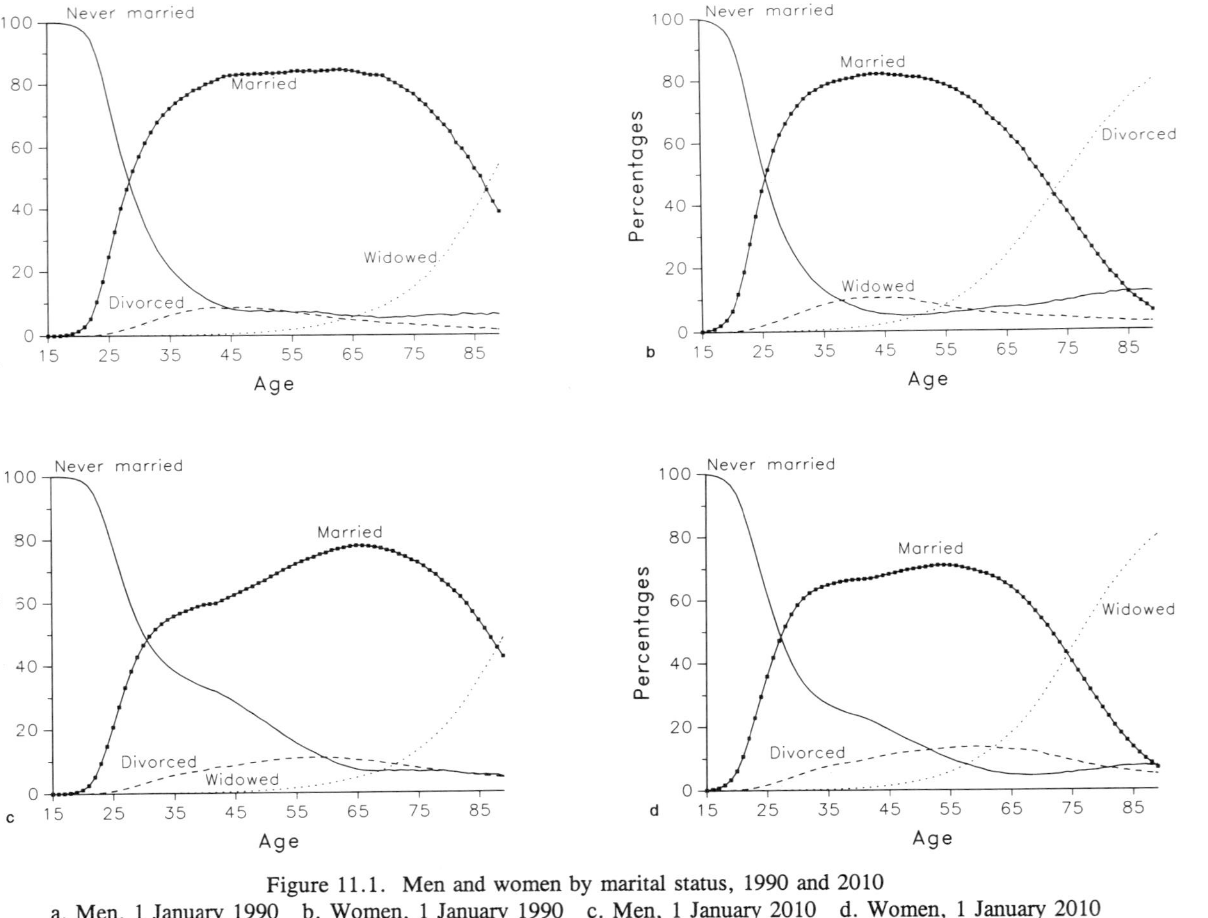

Figure 11.1. Men and women by marital status, 1990 and 2010
a. Men, 1 January 1990 b. Women, 1 January 1990 c. Men, 1 January 2010 d. Women, 1 January 2010

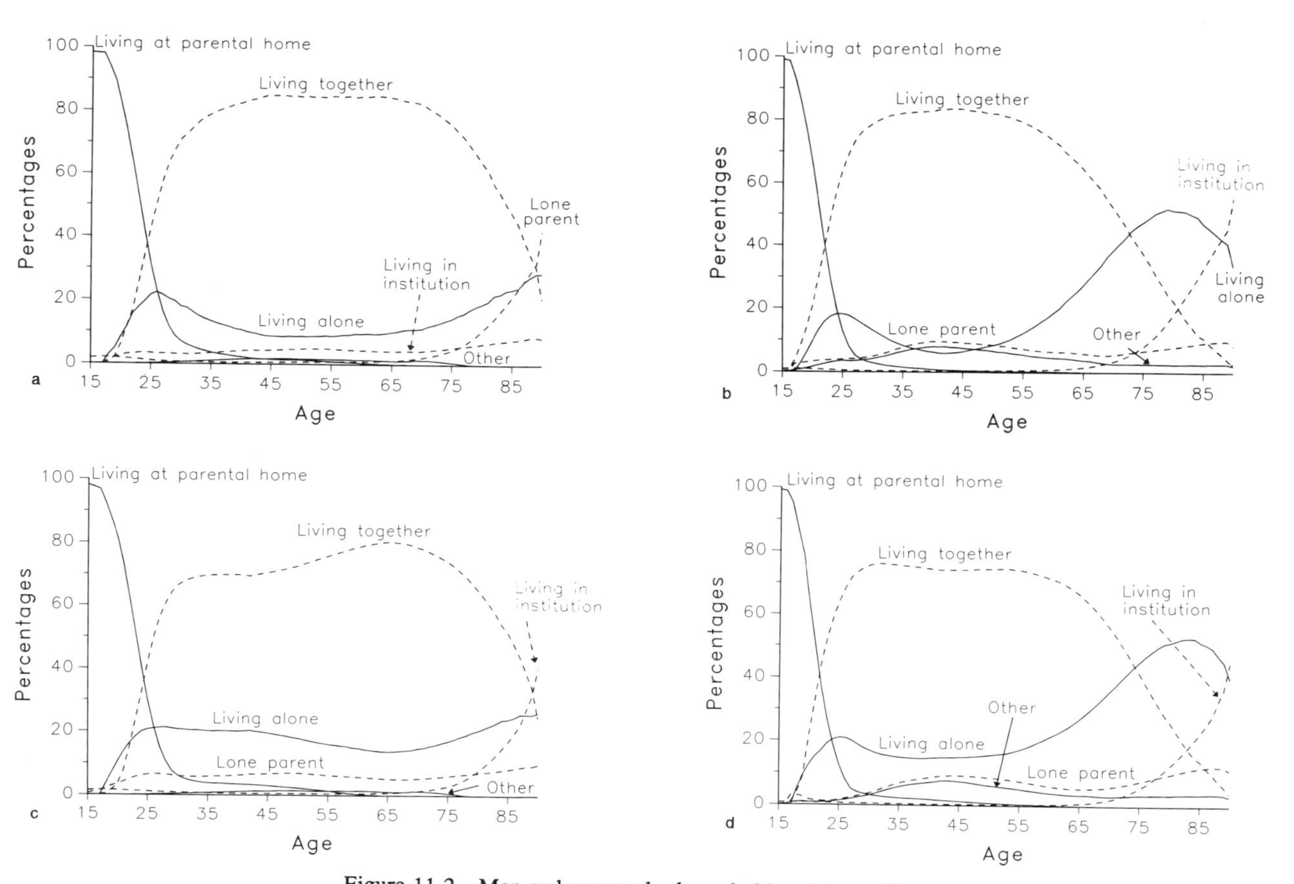

Figure 11.2. Men and women by household position, 1990 and 2010
a. Men, 1 January 1990 b. Women, 1 January 1990 c. Men, 1 January 2010 d. Women, 1 January 2010

is assumed to increase due to the decline of the average number of children per couple, resulting in a decline of the percentage of lone parents.

Of men aged 25 or older in 2010, about one out of five will live alone, whether or not temporarily. For women the proportion of one-person households will be high at old ages. Of women aged 75 or over, more will live alone than together with a partner. The reason is that life expectancy of women is six years higher than that of men, while they marry men who are on average two years older. Consequently, much more women than men are widowed.

The decline of the percentage of couples living together is due to the assumption that although non-marital cohabitation is assumed to increase, the increase is smaller than the decrease of the propensity to marry. For men the percentage living together in 2010 is highest around age 65. This is due to a cohort effect: a larger percentage of older cohorts than of younger cohorts is married.

The percentage of persons living in an institution decreases because the capacity of old people's homes is not assumed to increase, while the number of elderly persons will rise.

Taking into account changes in the age distribution of the population, the development of the total number of persons in the various household positions can be determined (see Figure 11.3). The number of children living at home will decrease slightly for some years and then rise again. This can be explained by changes in the number of births: a decrease until 1983, followed by an increase which is expected to continue until 1998. The number of one-person households will rise strongly: from 1.7 million in 1990 to 2.6 million in 2010. While 11.7% of the population lived alone in 1990, this will be 15.5% in 2010. The number of persons living together will increase until the early 21st century and then stabilise. The large majority of them will be married. The number of unmarried cohabitees will continue to increase until 2010.

On the basis of the numbers of persons in various household positions, a forecast can be made of the numbers of households. The total number of private households is expected to increase from 6.0 million in 1990 to 7.2 million in 2010 (Figure 11.4). According to the Low variant, the number will increase also, although much more slowly: to 6.8 million in 2010. The High variant expects 7.7 million households in 2010.

The increase in the number of households is mainly caused by the increase in the number of one-person households. Consequently, average household size will decline: from 2.45 in 1990 to 2.27 in 2010. If household size were not to decline, the increase in the number of households would be half the forecasted increase.

Broadly speaking, the results of the NCBS Household Forecasts are in line with the PRIMOS forecasts, which are based on the 1991 NCBS Population Forecasts. According to the PRIMOS forecasts, the number of households in

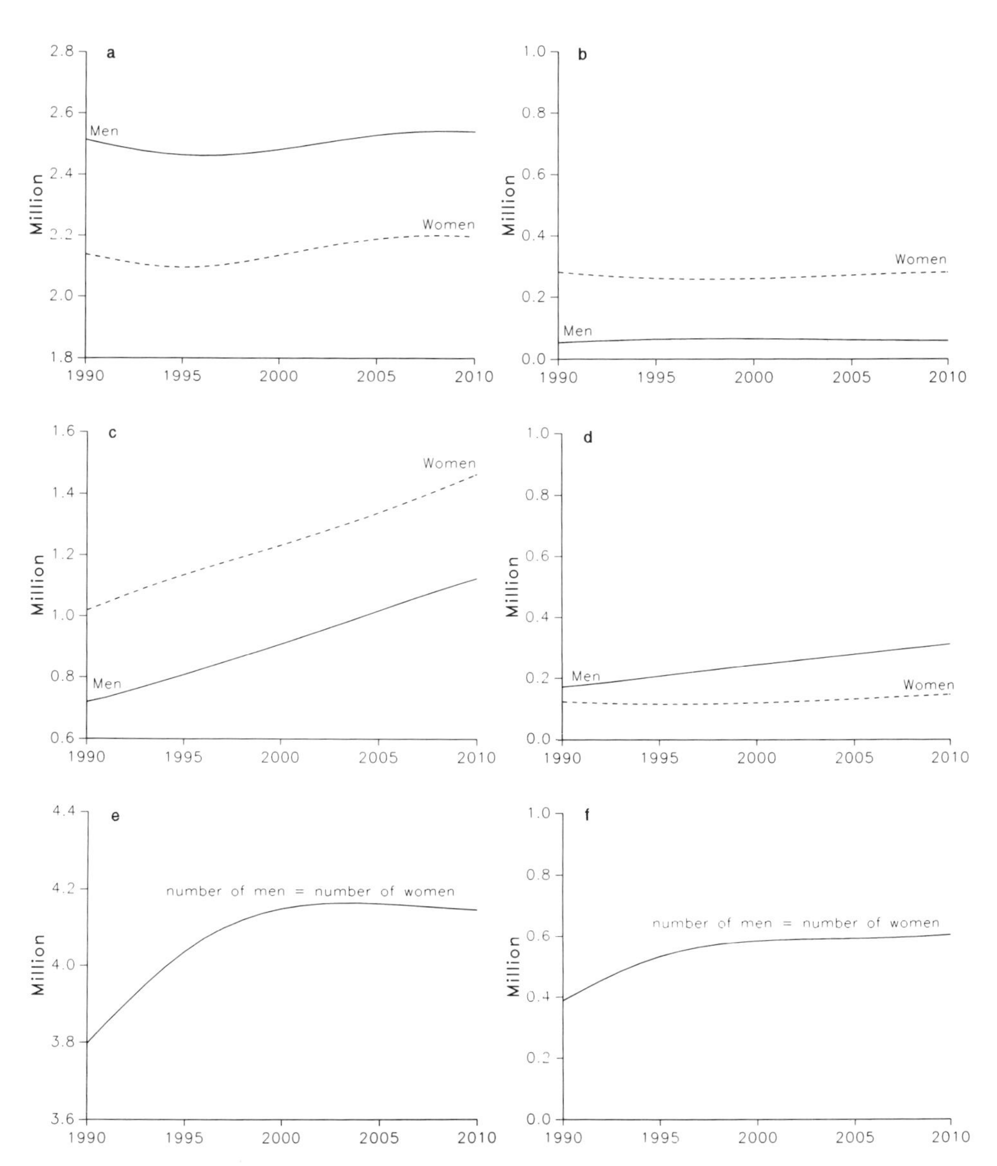

Figure 11.3. Population by household position, 1990-2010
a. Living in parental home b. Lone parent c. Living alone
d. Other e. Living together f. Non-married cohabitees

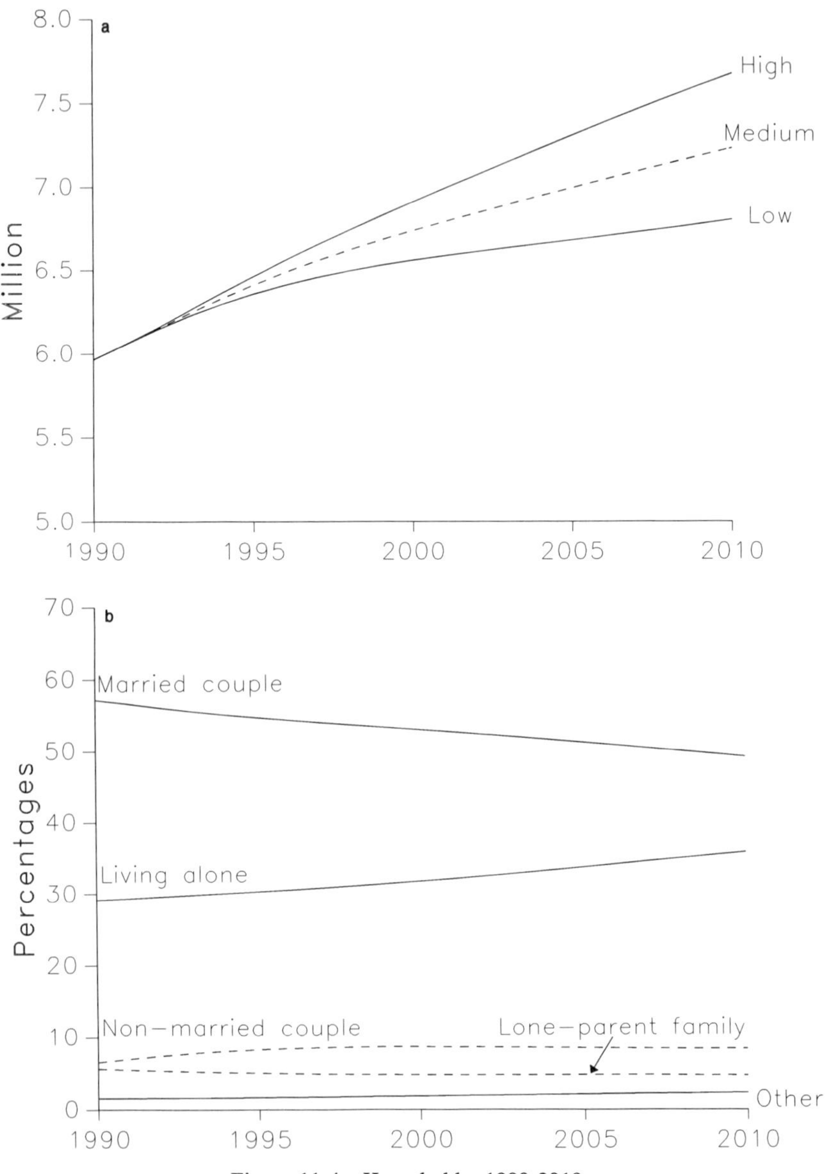

Figure 11.4. Households, 1990-2010
a. Total number of households
b. Households by type.

2010 will be 7.18 million, compared with 7.24 million according to the NCBS. The difference is considerably smaller than the width of the margin between the variants of the NCBS forecasts. However, the underlying differences for separate types of households are larger. For 2010 the PRIMOS model forecasts 4.38 million couples, and the NCBS 4.15 million. In combination with the NCBS forecast of the population by marital status, the PRIMOS forecast would imply that the rise in non-marital cohabitation is larger than the drop in marriage. However, as the PRIMOS forecast does not distinguish married and unmarried couples, it is not based on the NCBS forecasts of the population by marital status.

As for the number of one-person households, it is important to note that the definitions between the PRIMOS and the NCBS forecasts differ. If two or more persons in the household position 'other' form one household together, one of those is labelled 'single' in the PRIMOS forecast, whereas all persons are labelled 'other' in the NCBS forecast. Taking this into account, the number of households without partners in the NCBS forecast is about 300 thousand higher than in the PRIMOS forecast.

11.10. CONCLUDING REMARKS

The NCBS Household Forecasts will be updated annually. The updates of the Household Forecasts will be consistent with future revisions of the Population Forecasts, thus taking the most recent demographic trends into account. In addition, from 1993 the NCBS will publish annual household statistics (based on the labour force survey). Hence in updating the household forecasts, recent information on changes in household formation can also be taken into account.

The rather simple methods used for the first Household Forecasts should not be considered as 'definitive'. For future forecasts, it will be examined to what extent more sophisticated methods may result in better forecasts. This depends to an important extent on the availability of data of good quality and enough detail.

It is expected that in future issues of the Household Forecasts the number of household positions will be expanded, e.g. by relating children to parents. This will provide the possibility to forecast the average size of various types of households. Furthermore, it will be examined to what extent the model can be made more dynamic by means of introducing transition probabilities for several household positions.

ACKNOWLEDGEMENTS

The views expressed in this paper are those of the author and do not necessarily reflect the views of Statistics Netherlands. The author thanks Pieter Everaers, Hans Heida, Pieter Hooimeijer, Wim van Hoorn, Evert van Imhoff, Andries de Jong, Jan Latten, Kees Prins, and Huib van de Stadt for their comments on earlier versions.

REFERENCES

DE BEER, J. (1991), Bevolkingsprognose 1991: Methode, Veronderstellingen en Uitkomsten, *Maandstatistiek van de Bevolking*, Vol. 39(12), pp. 29-40.

HEIDA, H.R. (1991), *Primos Huishoudenmodel. Dynamische Simulatie van de Huishoudens-ontwikkeling*, INRO-TNO: Delft.

NETHERLANDS CENTRAL BUREAU OF STATISTICS (1984), *Prognose van de Bevolking van Nederland na 1980. Deel 2. Modelbouw en Hypothesevorming*, SDU: The Hague.

NETHERLANDS CENTRAL BUREAU OF STATISTICS (1991), *Netherlands Fertility Survey 1988. Cohabitation, Marriage, Birth Control, Employment and Fertility*, Demographic Working Paper, Department of Population Statistics, Voorburg.

VAN IMHOFF, E., & N.W. KEILMAN (1991), *LIPRO 2.0: an Application of a Dynamic Demographic Projection Model to Household Structure in the Netherlands*, Swets & Zeitlinger: Amsterdam/Lisse.

APPENDIX: Age Profile Described by Cubic Spline

An age pattern can be described by a cubic spline as follows:

$$y_x = \sum_{i=0}^{I} D_i F_i(x) + e_x , \tag{1}$$

where

$$F_i(x) = a_i + b_i(x-x_i) + c_i(x-x_i)^2 + d_i(x-x_i)^3,$$

$$D_i = 1 \text{ if } x_i \leq x < x_{i+1} \qquad \text{for } i = 0, 1, \ldots, I\text{-}1$$

$$D_i = 0 \text{ if } x < x_i \text{ or } x \geq x_{i+1} \qquad \text{for } i = 0, 1, \ldots, I\text{-}1$$

$$D_I = 1 \text{ if } x_I \leq x \leq x_{I+1}$$

$$D_I = 0 \text{ if } x < x_I,$$

$$F_i(x_{i+1}) = F_{i+1}(x_{i+1}) \qquad \text{for } i = 0, 1, \ldots, I\text{-}1 \tag{2}$$

$$F'_i(x_{i+1}) = F'_{i+1}(x_{i+1}) \qquad \text{for } i = 0, 1, \ldots, I\text{-}1 \tag{3}$$

$$y_x = \text{observation for age } x,$$

$$x_0 = \text{starting age},$$

$$x_i = \text{age at knot } i \ (i = 1, 2, \ldots, I),$$

$$x_{I+1} = \text{oldest age},$$

$$F'_i(x_{i+1}) = \text{first derivative at } x = x_{i+1},$$

$$e_x = \text{deviation between observation and spline.}$$

The coefficients a_i, b_i, c_i and d_i are determined by fixing the function values $F_i(x_i)$ and the first derivatives $F'_i(x_i)$. Taking into account the condition that the function values and the first derivatives of the adjacent cubic curves are equal at the knots, this yields $4(I+1)$ equations, from which the $4(I+1)$ unknown coefficients, a_i, b_i, c_i and d_i ($i = 0, 1, \ldots, I$), can be solved. The $4(I+1)$ equations are (2) and (3) together with (4)-(7):

$$F_i(x_i) = a_i \qquad \text{for } i = 0, 1, \ldots, I \tag{4}$$

$$F'_i(x_i) = b_i \qquad \text{for } i = 0, 1, \ldots, I \tag{5}$$

$$F_I(x_{I+1}) = a_I + b_I(x_{I+1}-x_I) + c_I(x_{I+1}-x_I)^2 + d_I(x_{I+1}-x_I)^3 \tag{6}$$

$$F'_I(x_{I+1}) = b_I + 2c_I(x_{I+1}-x_I) + 3d_I(x_{I+1}-x_I)^2. \tag{7}$$

Solving these equations yields the values of the coefficients. The values of a_i and b_i are given by (4) and (5). The values of c_i and d_i equal:

$$c_i = 3(a_{i+1}-a_i)/L_i^2 - (2b_i+b_{i+1})/L_i \qquad \text{for } i = 0, 1, \ldots, I \tag{8}$$

$$d_i = 2(a_i-a_{i+1})/L_i^3 + (b_i+b_{i+1})/L_i^2 \qquad \text{for } i = 0, 1, \ldots, I \tag{9}$$

where

$$L_i = x_{i+1}-x_i+1,$$

$$a_{I+1} = F_I(x_{I+1}),$$

$$b_{I+1} = F'_I(x_{I+1}).$$

12. LIPRO: A MULTISTATE HOUSEHOLD PROJECTION MODEL

Evert van Imhoff

Netherlands Interdisciplinary Demographic Institute (NIDI)
P.O. Box 11650
2502 AR The Hague
The Netherlands

Abstract. This chapter describes the construction of a very general, dynamic household model called LIPRO, based on the insights of multidimensional demography, and applies this model to trace the present and the future household situation in the Netherlands. The model classifies individuals by eleven household positions, corresponding to seven household types. These household positions are specified in such a way that a projection of individuals by household position can easily be transformed into a projections of households by household type.

12.1. INTRODUCTION

Household projection models developed in demography over the past few decades are predominantly of the headship rate type. The development of households by number and type is described by the development of number and type of their heads, on the basis of analysing trends in proportions of individuals who occupy the position of household or family head within some broader defined population categories. Like the labour force participation rate in labour force studies, the headship rate is not a rate in the demographic ('occurrence-exposure') sense, and its analytical use reflects a focus on changes in *stocks*, rather than a focus on *flows*.

In spite of this disadvantage, the headship rate method is often used. The method is easy to apply and its data demands are modest. In addition, the method can relatively easily be generalized in several directions. In the basic headship rate method, fractions of household heads are specified for each combination of age and sex. One possible generalization is to refine the population categories in terms of which the headship rates are defined, for

Household Demography and Household Modeling
Edited by E. van Imhoff *et al.*, Plenum Press, New York, 1995

instance, as in models with headship rates by age, sex, and marital status. Another possibility is to distinguish headship rates for different types of households: one-person households, couples, one-parent families, and so on. The household projection model used by Statistics Netherlands (NCBS), discussed more fully in chapter 11 by Joop de Beer, combines these two generalizations. Since the NCBS projects the population by marital status with the use of a dynamic model, one might call the NCBS household projection model *semi-dynamic*.

However, the *static* nature of the headship rate method and its extensions have led to a number of attempts to develop *dynamic* household models, as it is the dynamic processes of household formation and dissolution that really cause changes in the cross-sectionally observed stock numbers of households, and not the other way around. Early attempts to construct dynamic household models were largely of an *ad hoc* nature. Data availability and certain practical solutions to modelling problems were largely responsible for this *ad hoc* situation.

In recent years, better data on household dynamics became available for some countries. At the same time, a coherent and very general methodology for the modelling and projection of dynamic demographic aspects of a population was developed: multidimensional (or multistate) demography. It is the purpose of this chapter to describe the construction of a very general, dynamic household model called LIPRO, based on the insights of multidimensional demography, and to apply that model to trace the present and the future household situation in the Netherlands. What are the consequences of increasing divorce rates for the number of one-parent families? How many additional one-person households would there be if young adults were to leave the parental home at a lower age? These and similar questions may be answered by using a dynamic household model, whereas a static model would provide little insight into these matters.

The organization of this chapter is as follows. Section 2 gives an outline of the LIPRO methodology. Section 3 describes how the parameters of the underlying household dynamics have been estimated from empirical data. Some projection results are presented in section 4. The final section offers a summary and conclusions.

12.2. METHODOLOGY

12.2.1. Multidimensional Projection Models

Dynamic demographic projection models describe the development over time of a population. A population consists of a number of *individuals*, broken down by certain demographic characteristics (e.g. age, sex, marital status, geographic location, household status, and the like). This multidimensional breakdown of the population defines a *state space*. A vector in the state space is called a *state vector*; its elements consist of numbers of individuals at one point in time, broken down by demographic characteristics.

Theory and applications of multidimensional models (sometimes called multistate models) have appeared in the demographic literature since the mid-1970s. Most applications have focussed on multiregional models and marital status models. Other applications include working life tables and fertility by parity. The application to household dynamics is relatively new.

The development over time of the population can be described in terms of *events*: immediate jumps from one cell in the state vector to another. Examples of events are marriage, divorce, leaving the parental home, and internal migration. It is possible for an individual to experience several events within one single projection interval; of course, the probability of multiple events increases with the length of the projection interval.

The population under consideration is not closed: some individuals leave the population (death, emigration), others enter the population (birth, immigration). Such jumps into, or out of, the population are also termed events. In order to distinguish this latter type of events from the type discussed in the previous paragraph, jumps across the boundaries of the population will be termed *external events* as opposed to *internal events*. External events comprise *exits* and *entries*. Exits can be subdivided according to destination, entries according to origin.

A different classification of events is by *endogenous* and *exogenous* events. An endogenous event is an event that is 'explained' within the demographic model itself. In purely demographic models, the occurrence of events is explained by, first, the number of individuals occupying a certain state during a certain interval of time, and, second, the probability that a given individual will experience some event. Consequently, events are endogenous whenever the number of events is dependent on the distribution of individuals within the population over the various characteristics. All internal events and all exits are endogenous. Entries are in part endogenous (births), in part exogenous (immigration).

These various types of events have been illustrated in Table 12.1.

Before this general multidimensional projection model can become operational, several steps have to be taken. First, the state space has to be specified, i.e. the number of age groups, sexes, and the number of internal positions as well as their names. Second, for each event in the table, transition probabilities, occurrence-exposure rates, or jump intensities have to be specified, for each combination of age and sex, with the exception of the exogenous entries (immigration) where absolute numbers are used. Third, an initial population, classified by age, sex, and internal position is required. Finally, depending on the particular application, several additional model features may be necessary, such as a specification of consistency requirements, to be discussed in section 12.2.3.

Table 12.1. Classification of events

<table>
<tr><td colspan="3" rowspan="3"></td><td colspan="2">position after the event</td></tr>
<tr><td>internal positions</td><td>external positions</td></tr>
<tr><td>#1 #2 #3 etc.</td><td>dead rest world</td></tr>
<tr><td rowspan="3">position before event</td><td>internal positions</td><td>#1
#2
#3
.
etc.</td><td>internal events</td><td>exits</td></tr>
<tr><td rowspan="2">external positions</td><td>not yet born</td><td>endogenous entries</td><td>irrelevant</td></tr>
<tr><td>rest of the world</td><td>exogenous entries</td><td>irrelevant</td></tr>
</table>

It should be pointed out that the term 'model' is used here in two different senses. On the one hand, we have a model (such as the LIPRO model) as a *calculation device*. The computer program LIPRO can be used for any kind of multidimensional demographic projection. The program itself does not care whether the state space defines household types, regions, or marital states.

The model only becomes a household projection model (and this is the second sense in which the term 'model' is used here) if the state space defines household types. For the computer program, the definition of the state space is part of its input. Thus, the LIPRO household projection model is just a particular *specification* of the more general LIPRO multidimensional projection model.

12.2.2. Specification of the Household Model

In the LIPRO household model, the population is broken down by age, sex, and *household position*. This breakdown defines a state vector, the elements of which consist of numbers of *individuals* at one point in time, of a certain age, male or female, occupying a particular position in a particular household type. One important consideration in the choice of household positions is that the classification should facilitate the transformation of a projection of individuals into a projection of households, preferably of households *by type*. To date, the application of LIPRO has been limited to *private* households, thus excluding the population in nursing homes, homes for the elderly, and similar collective households.

The considerations that gave rise to the specification of the LIPRO household model are discussed extensively elsewhere (Van Imhoff & Keilman 1991, pp. 59-61) and will not be repeated here. After weighing all the arguments, it was decided to use the following classification of *household positions*:

1.	CMAR	child in family with married parents
2.	CUNM	child in family with cohabiting parents
3.	C1PA	child in one-parent family
4.	SING	single (one-person household)
5.	MAR0	married, living with spouse, but without children
6.	MAR+	married, living with spouse and with one or more children
7.	UNM0	cohabiting, no children present
8.	UNM+	cohabiting with one or more children
9.	H1PA	head of one-parent family
10.	NFRA	non-family-related adult (i.e. an adult living with family types 5 to 9)
11.	OTHR	other (multi-family households; multiple single adults living together)

These eleven household positions together define seven *household types*. Numbers of households of these various types (H) may be inferred easily from numbers of persons in the eleven household positions (P). Thus a household projection which reads in terms of individuals may be translated into one which reads in terms of households. The seven household types, as well as how their number may be derived from numbers of individuals, are the following:

1.	SING	a one-person household. H(SING) = P(SING).
2.	MAR0	a married couple without children, but possibly with other adults. H(MAR0) = ½·P(MAR0).
3.	MAR+	a married couple with one or more children, and possibly with other adults. H(MAR+) = ½·P(MAR+).
4.	UNM0	a couple living in a consensual union without children, but possibly with other adults. H(UNM0) = ½·P(UNM0).
5.	UNM+	a couple living in a consensual union with one or more children, and possibly with other adults. H(UNM+) = ½·P(UNM+).
6.	1PAF	a one-parent family, possibly with other adults (but not as partner to the single parent!). H(1PAF) = P(H1PA).

7. OTHR multi-family households, or multiple single adults living together without unions. The number of households of type OTHR equals the number of persons in household position OTHR divided by the average number of persons in OTHR households. This average household size was 2.82 persons in 1985. It is assumed that the average size of OTHR household remains unchanged throughout the projection period. Thus: H(OTHR) = P(OTHR) / 2.82.

We feel that this classification offers a reasonable compromise between the conflicting objectives of completeness and feasibility.

The events matrix corresponding to this model specification is depicted in Table 12.2. For the event 'birth', the household position of the newly-born child is uniquely determined by the household position of the mother prior to giving birth. Two important assumptions underlying the events matrix are:

- a return to the status of child is possible only from the household position 'single';
- adults can change household only via the status 'single'. This excludes events like 'from MAR+ to UNM0' (leave spouse and children in order to directly move into cohabitation with a new partner) or 'from H1PA to MAR0' (leave children in order to directly move into marriage).

12.2.3. Consistency Relations

An important part of the model consists of the module that handles the *consistency problem*. Within the context of household models, the consistency problem can be considered a generalization of the well-known two-sex problem in marital status models. Unless the modeller includes a two-sex algorithm in such a marital status model, male marriages will *not* be equal to female marriages (nor will male divorces correspond to female divorces, or deaths of married persons to transitions to widowhood of the other sex). In household projection models, numbers of male entries into cohabitation have to correspond to numbers of female entries into cohabitation in a certain period, and the number of last children who leave a one-parent household must be equal to the number of heads of such a household who become single. These requirements are but a few of the many consistency relations that may appear in the framework of a household projection model.

Table 12.2. Events matrix of the household model

from: to:	1	2	3	4	5	6	7	8	9	10	11	dead	rest
1. CMAR	*	*	.	.	.	.	.	.	*	.	.	.	.
2. CUNM	.	*	.	.	.	.	.	.	*	.	.	.	.
3. C1PA	.	.	*	.	.	.	.	.	*	.	.	.	.
4. SING	.	.	.	*	.	.	.	.	.	.	.	.	.
5. MARO	*	*	*	.	*	.	*	*	*	*	.	.	.
6. MAR+	*	*	*	.	.	*	*	*	.	*	.	.	.
7. UNMO	*	*	*	.	.	*	*	.	*	*	.	.	.
8. UNM+	*	*	*	.	*	.	.	*	.	*	.	.	.
9. H1PA	*	*	*	.	*	.	*	.	*	*	.	.	.
10. NFRA	*	*	*	.	*	*	*	*	*	*	.	.	.
11. OTHR	.	.	.	.	.	.	.	.	.	.	*	.	.
a. birth from													
CMAR	*	*	*	*	*	*	*	*	*	*	.	*	*
CUNM	*	*	*	*	*	*	*	*	*	*	.	*	*
C1PA	*	*	*	*	*	*	*	*	*	*	.	*	*
SING	*	*	.	*	*	*	*	*	*	*	*	*	*
MARO	.	*	*	*	*	*	*	*	*	*	*	*	*
MAR+	.	*	*	*	*	*	*	*	*	*	*	*	*
UNMO	*	.	*	*	*	*	*	*	*	*	*	*	*
UNM+	*	.	*	*	*	*	*	*	*	*	*	*	*
H1PA	*	*	.	*	*	*	*	*	*	*	*	*	*
NFRA	*	*	*	*	*	*	*	*	*	*	.	*	*
OTHR	*	*	*	*	*	*	*	*	*	*	.	*	*
b. rest	.	.	.	.	.	.	.	.	.	.	.	*	*

* = impossible event.
. = possible event.

The current version of the LIPRO computer program contains a very flexible consistency module that automatically produces consistent numbers of events once the user has specified which sets of events are linked in linear combinations. The algorithm was developed by Van Imhoff (1992) and is based on weighted linear least-squares optimization. For the present chapter, we have used the (generalized) harmonic mean solution to the consistency problem which involves a proportional adjustment of age-specific numbers of inconsistent events to find age-specific numbers of consistent events. For consistency relations that involve mortality events, the restrictions can be (and have been) formulated in such a way that the projected numbers of deaths are unaffected, leading, for example, to mortality-dominant consistency for entry into widowhood.

It should be stressed that the consistency algorithm is more than merely a mechanical procedure to enforce consistency upon the projection results. As has been shown by Keilman (1985a), the properties of the harmonic mean method are such that the corresponding adjustments in the age-specific numbers of events have a satisfactory *behavioural* interpretation.

Most of the consistency constraints (e.g. the two-sex requirements) stem from the nature of the household classification chosen; this type of consistency is referred to as *internal* consistency. Other constraints may occur because of interrelationships between different models. For instance, numbers of events computed from models of a low aggregation level may be required to add up to the corresponding numbers in the national population forecasts, which are of a higher aggregation level. The latter type of constraints is referred to as *external* consistency (Keilman 1985b). LIPRO's consistency algorithm ensures that the projected numbers of events satisfy certain linear constraints, thus allowing for both internal and external consistency requirements. External consistency has played a role during the phase of model estimation, to be discussed in section 12.3.

The events matrix for the LIPRO household model, depicted in Table 12.2, contains 69 internal events, 22 exits, and 22 entries. For two sexes, this amounts to a total of 226. Formulation of consistency relations between these 226 events was a painstaking activity, finally leading to 38 restrictions in terms of 129 variables. The main assumptions are:

- divorced partners do not continue to live together;
- adoption can be disregarded for the entry of a first child into the household;
- the formation and dissolution of homosexual consensual unions can be disregarded as far as the two-sex requirement for cohabitation is concerned;
- only complete households can migrate (in the sense of moving to or from the state 'rest of the world').

12.3. FROM DATA TO INPUT PARAMETERS

12.3.1. The Housing Demand Survey of 1985/1986 (WBO 1985/1986)

A projection of future household positions requires two types of input data:

- an initial population at the start of the projection interval;
- data on jump intensities or, alternatively, data on jumps and exposed population from which jump intensities can be estimated.

The jump intensities are the central behavioural parameters of the LIPRO model. As will be discussed below, the intensities are estimated from observed transition probabilities. Since many projection models are formulated in terms of transition probabilities, some readers may wonder why LIPRO was not formulated directly in terms of transition probabilities rather than in terms of intensities (or rates). The reasons for preferring intensities to transition probabilities are the following:

- As a behavioural parameter, rates are easier to interpret than probabilities. Also, contrary to probabilities, rates are independent of the length of the unit time interval.
- Transitions only give the cross of the state at the beginning and at the end of the unit time interval. Intensities can be used to obtain all events during the time interval. Thus, intensities give a more complete description of household dynamics (the flows) than transition probabilities. This is especially true if the time interval is relatively long (e.g. five years), with a relatively high probability of multiple events within the interval.
- The behaviour of immigrants can be modelled more easily with intensities.
- The formulation of scenarios for future demographic behaviour is easier in terms of intensities than in terms of transition probabilities.

Our main source of demographic data is the so-called 'Woningbehoeftenonderzoek 1985/1986' ('Housing Demand Survey 1985/1986') or simply WBO 1985/1986. This massive set of data, containing 46,730 households, was collected by the NCBS during the last few months of 1985 and the first few months of 1986.[1] It contains detailed information on, among other things, the household situation of the respondents at the time the survey was taken and their household situation one year earlier. A slight drawback of the WBO 1985/1986 is that the questionnaire focusses on private households. For persons in collective households, only a few basic questions are included.

The WBO 1985/1986 gives us the household position of all individuals in the sampled private households at a single point in time. This information was used, together with data on the distribution by age, sex, and marital status of the WBO respondents living in collective households, to construct the initial population for the simulation (see section 12.3.2). The WBO data were corrected so as to correspond to the observed population structure by age, sex, and marital status as per December 31, 1985.

Information on jumps between the eleven household positions can be obtained from variables indicating the household position of each person one year earlier, to be reconstructed from a small number of 'retrospective' questions included in the questionnaire. Unfortunately, this 'retrospective'

---

[1] In the course of 1994, the LIPRO calculations will be updated using data from the 1989/1990 version of the NCBS Housing Demand Survey. Then an assessment will also be made of the accuracy of the 1985-based projections.

information is incomplete, requiring the use of simplifying assumptions and approximation methods.

12.3.2. The Initial Population

The starting point of our projections is the situation as per December 31, 1985. From the WBO 1985/1986, we can calculate (using the weighting factors provided by the NCBS) the number of persons in each of the eleven household positions, by age and sex. Since these WBO 1985/1986 figures are subject to sampling error, we have adjusted them to bring them into line with the official NCBS population statistics for December 31, 1985. These population statistics are broken down by age, sex, and marital status.

First, the population statistics were adjusted to eliminate the population living in collective households. Next, the age- and sex-specific numbers from the WBO 1985/1986 were adjusted proportionally over the eleven household positions, equalizing the sum of the numbers in positions mar0 and mar+ to the numbers in marital state 'married', and equalizing the sum of the numbers in the other nine household positions to the sum of the numbers in marital states 'never married', 'widowed', and 'divorced'.

12.3.3. Transition Probabilities

If we knew the household position for each individual in the sample, at some previous point in time, we would be able to calculate transition probabilities from the simple age- and sex-specific cross-tables of past versus present household positions. Unfortunately, the variables in the WBO 1985/1986 do not allow an exact reconstruction of past household positions, at least not in terms of our eleven-cell classification. Therefore, an approximation method had to be devised.

For each individual, the WBO 1985/1986 gives the following relevant variables on the household situation one year before the survey date:

- did the individual live in the same household?
- relation to head of household (RELTOHEAD) in which the individual lived at the time (whether or not this was the same household as the present one), coded as follows:
  1. head of one-person household
  2. head of multi-person household
  3. married spouse of head
  4. unmarried partner of head
  5. (step)child of head and/or partner
  6. other adult

7. other child
8. not yet born (i.e. born during last year)
9. living abroad (i.e. immigrated during last year)

- marital status
- how many individuals left the household during the past year?

The households in the sample fall into one of three categories:

- Households without entries and without exits. For these households, reconstruction of the state of its members one year ago is trivial. The only point to bear in mind is the possibility of a transition from an unmarried to a married couple.
- Households with entries. For the non-entrants (i.e. the members of the household who were already present one year earlier), the household position can be found by 'subtracting' the entrants from the present household composition. For the entrants, the previous household position has been approximated by considering their previous value on the variable RELTOHEAD and assuming that the age- and sex-specific cross-tabulation of RELTOHEAD versus our eleven-cell classification did not change between 1984 and 1985. Immigrants were not taken into account. Newly-born individuals were treated separately.
- Households with exits. Here too an approximation method had to be used, since the household position of the person(s) who left, and consequently the household position of the remaining household members before the departure, is unknown. For the stayers, their previous values on the variables RELTOHEAD, marital status, and household size were considered, assuming an unchanging age- and sex-specific distribution across the eleven household positions within each combination of RELTOHEAD, marital status, and household size. The departed persons do not need separate treatment. If they moved into another household, they can be assumed to be included in the entrants discussed in the previous paragraph. If they left the population (through death or emigration), the corresponding jump intensity is estimated from different sources.

For those individuals born in the year before the survey date, we tried to reconstruct the household position of the mother at the moment of childbirth. If necessary, we made simplifying assumptions. Once the household position of the mother has been determined, the household position into which the child is born follows automatically. If this household position at birth is crossed with the household position at the survey date, the transition matrix for age group 0 can easily be calculated.

12.3.4. From Transition Probabilities to Intensities

The computations thus far have yielded single-year transition matrices for internal events, for each age/sex group. As already explained in section 12.3.1, what we need are *intensities*, being the fundamental parameters of the exponential multidimensional projection model (Van Imhoff 1990). The mathematical relationship between a transition matrix $\boldsymbol{T}$ and an intensity matrix $\boldsymbol{M}$ is given by:[2]

$$\boldsymbol{T} = \exp[\boldsymbol{M}h] \tag{1}$$

where h is the length of the observation interval (here one year, except for age group 0 where h is approximately equal to ½). Then

$$\boldsymbol{M} = \log(\boldsymbol{T}) \;/\; h \tag{2}$$

where the log of a matrix is defined in terms of its Taylor power series. It may happen that the latter power series does not converge. In that case, the empirical transition matrix $\boldsymbol{T}$ is said to be non-embeddable, i.e. inconsistent with the assumptions of the exponential model (Singer & Spilerman 1976). In our application, only four out of 176 transition matrices turned out to be non-embeddable.

However, even for embeddable transition matrices, application of (2) leads to very unreliable estimates of the intensities. This is caused by the fact that the logarithm of a matrix is very sensitive to small changes in one of its elements. Since the empirical transition matrices $\boldsymbol{T}$ are subject to a high degree of sampling error, the resulting intensities exhibit a very irregular and unrealistic pattern when plotted as a function of age.

Therefore, it was decided to follow a different approach. This approach rests on the assumption that an observed transition can be identified with an *event*. That is, it is assumed that each individual experiences one event at most during the observation period. Since the observation period is rather short (one year), this assumption appears to be quite reasonable. The only exceptions were made for transitions that are impossible events according to the events matrix of Table 12.1. For example, if a woman was observed to be in position UNMO one year before being observed to be in state MAR+, it has been

---

[2] For some types of events (e.g. leaving the parental home, formation of marriage or consensual union), the WBO contains retrospective information that covers more than just the last year before the interview. For these events, an alternative method of obtaining intensities from retrospective data would be to estimate hazard rates for duration models. This line of research has, however, not been followed in the LIPRO project.

assumed that she experienced two events, namely first the event from UNM0 to MAR0 and then the event from MAR0 to MAR+.

From the events matrices constructed in this way, intensity matrices can be estimated using the moment estimator developed by Gill (1986). The computer program for estimating intensities from events matrices of any rank is described by Van Imhoff and Keilman (1991, Appendix to Chapter 4).

The intensity matrices obtained in this way refer to internal events only. In order to estimate household-position specific mortality and emigration intensities, we used marital status as a proxy. From the NCBS population statistics 1981-1985, marital status-specific exit intensities were estimated using the method of Gill and Keilman (1990). These intensities were subsequently transformed into intensities by household position, using the age- and sex-specific marital status distribution for each household position as weights. A similar approximation method was used to produce estimates of the immigrant population (absolute numbers) by household position from immigration statistics 1981-1985 by marital status.

Since the number of estimated intensities is very large compared with the number of observations, the resulting estimates are subject to large random variations. In order to reduce this variation, the one-year/single age-group intensities were transformed into five-year/five-year age group intensities. A secondary advantage of this transformation is that it reduces the number of computations for a given projection by a factor of 25.[3] The transformation involves a weighted average of the single-year intensities, using the average population (over the year) in each household position as weights, and taking into account the fact that a five-year age group over a period of five years involves 9 different one-year age groups. This procedure resulted in 38 sets of intensity matrices (namely two sexes and 19 age groups, 18 ranging from 0-4 to 85+ and one for the age group born during the five-year period).

12.3.5. Adjusting Intensities to Achieve Internal and External Consistency

The five-year intensities were used to make a household projection over a single projection interval, i.e. the five-year period 1986-1990. The projected numbers of events, not surprisingly, failed to satisfy the conditions for internal consistency. In addition, the results on vital events diverged in several respects from the official numbers of the NCBS. Using the consistency algorithm, the numbers of events projected by LIPRO were adjusted to yield internal con-

[3] Fitting parametrized functions to the age-intensity profiles, which is an alternative method for removing sample fluctuation from the estimated intensities, does not possess this latter advantage.

sistency, as well as external consistency with the official numbers on seven vital events:

- number of births;
- number of deaths;
- number of marriages;
- number of marriage dissolutions;
- number of male entries into widowhood;
- number of female entries into widowhood;
- net international migration.

The intensities were reconstructed from these both internally and externally consistent numbers of events. It is this adjusted set of jump intensities that constitutes the basis of the projections to be discussed in the next section.

12.4. PROJECTION RESULTS FOR THE NETHERLANDS, 1985-2050

The final step in preparing a household projection consists of formulating assumptions with respect to the future trends in household formation and dissolution behaviour. One scenario was created which we have rather optimistically termed our 'Realistic Scenario'. In this scenario we try to adhere, as closely as possible, to the assumptions underlying the official national population forecasts prepared by the NCBS (we used the medium variant of the 1989 forecast; see Cruijsen 1990). Extreme trends are not present in this scenario. The following 'reasonable' assumptions[4] were made:

- a slight further increase in life expectancy at birth, being somewhat larger for males than for females;
- a slight fertility increase, but remaining significantly below replacement level;
- an increase in the proportion of extramarital births;
- a drop in marriage rates, fully compensated by a simultaneous increase in the propensity to form consensual unions;
- an increase in divorce rates, for both marriages and consensual unions;
- a modest decline in international migration.

All these changes are assumed to occur gradually over time.

---

4 This "translation" of NCBS assumptions into LIPRO assumptions should not be confused with imposing external consistency. Assumptions are in terms of intensities; consistency relations are in terms of numbers of events.

In setting scenarios, one has to take into account the fact that many of the intensities are linked together by the consistency relations. If one fails to adjust these related intensities, then the net effect of setting the scenario is smaller than intended. For example, if male mortality is reduced by ten per cent without reducing the rate at which married women are widowed, then the consistency algorithm will partially offset the ten per cent reduction in male mortality. Adjusting related intensities is rather complex, and the exact way in which it has been done will not be discussed here in detail.

It should be pointed out that although our Realistic Scenario was set in such a way that the increase in life expectancy exactly follows the life expectancy in the NCBS forecast, the corresponding projected number of deaths are much smaller in the LIPRO projection as compared to the NCBS forecast. This rather peculiar phenomenon can be explained from the fact that LIPRO and the NCBS have a different way of handling mortality differentials between household positions or marital states (Van Imhoff 1993). Although the NCBS model recognizes mortality probabilities that differ across marital states, the model calculates both life expectancy and the projected number of deaths *without* taking the composition of the population by marital state into account; it is only in the second phase of the projection process that the projected aggregate number of deaths is subdivided into number of deaths by marital status. In the LIPRO model, on the other hand, deaths are immediately attributed to the relevant household positions. The result is that, with constant mortality rates, changes in the marital composition of the population do *not* affect the life expectancy in the NCBS model, while changes in household structure *do* affect the life expectancy in the LIPRO model.

The development of the total population by household position is illustrated in Figure 12.1 for the Realistic Scenario. Total population size reaches its maximum in 2025 (16.4 million), after which a gradual decline sets in. Developments of the population by household position are characterized by:

- a decrease in the number of children;
- a strong decline in the number of couples with children;
- an increase in the number of lone parents which is modest in the absolute sense, but much stronger in the relative sense;
- a diminishing average household size;
- an enormous growth in the number of individuals living in a one-person household.

The tremendous increase in the number of persons living alone is the most striking result of the application of the LIPRO model to household projections in the Netherlands, irrespective of the scenario chosen. This trend is even stronger when the number of households is considered, instead of the number

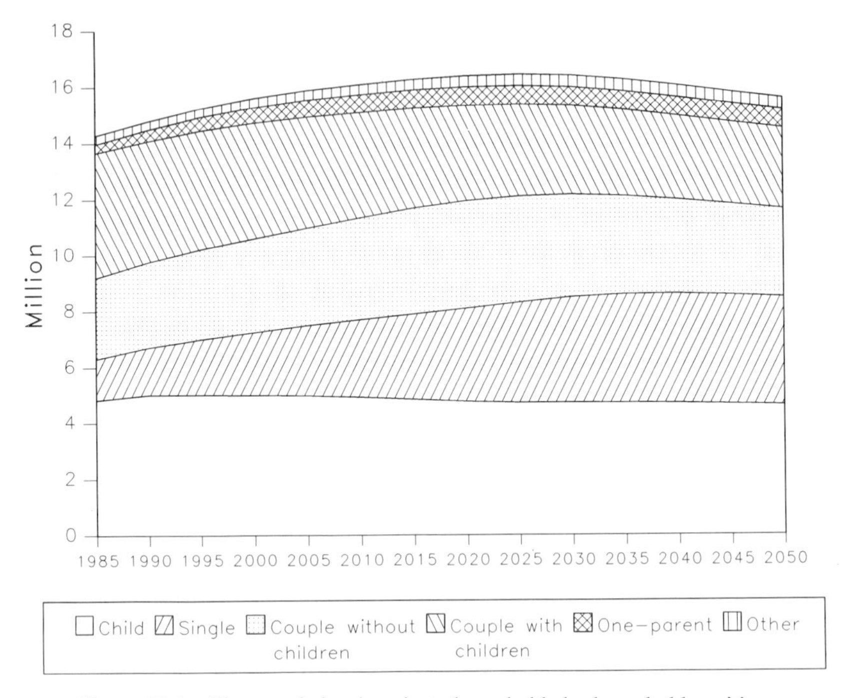

Figure 12.1. The population in private households by household position, 1985-2050 (Realistic Scenario)

of persons by household position. Figure 12.2 shows the development in the number of households of various types for the Realistic Scenario. The increase in the proportion of one-person households is dramatic: from 27 per cent in 1985 to no less than 51 per cent in 2050.

The rise in the number of persons living alone goes hand in hand, to a large extent, with the general aging of the population. In 1985, 34 per cent of the persons living alone was 65 years of age or older; in 2035 and in 2050 this will be 49 per cent. Also note, in Figure 2, the diminishing share of the traditional family, i.e. the married couple with one or more children. Changes in the age structure explain this trend to a small extent only: elderly couples are more frequently in the 'empty nest' stage than younger couples. However, changes in household formation patterns are more important: more couples remain childless, less persons marry, and more marriages are dissolved at a relatively early stage.

Age pyramids, not shown in this chapter, further illustrate the age-specific developments in household structure between 1985 and 2035, the latter being the peak year for the aging process. The strong growth in the number of elderly persons comes out very clearly. Improved longevity is responsible for the rising

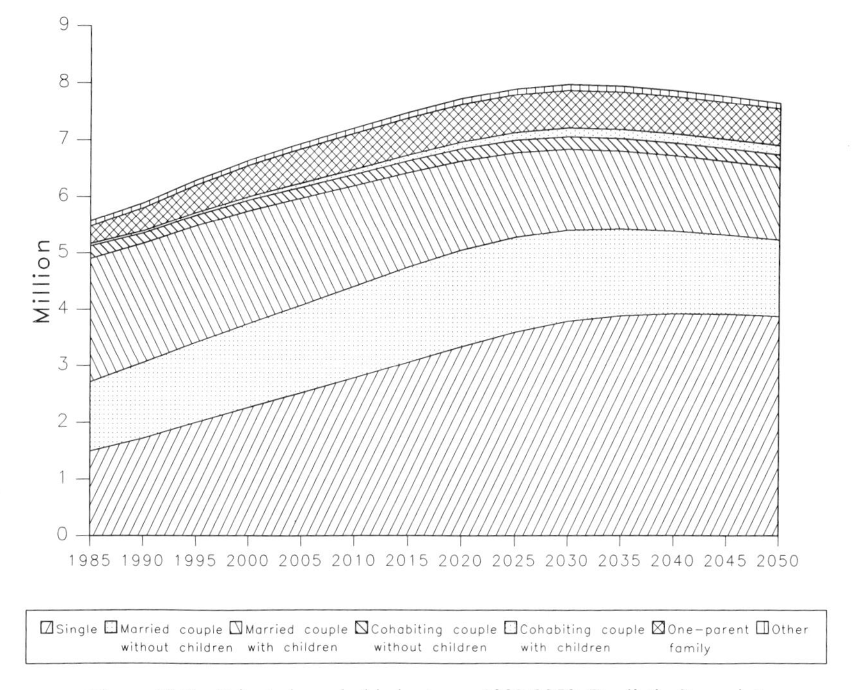

Figure 12.2. Private households by type, 1985-2050 (Realistic Scenario)

numbers of elderly couples (both married and in consensual union) without children. Unmarried cohabitation will slowly become more popular among the elderly in the Realistic Scenario, although the numbers involved will remain low. Stated differently, the mean age of cohabiting persons will rise during the first half of the 21st century, especially among males.

12.5. CONCLUDING REMARKS

This chapter has shown how a multistate projection model can be used to trace current and future dynamics in living arrangements. We have used a household classification of eleven distinct positions that an individual person may occupy at a certain point in time. Our model is able to describe household events for the individuals concerned as they jump from one household position to another.

The model was applied to data for the Netherlands for the years 1985/1986. The most striking outcome of our illustrative projections is a major shift towards one-person households during the next half century.

LIPRO is a dynamic household projection model, contrary to the traditional headship rate models which are essentially static. In the basic headship rate method, it is assumed that fractions of household heads are constant within specific categories of age and sex; these constant rates are then applied to a standard projection of individuals by age and sex in order to yield a projected number of households. One can refine this method by increasing the number of demographic categories for which the headship rate is assumed to be constant; e.g. headship rates by sex, age, and marital status.

The eleven-state LIPRO application can be interpreted as a further refinement, in which headship rates are defined for each combination of age, sex, and *household position*. For this 'ultra-refined' headship rate model, the (implicit) headship rates have specific values only, namely 1 (for states SING and H1PA), ½ (for states MAR0, MAR+, UNM0, and UNM+), 0 (for states CMAR, CUNM, C1PA, and NFRA), and 1/2.82 (for state OTHR). However, there are at least two crucial differences between a 'mere' headship rate model and the LIPRO household model:

- the former yields numbers of households only, while the latter in addition yields a distribution of households by *type of household*;
- LIPRO not only produces the number of households at particular points in time, but also the numbers of the various household *events* that give rise to this projected number of households. That is, LIPRO explicitly opens the black box of the underlying household formation and dissolution processes.

The 'semi-dynamic' NCBS household projection model (De Beer, chapter 11 in this volume) is somewhere in between. It shares the first feature with LIPRO, but not the second; the NCBS model recognizes nuptiality events, which only constitute a subset of all events underlying household formation and dissolution.

ACKNOWLEDGEMENTS

This paper draws heavily on earlier publications, especially (Van Imhoff & Keilman 1991) and (Van Imhoff 1995). The research reported here was financed by the Netherlands Ministry of Social Affairs and Employment. Most of the data used for the calculations presented in this paper were collected by Statistics Netherlands.

REFERENCES

CRUIJSEN, H. (1990), Bevolkingsprognose 1989: Meer Buitenlandse Migratie, *Maandstatistiek van de Bevolking*, Vol. 38(1), pp. 6-7.

GILL, R.D. (1986), On Estimating Transition Intensities of a Markov Process with Aggregate Data of a Certain Type: Occurrences but no Exposures, *Scandinavian Journal of Statistics*, Vol. 13, pp. 113-134.

GILL, R.D., & N.W. KEILMAN (1990), On the Estimation of Multidimensional Demographic Models with Population Registration Data, *Mathematical Population Studies*, Vol. 2(2), pp. 119-143.

KEILMAN, N.W. (1985a), Nuptiality Models and the Two-Sex Problem in National Population Forecasts, *European Journal of Population*, Vol. 1(2/3), pp. 207-235.

KEILMAN, N.W. (1985b), Internal and External Consistency in Multidimensional Population Projection Models. *Environment and Planning A*, Vol. 17, pp. 1473-1498.

SINGER, B., & S. SPILERMAN (1976), The Representation of Social Processes by Markov Models. *American Journal of Sociology*, Vol. 82(1), pp. 1-54.

VAN IMHOFF, E. (1990), The Exponential Multidimensional Demographic Projection Model, *Mathematical Population Studies*, Vol. 2(3), pp. 171-182.

VAN IMHOFF, E. (1992), A General Characterization of Consistency Algorithms in Multidimensional Demographic Projection Models, *Population Studies*, Vol. 46(1), pp. 159-169.

VAN IMHOFF, E. (1993), Leefvorm en Sterfte in Demografische Vooruitberekeningen, *Tijdschrift voor Sociale Gezondheidszorg*, Vol. 71(2), pp. M40-44.

VAN IMHOFF, E. (1995), Modelling the Impact of Changing Household Structure on Social Security in the Netherlands. *In*: J.P. GONNOT, N. KEILMAN & Ch. PRINZ (eds.) *Social Security, Household, and Family Dynamics in Ageing Societies*, Kluwer: Dordrecht.

VAN IMHOFF, E., & N. KEILMAN (1991), *LIPRO 2.0: an Application of a Dynamic Demographic Projection Model to Household Structure in the Netherlands*, NIDI CBGS Publications 23, Swets & Zeitlinger: Amsterdam/Lisse.

13. HOUSEHOLD PROJECTIONS AND HOUSING MARKET BEHAVIOUR

Pieter Hooimeijer[1] and Hans Heida[2]

[1]Faculty of Geographical Sciences
Utrecht University
P.O. Box 80115
3508 TC Utrecht
The Netherlands
[2]FOCUS BV
Jacoba van Beierenlaan 81-83
2613 JC Delft
The Netherlands

Abstract. Demographic behaviour with respect to household formation and dissolution is both motivational and situational. Very few household models deal with either motive or situation. This chapter sketches an approach to incorporate aspects of the situation in the modelling effort. At the intra-regional level the housing market has the most pervasive effects on processes of household evolution. In turn, household evolution is an important driving force behind housing market transactions. Some household events, like leaving the parental home, or a separation, imply a residential move of at least one of the persons involved and are therefore dependent on the availability and accessibility of a dwelling vacancy. Other household events, like the death of a single person, generate vacancies in the housing stock.
The SONAR model is an event-driven dynamic simulation model of both household evolution and housing market transactions at the local (municipal) level, which can be applied to any user-defined combination of municipalities in the Netherlands. The model supplies full detail on household events that either imply a move (event-dependence) or increase the propensity to move (state-dependence) and on the intensity and outcome of housing market search. The generation and allocation of vacant dwellings is part of the housing market search algorithm and feed-back effects on household evolution can be traced. Policy measures with respect to the number and composition of dwelling construction and housing allocation can be evaluated in terms of the redistributive effects of dwellings over households.

Household Demography and Household Modeling
Edited by E. van Imhoff *et al.*, Plenum Press, New York, 1995

13.1. INTRODUCTION

Household evolution processes are particularly sensitive to a wide variety of external influences. At the national level, economic and socio-cultural trends might generate shifts in household formation and dissolution. At the sub-national level, the interdependence of household change and migration will have a decisive impact on the size and structure of the population in households. At the regional level, changes in employment will redirect household change. At the intra-regional level, the housing market will have a decisive impact on residential mobility, not only causing a redistribution of households, but also having an impact on household formation rates. In short, household behaviour is highly dependent on the context and the relevant aspects of the context will differ, depending on the spatial level at which household evolution is being analysed and modelled. For research on household modelling, this assertion has a number of major implications.

The first implication refers to the substantive properties of the model. The temporal and spatial instability in household forecasts, which show up when results of the models are compared with observed data, are often assumed to be a result of population heterogeneity (De Vos & Palloni 1989). However, this instability could be an expression of temporal and spatial non-stationarity, arising from shifts in the context rather than from a change in unobserved characteristics of the population. Therefore, it might be rewarding to direct the research effort towards endogenizing relevant aspects of the context, rather than towards a quest for sources of unobserved heterogeneity.

The second implication refers to the structural properties of the model. Variation in the context over time and space will not affect each (potential) household in the same way. Variation in housing shortages might affect dissolution rates caused by separation, but certainly will not affect dissolution rates due to widowing. Contextual change is linked to dynamics in household structure through the way it affects the underlying household processes. This implies that the modelling effort should be based on the dynamics in household formation and dissolution in terms of the underlying processes of leaving the parental home, union formation, separation, etc. Many household models use comparative-statics methods (e.g. headship rates) to produce household forecasts. Examples can be found in the forecast of Statistics Netherlands (De Beer, this volume) and in the forecast of the Department of the Environment in the UK (DoE 1988). The link between contextual change and changes in headship rates can only be inferred statistically. Extending the method to include multiple household compositions (Akkerman 1985) does not solve this problem. The introduction of multistate demographic concepts, employing transition matrices, does not offer a solution either, if the transitions between states are not defined in terms of the processes that direct these transitions. For instance, the multistate models (among others Hårsman & Snickars 1984, Rima & Van Wissen 1987) that model household dynamics on the basis of transitions between households of various sizes do not distinguish between separation and widowing in the transition from a two-person to a one-person household. To incorporate contextual change, it is required that household dynamics are modelled on the basis of events, rather than on transitions from one state to another.

The third implication refers to the time dimensions used in the model. Contextual change is situated along the calendar time dimension. Life events are situated along the age dimension. Linking contextual change to household dynamics implies that the intersection of age and calendar time should be dealt with. For instance, a change in attitudes with respect to living arrangements will have a different impact on those who cohabit with a partner compared to those who have not yet entered a consensual union. Since contextual change can lead either to a postponement of an event or to a decreased participation in a process, it is necessary to adopt a cohort-component framework in incorporating contextual change.

The purpose of this contribution is twofold. First it will be argued that the development of dynamic event-driven models (see also Van Imhoff, and Nelissen, this volume) using the proper time dimension is a precondition to improving the substantive properties of household models by opening up the model to relevant aspects of the context. This might either take the form of improving the way in which hypotheses about contextual change can be translated into model parameters, or by endogenizing that part of the context which is relevant for the process at hand. Second, it will be shown that multi-dimensional household models can be extended to incorporate housing market behaviour, providing useful policy information on the distributive effects of planned construction of housing.

This argument will be developed on the basis of the experience that arises from the cooperation of the authors in developing an operational household projection model at the *local* level. In producing small-area household projections, the *regional* housing market is the most relevant aspect of the context. The interactions between (spatial) household dynamics and the operation of a regional housing market are manyfold and complicated. Small-area household projections therefore require the design of a fully integrated household and housing market model in which these interactions can be simulated.

Before turning to the model design at the local level, we will first introduce the household model at the national level, which serves as a basis for the local model. In section 13.2, the structural properties of the PRIMOS Household Model will be discussed. It will be shown that the structural properties of this multi-state model, an event-driven model using a cohort approach, allow for rigid hypothesizing with respect to the effect of future contextual change on household evolution. These structural properties are also a precondition for the development of the SONAR Household and Housing-Market Model, which is described in sections 13.3 to 13.7. First, the general structure of the model will be introduced. Then the various components of the model: the state-space, the demographic module, the housing market module, and the feed-back mechanism between the two, are discussed in more detail.

13.2. THE NATIONAL LEVEL: THE PRIMOS HOUSEHOLD MODEL

The PRIMOS model has been developed to provide national (and sub-national) estimates of the number of households in order to gain insight into future housing needs. The forecasts produced by the model have been used to provide a quantitative basis for national policy targets with respect to new construction of dwellings.

The original formulation of the PRIMOS model (Heida & Gordijn 1985) was tailored to this aim. A limited number of household positions was defined as states in the macro-simulation model: child living with parents, living single, living together, having lived together, and living in institutions. The model was person based, estimating transitions between states using transition probabilities. Some of these transitions between states were derived from events (e.g. the transition from living together to living single were derived from divorce rates and mortality quotients of the opposite sex). Others were derived cross-sectionally (e.g. the transition probabilities of leaving the parental home were derived by calculating first differences from the age distribution of people living with their parents). The output consisted of the number of individuals in each household position for each sex and one-year age groups. A number of improvements have been made on the basis of an evaluation of model structure and model performance (Hooimeijer & Linde 1988) two of which are pertinent to the discussion raised in the introduction of this contribution.

13.2.1. Context, Events, and Time Dimension

At the national level, contextual change determining rates of household formation and dissolution is often supposed to be economic or socio-cultural. It is often assumed that economic circumstances determine fluctuations in rates of household formation (Van Fulpen 1985). Haurin, Hendershot & Kim (1991) found some evidence in the United States that the age of household formation was correlated to (regional) income levels among the young and to house price levels. Research in the Netherlands provided contradictory proof of this relation (Ferment 1986, 1990). The general hypothesis is that it is not the income level as such, but the distribution of sources of income that determine whether children will be able to leave their parents. Welfare provisions in the Netherlands guarantee an independent income for everyone aged eighteen and over. Recent panel analysis (Klaus & Hooimeijer 1993) shows that 98% of all young people leaving their parental home had a source of income independent of their parents before the moment of nest-leaving.

The major contextual change that has led to dynamics in household evolution over the last decades is therefore probably socio-cultural and has been described as the individualization process. This is a very general concept, referring to a major shift in norms and attitudes towards living arrangements (Van de Kaa 1987). The process has affected the timing and occurrence of each event in household evolution, although in different time paths. It led to a decreased age at the moment of leaving the parental home in the fifties and sixties, to a drop in fertility in the late sixties and seventies, to increased separation (divorce) rates in the seventies, and to decreased rates of union formation in the eighties (Hooimeijer & Linde 1988).

Even though the changes in attitudes have touched upon each segment of the population (De Feijter 1991), the effect of the changing norms and values on living arrangements has been very different for successive birth cohorts, depending on the past level of participation in demographic processes. Figure 13.1 depicts the 'survival' function of two birth cohorts of females living with their parents.

This figure illustrates the 'mistake' that is being made in life table analysis, if transition probabilities are derived form a cross-sectional age distribution, treating the cross-section as a synthetic cohort. The cross-section is a mixture of survival functions of successive cohorts. If age-specific intensities change over time, as in Figure 13.1, one mixes the high intensities at younger ages of the later cohorts with the high intensities at older ages of the earlier cohorts (see also De Beer, this volume). This well-known problem in demography can be solved in dynamic models by specifying proportions longitudinally. Another advantage of the longitudinal approach is that explicit hypotheses have to be formulated with respect to future behaviour of unfinished cohorts. It should be decided whether a drop in age-specific intensities at younger ages is a matter of postponement or of decrease in overall intensity.

However, hypotheses on the effects of contextual change can only be formulated in terms of the processes at hand. The model should preferably be event-driven, to ensure that a distinction can be made between various processes. The PRIMOS model has been reformulated accordingly. Table 13.1 shows the various household states of individuals in the model and the events that mark the jumps from one state to another.

Apart from mortality, it is assumed that only one event takes places in a single period of one year (unless events occur simultaneously, like leaving the parental home to cohabit), and that rates are piecewise constant over that period

Table 13.1. States and events in the household model

	child	single	one-parent	couple	family	institution
child	*	1	-	1/2	1/2	6
single	-	*	-	2	2	6
one-parent	-	1	*	-	2	6
couple	-	4/5	-	*	3	6
family	-	4/5	4/5	1	*	6
institution	-	-	-	-	-	*

1. Leaving parental home; 2. Cohabitation; 3. Birth of child; 4. Separation; 5. Widowing; 6. Moves to institutions.

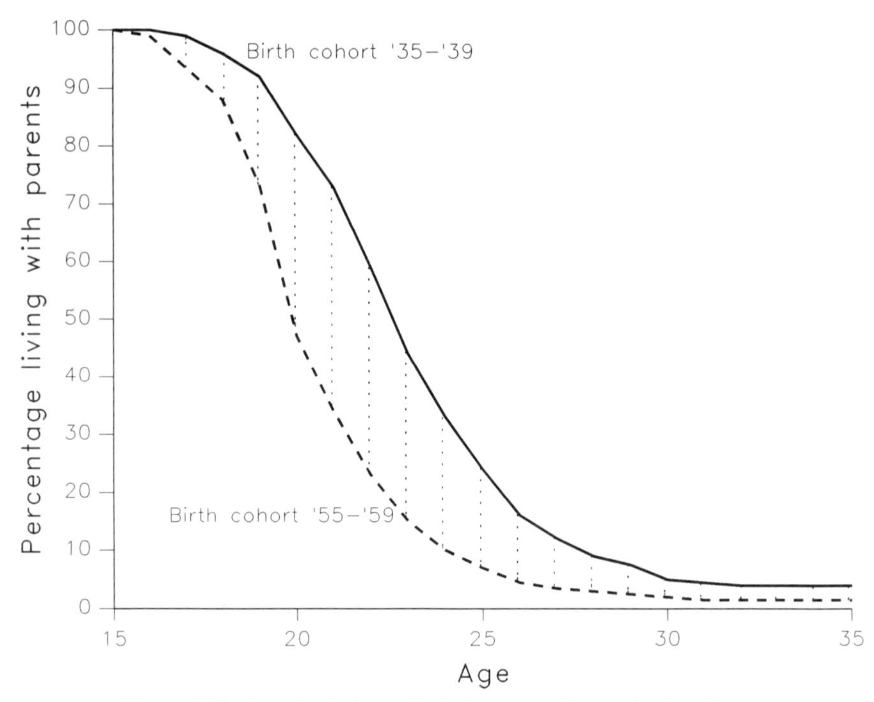

Figure 13.1. Females living with their parents for two birth cohorts

(see Van Imhoff, this volume, for a different specification). The table only provides a summary of all the states and events in the model. For instance, families are decomposed according to their number of children. Changes in this number are estimated using local parity progression ratio's. Three different types of institutions are also distinguished. International migration is exogeneous to the model (numbers are derived from the national population forecast of Statistics Netherlands).

As an example of how future changes in household processes are dealt with in the model, we will show some results from the 1990 forecast. The data for this forecast were derived from retrospective questions on household formation and dissolution in the Housing Demand Survey (WBO) 1985/1986, in the Fertility Survey 1985, and from population statistics on fertility and on moves to homes for the elderly and to nursing homes. All data have been collected by Statistics Netherlands (NCBS). As most processes had stabilized during the eighties or before, no quantitative hypotheses were needed to estimate future behaviour of new cohorts. The parameters of the last cohort could be used for producing the forecasts. The rates of first union formation provide an exception. A recent trend, starting with the birth cohort of 1960-1964, shows decreasing rates at younger ages (in the early eighties). A number of hypotheses with respect to the future rates of this and subsequent cohorts have been formulated:

- The first hypothesis states that the decrease in rates is an expression of an overall decreasing tendency to form unions (cohort effect). Age-specific rates of union formation are lowered to such an extent that the completed cohort intensity drops to a level of 75%.
- The second hypothesis claims that the decrease in rates at younger ages is a matter of postponement of the moment of first union formation (period effect). Rates at higher ages are raised to such an extent that the overall intensity remains at the level of 90%.
- The third hypothesis depicts a change in age-specific participation. It is assumed that the present age-specific occurrence/exposure rates will correctly describe cohort union formation in the future.

These hypotheses correspond to a high, low, and middle scenario, respectively, regarding the number of households in the future. Running the model for the period 1986-1990 allows for a comparison of the model outcomes with data from the Housing Demand Survey 1989/1990 which have recently become available. The results of the middle scenario are depicted in Table 13.2.

Even though the forecast of the total number of households is right on target, household states are predicted less accurately. A possible explanation for the deviations is given by Mulder (1993). On the basis of event history analyses of nest-leaving and union formation in the period 1978-1990, she shows that the early eighties witnessed a postponement in nest-leaving among the young that start cohabitation at the moment of leaving the parental home (catching up in the second half). The same trend did not occur among the young that left their parents to live alone. As a result, the parameters in the model underestimated the total number of nest-leavers in the second half of the eighties and

Table 13.2. Performance test of the national model (number * 1.000)

	WBO-86	Forecast	WBO-90	Diff.
Single male	388.7	489.1	470.2	+18.9
Single female	393.6	452.9	416.8	+36.1
Couples	3692.1	3805.3	3877.5	−72.2
Male sep. or wid.	287.1	318.9	297.9	+21.0
Female sep. or wid.	804.0	889.2	892.2	− 3.0
Total	5565.4	5955.4	5954.7	+0.7

overestimated the share of people living alone. The delay in union formation might be the result of a lack of housing opportunities in the early eighties (a period of recession). It has been shown for the Netherlands (Goetgeluk *et al.* 1991) that new one-person households are less sensitive to housing shortages, because they more often decide to take up temporary accommodation (furnished rooms, caravans, etc.).

13.2.2. From Individuals to Households: Consistency Relations and Feedback

To apply the model at the sub-national level, endogenizing housing market processes, implies that the model should provide consistent output on households rather than individuals. In any housing market model, the household is the basic unit of analysis. In a micro-simulation approach, this could be solved by including pointer variables, linking individual records in the data-base to each other. In a macro-simulation approach, comparable results can be generated by introducing age-relation matrices. In the present formulation of the model, the female has been chosen as the marker of the household. By introducing matrices describing the age-relation between the female and her partner, and the age-relation of the female and her children (or the male in the case of male-headed, one-parent families), changes in the households can be derived from events occurring to individuals. Both matrices are updated each year on the basis of the relevant processes at the individual level. Changes in male life expectancy, for instance, lead to adjustment of the age relation between partners at older ages. The age-relation matrix of the parent with the children is updated on the basis of parity progression and on the basis of nest-leaving. This is an efficient way of introducing feedback relations between various processes in the model. A postponement of nest-leaving will result in a prolonged stay in the position of family by the mothers.

The reasons for choosing this solution to ensure consistency relations have been largely pragmatic. The best alternative would be to specify a full partner-market model, analogous to the housing search model described in the following sections. Certainly, this would mean the introduction of non-demographic variables like education, etc. into the model. Alternative approximations described by Keilman (1985) and by Van Imhoff (1992) are mathematically more elegant, and particularly suitable if no a priori dominance can be given to an event. This is clearly the case in the two-sex problem in union formation. The reason for opting for female dominance in the PRIMOS model has been a matter of data availability. Reliable retrospective data on female behaviour with respect to union formation is available from the Fertility Survey. Comparable information on male behaviour is lacking. The Housing Demand Survey gives data on both sexes, but only provides retrospective information on union formation of people living single before, for a period of twelve months. Due to the sample design, recent movers (in case of union formation, at least one person will have to change residence) are underrepresented in the response. This not only leads to small numbers, but also to an underestimate of events.

Even at the national level, indications arise that housing shortages affect rates of household formation. Static models are inadequate for representing these effects of the context on household evolution in terms of behavioural responses. It has been shown above that dynamic event-driven models allow for the formulation of hypotheses that incorporate these responses. At the regional and the local level, the housing market will have a far more decisive impact on household evolution. To evaluate the effects of changes in housing opportunities on household evolution at this level, it is insufficient to formulate hypotheses. The housing market itself should be brought into the model. In the following sections, it will be shown that this is a viable alternative, even within the mutidimensional approach.

13.3. THE SONAR MODEL: GENERAL STRUCTURE

Housing capacity models have been used before to generate population projections at the local level. In the Netherlands, several single-region household models have been produced which explicitly link household dynamics with housing market processes (e.g. the RIWI model of Rima & Van Wissen 1987, and the QUATRO model of Heida & Den Otter 1988). The need for developing the SONAR model stems from a shift in housing policy in the Netherlands. In a process of rapid decentralization, the prime responsibility for housing has shifted from the national government to the local municipalities. The municipalities are obliged to cooperate in regional housing authorities which are responsible for the planning of new construction and the allocation of housing.

This calls for new planning information at two levels. The first is an assessment of housing needs at the regional level, to decide on the number of additions to the stock. The second is an evaluation of the effect of new construction on the redistribution of households over dwellings and over municipalities. In particular for the local government, the effects of the new policies on the total number and structure of the population in households is decisive in the planning of policy measures with respect to housing, employment, and the provision of public services.

In developing the local model, some new requirements have been met in order to obtain full integration between demography and housing. The result can be regarded as a housing market model, or as a household model, depending on the type of output the user is interested in. These requirements are the following:

- The model should provide full information on the effects of household formation, expansion, reduction, and dissolution on housing demand and supply, in quantitative and qualitative terms.
- It should provide projections of the number and composition of households at the local level, both as a result of household evolution processes and of the operation of the regional housing market.
- In doing so, the model should distinguish between housing-related moves within the region and migration for other reasons (henceforth called structural migration), both within and across the regional boundaries.
- The housing market adjustment mechanism, removing sectoral and spatial mismatches between demand and supply, should be explicitly modelled.

The overall design of the model is completely in line with the national household model. It can be characterized as a multi-dimensional, deterministic, demographic accounting model in which individuals occupy various states and are exposed to jump probabilities which are fixed during each projection interval (one year). However, endogenizing the housing market does have far-reaching effects on the model. The original state space of the national household model consists of one-year age groups for each household position — for each sex. Including dwelling types and municipalities would enlarge the state space to such an extent that this would become infeasible. The household model has therefore been simplified.

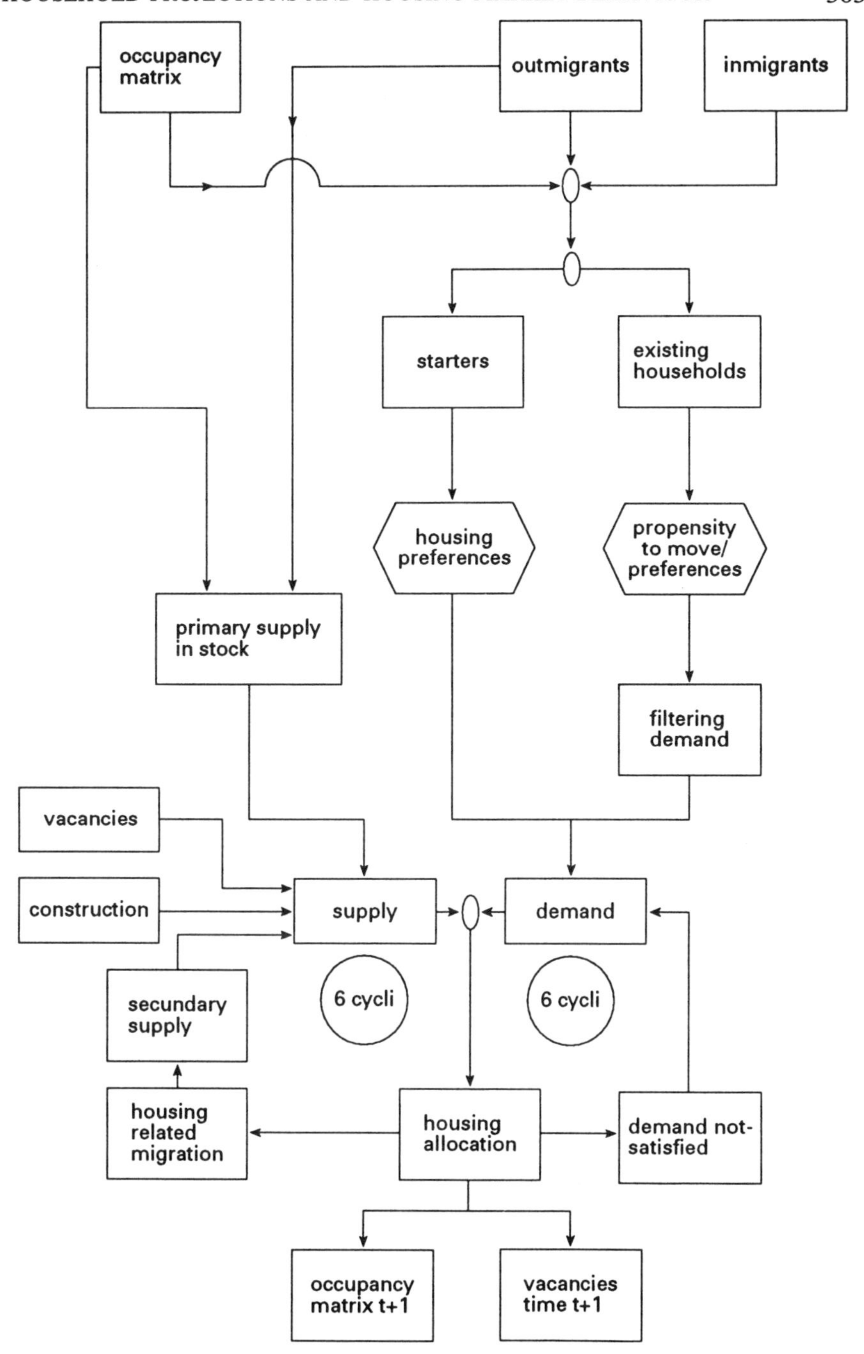

Figure 13.2. Generalized flow chart of the model.

Figure 13.2 depicts a generalized flow chart of the model. The starting point of the model is the occupancy matrix (which classifies households by age of the marker, household type, income group, and dwelling type) for each municipality selected at the start of the projection period.

The first step in the model is the demographic module in which household events are being estimated. A number of demographic events have a direct effect on the housing market; household formation of people living with their parents results in new housing demand, as does separation of couples and immigration of households. Other demographic events, like household expansion and household reduction, will affect the housing preferences of the household and therefore its propensity to move to another dwelling (indicated as 'filtering' demand). Housing demand is largely demography-driven.

A large share of the housing supply is also generated by demographic events; household dissolution as a result of mortality among singles, moves to institutions, and outmigration generates primary supply on the housing market. The rest of the primary supply (i.e. supply that starts a vacancy chain) is generated by new construction. The remaining supply is generated in the filtering process as people move from one dwelling to another.

In the housing-market module, demand and supply are matched by a housing allocation algorithm. Within each projection year, this confrontation between demand and supply takes place six times, reflecting an average vacancy duration of two months. In each allocation cycle, vacancies are being occupied and other vacancies released if the new occupant leaves a dwelling behind.

Each step in the model is based on jump probabilities, producing numbers of events which define the jump from one state to another. This dynamic approach is superior to comparative-statics approaches in three ways:

- It offers full detail on the processes of household and housing market change and the interaction between the two (e.g. using headship rates to estimate levels of household formation and dissolution only provides information on the net outcome of these processes, while the gross number of events is of importance to the operation of the housing market).
- It simulates the feedback mechanisms between household behaviour and the housing market context (e.g. the rates describing residential mobility are produced by the model on the basis of the availability and accessibility of supply on the housing market, constrained by the demand for this type of supply).
- Scenarios can be built efficiently (e.g. the effects of changing the level and composition of new construction or of restricting access to certain types of dwellings on the number and structure of households in various municipalities can be evaluated directly).

To illustrate these points, we will describe each step in the model in some detail in the following sections. In section 13.4, the state space will be defined and it will be shown that in spite of the absence of detailed data at the local level, the necessary data can be estimated from known marginal distributions. In section 13.5, the demographic module will be introduced, dealing with migration and household evolution at the local level. In section 13.6, the housing market module will be described in some detail, to provide insight into the reciprocal relations between household change and housing market change.

13.4. DEFINING THE STATE SPACE: THE OCCUPANCY MATRIX

The spatial unit in the model is the municipality (647 in the Netherlands in 1991). The reason for choosing this administrative unit has been stated above. The municipality is the level at which housing policies are formulated and executed. In the absence of census material on households and the housing stock, an approximation of the occupancy matrix at the level of the municipality had to be made.

Table 13.3. The typology of dwellings used in the model

1. Dwellings n.e.c. (bed-sitters, house-boats, etc.)
2. Rental, multi-family, <= 3 rooms, < *f* 490
3. Rental, multi-family, <= 3 rooms, *f* 490 - *f* 650
4. Rental, multi-family, <= 3 rooms, > *f* 650
5. Rental, multi-family, > 3 rooms, < *f* 490
6. Rental, multi-family, > 3 rooms, *f* 490 - *f* 650
7. Rental, multi-family, > 3 rooms, > *f* 650
8. Rental, single-family, <= 3 rooms, < *f* 490
9. Rental, single-family, <= 3 rooms, *f* 490 - *f* 650
10. Rental, single-family, <= 3 rooms, > *f* 650
11. Rental, single-family, > 3 rooms, < *f* 490
12. Rental, single-family, > 3 rooms, *f* 490 - *f* 650
13. Rental, single-family, > 3 rooms, > *f* 650
14. Owner-occupied, multi-family, < *f* 120,000
15. Owner-occupied, multi-family, *f* 120,000 - *f* 175,000
16. Owner-occupied, multi-family, > *f* 175,000
17. Owner-occupied, single-family, <= 4 rooms, < *f* 120,000
18. Owner-occupied, single-family, <= 4 rooms, *f* 120,000 - *f* 175,000
19. Owner-occupied, single-family, <= 4 rooms, > *f* 175,000
20. Owner-occupied, single-family, > 4 rooms, < *f* 120,000
21. Owner-occupied, single-family, > 4 rooms, *f* 120,000 - *f* 175,000
22. Owner-occupied, single-family, > 4 rooms, > *f* 175,000

Rents in guilders (*f*) per month, market value for owner-occupied units.

The categories in the typology (Table 13.3) reflect the basic features of the segmentation of the housing market in the Netherlands. Price levels in the rental sector are based on the allocation rules in the Netherlands. Housing with a monthly rent below *f*490 is inaccessible to higher income households, while housing with a rent over *f*650 is inaccessible to lower income households. The official distinction between higher and lower incomes is a net annual income of *f*22.000 for singles and of *f*30.000 for multi-person households. This demarcation is also used in the household typology.

The 72 household types represent the multiplication of the following household characteristics:

- household composition: single, one-parent, couple, family (see also Table 1);
- age of the marker (the female): 18-24, 25-29, 30-34, 35-44, 45-64, 65+;
- net annual household income (in *f*1,000): <22, 22-30, >30.

Until now, the NCBS only provides statistics on age, gender, and marital status. Household statistics at the level of the municipality may become available in the years to come, but cannot be provided at this moment. Two basic spatial differentials exist with respect to the age-specific household composition of municipalities in the Netherlands. Inter-regional differences occur due to the uneven distribution of educational and employment opportunities. Intra-regional differences occur as a result of the uneven distribution of housing opportunities within each region. The general equation to estimate the number of households in each category is as follows (see Heida 1991 for more detail):

$$Hh_{h,a,m} = \sum_g (Pop_{a,m,g} - Pin_{a,m,g} - Ed_{a,m,g}) \times \times (Pnah_{h,a} \times Ufac_{h,a,m} \times Rfac_{h,a,m}) \quad (1)$$

where:

$Hh_{h,a,m}$ is the number of households of type h in age group a in municipality m
$Pop_{a,m,g}$ is the population in age group a in municipality m of gender g
$Pin_{a,m,g}$ is the population in institutions
$Ed_{a,m,g}$ is the number of university students[1]
$Pnah_{h,a}$ is the national proportion of households type h in age group a

1 University students are treated separately due to their very specific household structure.

$Ufac_{h,a,m}$ is the relative deviation from the national proportion for municipality *m*; this factor depends on the degree of urbanization

$Rfac_{h,a,m}$ is the relative deviation from the national proportion for municipality *m*; this factor depends on the region in which *m* is located.

The next step involves the addition of the income distribution per household category, using the age-specific statistics on personal income for each municipality from the NCBS (for more detail, see Leering & Relou 1991).

After having established the distribution of the dwellings over the dwelling typology and the population over the household typology, the combination of both typologies in an occupancy matrix will have to be estimated. This has been done using Iterative Proportional Fitting on known distributions, according to region and degree of urbanization, that were fitted to the marginal distributions of households and dwellings in each municipality.

13.5. THE DEMOGRAPHIC MODULE: STRUCTURAL MIGRATION AND HOUSEHOLD EVOLUTION

The basic link between the housing market and household evolution is twofold. The residential mobility process, the interaction of demand and supply on the regional housing market, determines the spatial location of households, thereby changing the local household structure. A second aspect of this link is provided by the fact that some household events (like leaving the parental home) are dependent on a residential move. Both aspects are dealt with in the housing market module (see section 13.6). However, not all moves are housing related. Many migrations (in particular, those occurring over longer distances) are made for other reasons, like accepting a job, entering higher education, etc. This type of migration, henceforth called 'structural migration', is treated exogenously. Housing-related migration is endogenous to the model. Before turning to the household model, we will describe the way in which structural migration is isolated.

13.5.1. The Migration Algorithm

Migration patterns are very much age- and household-specific. The reason is that the motives for migration vary over the life course of the individual. At a relatively young age, many migrations are motivated by engaging in higher forms of education, accepting a job at a large distance from the present residential location, or union formation (which might occur both at short and at long distances). Later in life these motives become less important and most moves are made to improve the housing situation. In terms of the modelling exercise, this distinction is crucial. Housing-related moves will be affected by the development of the marginal supply on the regional housing market, and people might easily postpone the move if this supply does not meet their demand. Migration for other reasons is less prone to housing market conditions.

The isolation of structural migration is done in two steps. The first step is to define a region which corresponds to a regional housing market. In this contribution, we will not go into detail on how this can be done. All migration flows of the municipalities in the region with municipalities outside the region are treated as structural migration. Projections of this type of migrations are made on the basis of the average flows over the last five years, which can be derived from vital statistics.

The second step is to decompose the migration between municipalities within the region into housing-related and structural migration. This is done using age- and household-specific distance deterrence functions that describe the proportion of housing-related moves in the total migration flows between municipalities. By multiplying the registered flows from vital statistics with these proportions, structural migration (the complement of housing-related moves) is isolated. Projections of the structural intra-regional migration are made in the same way as described above for the inter-regional migration.

13.5.2. Household Evolution Processes

The household module serves two very important purposes in the overall model. The first is to estimate the changes in household structure which occur as a result of household evolution processes. The second is to 'translate' demographic events that are directly related to the housing market into the quantitative and qualitative effects on housing supply and/or demand within the region. The application of the model to municipalities necessitated further adjustments.

The age-specific jump probabilities from the national model have been aggregated into jump probabilities for the six age groups by taking simple weighted averages. Although some heterogeneity is introduced by this aggregation scheme, it has been shown to be rather limited.

Leaving the Parental Home. Apart from inmigration, leaving the parental home is the most dominant cause of sustained additions to quantitative housing needs. Whether or not this event leads to extra demand on the housing market depends on the destination. If the person leaves to live alone, an extra dwelling is needed. If the persons leaves to live with someone who also lived with the parents before, only half a dwelling is needed. If the person moves in with someone who lived independently before, no extra demand is generated and the household event will not depend on the availability of a suitable vacancy. The present model distinguishes between these destinations in the estimation of the number of events, but only assumes age-specificity in partner selection. The housing situation before the event does not influence partner choice, apart from the fact that this situation will be different in various age groups. An indirect effect of leaving the parental home is that the number of children living with their parents decreases. When the last child leaves, housing preferences of the parents will change. This change in the parental household is estimated using the mother-child age-relation matrix.

Cohabitation. Housing market effects of entry into cohabitation again depend on the housing situation before the event. The model estimates cohabitation for females on the basis of their age and household position (living with parents, living alone, being divorced or widowed). Using the age-relation matrix, a male is selected from the population at risk for each female. Housing market effects depend on the housing market situation of both. This is known from the occupancy matrix and varies with age and household position. As the distribution over the various dwelling types is known for both sexes within each age group, this introduces the possibility of releasing the worst (smallest) vacancy of the two on the market. The procedure is simple. For each group of women (per age group, household position, and dwelling type), the age relation matrix specifies the full distribution of the men (according to age group, household position, and dwelling type). For the future dwelling situation, the highest number in the dwelling typology can be chosen. Stated differently, the choice of a partner is considered independent of the housing situation, but the choice of housing at the moment of matchmaking is not.

Obviously this procedure is a simplified representation of partner choice. It might very well be that people who live in owner-occupied property are considered more attractive partners, or have moved into owner-occupation in anticipation of cohabitation. As has been stated before (part two), the introduction of age-relation matrices is a rather crude approximation of partner selection. In our opinion, further research into matchmaking deserves a high priority.

Separation. This always leads to extra demand on the housing market and changes the housing preferences of the partner who remains behind. If there are children present in the household, 90% remain with their mother without moving. For the other 10%, the house and the children are allocated to the father. Separation always initiates a housing market search by one of the partners.

Mortality and Moves into Institutions. The death of a one-person household releases a vacancy onto the market, as does the move into a home for the aged. The type of dwelling released can be derived directly from the occupancy matrix, and will depend on the age, household, and income position of the person(s) exposed to the event. No interaction is assumed between the housing situation and the occurrence of the event. In a multi-person household, mortality will cause a transition to widowhood, changing the housing preferences of the partner that remains behind. Access to homes for the aged is restricted to its present capacity, the capacity itself being dependent on the mortality of the population in institutions.

The Birth of a Child. This has no direct housing market effect, but will affect the housing preferences of the parents.

In summary, the household model generates three types of input for the housing market model:

- a full specification of the housing supply that is generated as a result of outmigration, cohabitation, the death of one-person households, and moves to institutions;
- a full specification of new demand on the housing market as a result of inmigration and household formation;
- an update on household composition of the local population in households and in dwellings, which serves as a starting point to estimate the propensity to move and the housing preferences.

13.6. THE HOUSING MARKET MODULE

The housing market module consists of two elements: a housing search model and a housing allocation model. The description below sketches the various steps in the algorithm.[2] The housing market module changes the household structure of the municipalities in two ways. The first is through a feedback mechanism to the household module. If the supply of dwellings is insufficient to accommodate the number of new entrants (starters and inmigrants) on the regional housing market, the resulting vector of new entrants that have not been allocated a dwelling are returned to the household module (implying that the demographic event could not take place). Second, if households are allocated a dwelling, the occupancy matrix is updated, including the moves from

[2] A prototype of this algorithm was devised and programmed by Drost, Hooimeijer, and Kuijpers-Linde of the Faculty of Geographical Sciences.

one dwelling and municipality to another. The whole process is steered by the balance between the (qualitative) demand of various types of households and the supply of these dwellings in the municipalities that make up the region. Therefore, the first step is to estimate demand.

13.6.1. Estimating Housing Demand

At the start of the model run, a pool of housing demand is established. This pool is estimated by multiplying each cell in the occupancy matrix with the probability that a household will be active in housing-market search in the projection interval. The probabilities were derived from the Housing Demand Survey and are estimated using a binary logit model. The starting households and the (structural) inmigrant households generated by the household model are added to this pool.

From this pool of potential movers, the number of households in search of a specific dwelling type is derived using stated preferences (derived from the Housing Demand Survey) that depend on age, household type, type of present dwelling, and type of municipality.

13.6.2. Allocating Supply

Once the sectoral demand has been estimated, the housing allocation module starts. Supply is allocated to demand in six cycles for each year. A (short-term) vacancy rate is needed to enable people to move from one dwelling to another. The estimated duration of vacancies is about two months. Allocation proceeds in a number of steps in each cycle. The main loop refers to the preferred dwelling. Within this loop, supply in this sector is allocated in each municipality. The steps in the allocation can be summarized as follows:

Localizing demand in municipality s. The total sectoral demand is localized by estimating the number of potential movers resident in each municipality that will exert demand for the dwelling type at hand in the municipality where supply is offered (2):

$$SQdem_{p,s,h,d,m} = [Qdem_{p,h,d,m}] \quad \text{for } m=s$$

$$= [Qdem_{p,h,d,m}] + [Qdem_{p,h,d,m} \times Pout_{p,h,d,m} \times Pdis_{p,s,m}] \text{ for } m \neq s \quad (2)$$

where:

$SQdem_{p,s,h,d,m}$ is the number of households of type h in present dwelling type d from municipality m that searches for a dwelling of type p in municipality s

$Qdem_{p,h,d,m}$ is the number of households of type h in present dwelling d living in municipality m that searches for a dwelling of type p

$Pout_{p,h,d,m}$ is the probability that a household h in dwelling d in municipality m searches outside the municipal boundaries

$Pdis_{p,s,m}$ is the probability that a household searching outside the municipal boundaries from municipality m searches in municipality s (these probabilities depend on the distance from s to m)

As can be seen from equation (2), a household can search for a dwelling in more than one municipality. Every household is assumed to consider supply in the present municipality of residence. The structural migrants form an exception. By definition, they only search in the municipality to which they move. Once they have become a resident, they may search for a dwelling elsewhere. An interesting option is that constraints can be imposed on the cells of this matrix, introducing housing allocation rules into the model. This way it can be simulated that low-priced rental dwellings are not allocated to higher income households and vice versa, which is standard practice in the Netherlands.

Calculating Supply in Municipality s. For each dwelling type p and each municipality s, total supply is calculated by summing the vacancies that have been released in the past, the vacancies that arise as a result of construction, and the vacancies that arise in the stock due to household dissolution (outmigration, cohabitation, deaths of one-person households moves into institutions).

Allocating Supply. For each dwelling type p in every municipality s, an allocation factor is determined by dividing supply by the sum of demand. Allocation is skipped if demand or supply equal zero. The allocation factor is set to 1 if supply is larger than demand. The factor is used to estimate the characteristics of the households that have been allocated a dwelling, and to calculate the number of vacancies left behind as a result of this allocation. After each allocation, the number of households in demand for a dwelling type from each municipality is reduced by the number that has been allocated a dwelling.

Updating supply and demand. After each cycle the supply of dwellings is being updated by calculating the number of vacancies that have not been occupied or have been released:

$$Vac^{t+1}_{p,s} = Vac^{t}_{p,s} - [\Sigma_h \Sigma_d \Sigma_m Afac_{p,s} \times SQdem_{p,s,h,d,m}] +$$

$$+ [\Sigma_p \Sigma_s \Sigma_h Afac_{p,s} \times SQdem_{p,s,h,d,m}] \quad \text{for each } d=p \text{ and each } m=s \quad (3)$$

where:

$Vac^{t}_{p,s}$	is the number of vacancies in dwelling type p in municipality s at time t
$Afac_{p,s}$	is the allocation factor for dwelling type p in municipality s
$SQdem_{p,s,h,d,m}$	is the number of households of type h in present dwelling type d from municipality m that searches for a dwelling of type p in municipality s

The number of vacancies for the next cycle is thus estimated as the supply at the beginning of the cycle minus the dwellings which have been allocated, plus the dwellings which have been vacated by movers from this particular dwelling type in this municipality. The resulting $Vac_{p,s}$ is used to estimate supply at the start of the next cycle.

The demand for each type of dwelling decreases during the allocation loop in each cycle. This could eventually lead to a situation where the balance of supply and demand becomes distorted. It might be possible that there is a high demand for 4-room, single-family dwellings with a rent of *f*490-*f*650 and a low supply in this sector, while the same type of dwelling in rent categories <*f*490 and >*f*650 are abundant. This is unrealistic. To some extent, households will adjust their preferences and decide to accept a cheaper or more expensive dwelling. To solve this problem, the following procedure has been followed. The number of potential movers decreases as a result of allocation, but the preferences of the remaining group are recalculated, using the same vector of (stated) preferences per household group that was applied in estimating the preferences in the first cycle. This implies a substitution of preferences towards sectoral and spatial sub-markets which are more abundant.

Feedback to the household module. This feedback occurs once a year. At the end of the year, the original occupancy matrix is updated. The housing characteristics and the place of residence of the households in the occupancy matrix are updated on the basis of the allocation of the dwellings, effectuating a change in household structure of both the spatial sub-markets (the municipalities) and the sectoral sub-markets (the dwelling types). This includes the allocation of dwellings to starters and inmigrating households. This new occupancy matrix is the starting point for the next round of demographic changes in the households.

The new entrants who have not been allocated a dwelling are also returned to the household module. These can be either starting households or inmigrating households. To return these households to the households module, implies that their original household situation should be known. This is done by adding two columns to the occupancy matrix (one for each group) each having 72 rows containing (the original) household positions.

This way a lack of housing opportunities on the regional housing market leads to a postponement of household events. The mechanism in its present formulation is still relatively straightforward. Only those household events are affected that imply a residential move, that is, household formation as a result of leaving the parental home (of both partners), as a result of separation and as a result of inmigration. Household dissolution as a result of moves to homes for the aged is also dependent on the opportunities provided for this type of event.

13.7. THE FEASIBILITY OF THE MODEL: A PERFORMANCE TEST

The algorithms described above have recently been programmed into a computer model[3] which is now in the stage of being tested. This section shows some preliminary tests on the performance of the model when applied to the region of Utrecht. This region was chosen because it is still characterized by a housing shortage, mainly due to a high level of inmigration.

Two tests have been carried out. The first refers to the correct estimation of the number of households. As household data become available only once every four years (from the Housing Demand Survey), the testing had to be done on the period 1986-1990. New information on household structures will become available in 1995, making another test possible. The second test refers to the prediction of the migration. Information on migrating households becomes available every year (with a time lag of about a year). The test on migration is therefore done in the period 1990-1992. The results of the first test are depicted in Table 4. Cell entries are households according to the age of the marker (the female).

[3] The authors are greatly indebted to the Ministry of Housing for providing funds to finance part of this research, and to Companen BV for doing the actual, complicated programming.

Table 13.4. Test of the model's performance on household structure

	18-29 year olds		30-64 year olds		65-plus		Total	
	One-Per	Multi-Per	One-Per	Multi-Per	One-Per	Multi-Per	One-Per	Multi-Per
START86	32,574	36,105	40,412	132,171	29,200	24,840	102,186	193,116
MODEL90	40,048	32,106	43,681	139,321	35,239	25,673	118,968	197,100
WBO90	38,147	35,829	48,440	140,074	32,935	24,038	119,522	199,941

Total household development is depicted rather well by the model. The total number of single households is underestimated by 650, the number of multi-person households by 1800. The age-specific prediction behaves less well. This is probably due to the fact that only six age groups were defined in the state-space of the model. The number of household ageing into the next age group is proportional to the size of the group as a whole. If one-year cohorts of *households* are not of equal size, this could lead to distortions. A further test, including a comparison with more simple, straightforward (and perhaps more robust) models, is being undertaken at this moment.

Another test on the model includes the effects of housing shortages on household formation and migration. Another run with the model, providing enough housing opportunities to alleviate the housing shortage, led to an increase in the estimated total number of households of about 5,000, indicating that the (regional) housing shortage does have an effect on household formation. Obviously this is not a test, as this cannot be compared to a real world situation. That is why the model is being run in various parts of the Netherlands at this moment, to evaluate whether the model does predict the number of households correctly in regions without a housing shortage, leading to fewer feedbacks into the household module. Results of these tests will become available in 1995. The general feeling at this moment is that the model slightly overestimates the effects of the housing shortage on household formation. It might very well be that young people 'opt' for a temporary housing situation, doubling up with friends or relatives.

The second test refers to the information at the level of the municipality. The most decisive demographic change in terms of changes in household numbers and structures is the effect of migration. In fact, the very elaborate housing market model is designed to make a correct simulation of migration effects at the household level. Migration statistics published by the NCBS for each municipality every year include details on the household position of the people who migrate into and from a municipality. Using some minor corrections, the number of migrating households can be derived from these numbers. Flows between municipalities, however, are not reported by the NCBS in the

Table 13.5. Test of the model's performance on municipal in- and outmigration

	Inmigration		Outmigration	
	Model	Actual	Model	Actual
Utrecht	19906	19849	17166	18025
Nieuwegein	3603	3564	3746	4055
Houten	1879	2127	939	1187
Maarsen	2603	2700	1942	2353
Vleuten	864	802	759	1138
IJsselstein	1566	1518	1152	1133
Bunnik	865	888	669	1035

same detail to prevent disclosure of the identity of the migrants. The test is therefore restricted to the number of inmigrating and outmigrating households in each municipality in a two-year period (1990-1991). The results are depicted in Table 13.5.

The general conclusion of this table is that inmigration is predicted very well by the model. Only in one municipality (Houten) the estimates of total inmigration deviates by more than 10% from the actual flows in households. Outmigration is generally underestimated by the model. In four out of seven municipalities, the deviation is more than ten per cent.

It is too early to tell whether the model performs better than comparative statics models that are more crude but also easier to make. A further comparison of the model outcomes with both real-world data and other model structures will be published in due course. However, if the model only simulates (changes in) stocks equally well, we feel that it is nevertheless superior in providing a wealth of information on flows which might be of particular relevance to policy-makers at the municipal and regional level. Even in its present state, the model might serve as a very helpful didactic device in generating housing policies and evaluating these ex ante with respect to their effects on household number, household structures, and quantitative and qualitative mismatches in the regional housing market.

13.8. CONCLUSION

In 1988, Keilman concluded that the multistate demographic approach holds considerable promise to generate sound household projection models. Most efforts in improving the models have been invested in the sound specification of the transition probabilities and in the analysis of unobserved heterogeneity. Relatively little effort has been devoted to improving the substantive properties of the models. Starting from the assertion that model performance is determined more by the correct specification of exogenous parameters than by the internal specification of the model, we have tried to show that the multistate demographic approach is sufficiently flexible to provide a sound design for models which actually endogenize relevant aspects of the context of household evolution processes. Even though the state space of the (general) model described here will contain a large number of empty cells in (small) municipalities in the Netherlands, it will be faster than a concomitant microsimulation model. An obvious drawback in the approach is the lack of insight into the effect of random fluctuations on the outcomes of the model.

Until a few years ago, a model of this complexity would hardly make any serious chance of being estimated and applied. Only mainframes could handle the large number of calculations involved. Also, the data requirements would prevent any user from applying this model in a local or regional context. However, times have changed. Elaborate compilers have been developed for PC's, and hardware has rapidly become much more sophisticated. The algorithms described in this contribution have been programmed[4] using Turbo Pascal and run in 4.5 hours on a PC with a 386 processor and a 387 co-processor for a region including 15 municipalities over a period of five years.

Data requirements are still a problem. Much energy had to be devoted to specifying the state space of the model. With the rapid development of geodemographic information systems, this might become redundant in the future.

ACKNOWLEDGEMENTS

The authors are greatly indebted to the Priority Programme on Population Research of the Netherlands Organization for the Advancement of Scientific Research and to the Ministry of Housing, Physical Planning, and the Environment for funding the research reported in this chapter.

REFERENCES

AKKERMAN, A. (1985), The household Composition Matrix as a Notion in Multi-Regional Forecasts of Population and Households, *Environment and Planning A*, Vol. 17(3), pp. 355-371.

---

[4] The actual programming has been carried out by Geert Rozenboom, Joost Drost, and Ingrid Ooms.

DE FEIJTER, H. (1991), *Voorlopers bij Demografische Verandering*, NIDI Report 22, NIDI: The Hague.

DE VOS, S., & A. PALLONI (1989) Formal Models and Methods of Kinship and Household Organization, *Population Index*, Vol. 55(2), pp. 174-198.

DOE (1988), *Estimates of Numbers of Households in England, the Regions, Counties, Metropolitan District and London Burroughs*, Department of the Environment: London.

FERMENT, B. (1986), *De Invloed van Economische Factoren op de Huishoudensvorming*, Strategisch Marktonderzoek: Delft.

FERMENT, B. (1990), The Influence of Economic Factors on Household Formation. *In*: R. CLIQUET, G. DOOGHE, J. DE JONG GIERVELD & F. VAN POPPEL (eds.) *Population and Family in the Low Countries VI*, NIDI CBGS Publications 18, NIDI: The Hague.

GOETGELUK, R., P. HOOIMEIJER & F. DIELEMAN (1990), Household Formation and Access to Housing, mimeographed, Faculty of Geographical Sciences, Utrecht University.

HÅRSMAN, B., & F. SNICKARS (1984), A Method for Disaggregate Household Forecasts, *Tijdschrift voor Economische en Sociale Geografie*, Vol. 74(4), pp. 282-290.

HAURIN, D., P. HENDERSHOT & D. KIM (1991), The Impact of Real Rents and Wages on Household Formation, Working Paper, Centre for Real Estate Education and Research, Ohio State University.

HEIDA, H. (1991), *Het PRIMOS Huishouden Model: Aanpassing en Uitbreiding*, Ministerie van VROM: The Hague.

HEIDA, H., & H. GORDIJN (1985), *PRIMOS Huishoudensmodel*, Ministerie van VROM: The Hague.

HEIDA, H. & H. DEN OTTER (1988), *QUATRO: Modellen voor de Vooruitberekening van de Woningvraag naar Type en Grootte*, INRO/TNO: Delft.

HOOIMEIJER, P., & M. LINDE (1988), *Vergrijzing, Individualisering en de Woningmarkt*, Ph.D. Thesis, Faculty of Spatial Sciences, Utrecht University.

KEILMAN, N.W. (1985), Internal and External Consistency in Multidimensional Population Projection Models, *Environment and Planning A*, Vol. 17, pp. 1473-1498.

KEILMAN, N.W. (1988), Dynamic Household Models. *In*: N. KEILMAN, A. KUIJSTEN & A. VOSSEN (eds.) *Modelling Household Formation and Dissolution*, Clarendon Press: Oxford.

KEILMAN, N.W. (1990), *Uncertainty in National Population Forecasting: Issues, Backgrounds, Analyses, Recommendations*, Swets & Zeitlinger: Amsterdam/Lisse.

KLAUS, J. & P. HOOIMEIJER (1993), Inkomen en Ouderschap: een Longitudinale Analyse, *Supplement bij de Sociaal-Economische Maandstatistiek*, 1993/4, pp. 15-24.

LEERING, D. & W. RELOU (1991), De Schatting van de Basismatrix, INRO/TNO: Delft.

MULDER, C.H. (1993), *Migration Dynamics: A Life Course Approach*, Thesis Publishers: Amsterdam.

RIMA, A., & L. VAN WISSEN (1987), *A Model of Household Relocation*, Ph.D. Thesis, Vrije Universiteit Amsterdam.

VAN DE KAA, D. (1987), Europe's Second Demographic Transition, *Population Bulletin*, Vol. 42(1).

VAN FULPEN, J. (1985), Volkshuisvesting in Economisch en Demografisch Perspectief, Staatsuitgeverij: The Hague.

VAN IMHOFF, E. (1992), A General Characterization of Consistency Algorithms in Multidimensional Demographic Projection Models, *Population Studies*, Vol. 46(1), pp. 159-169.

14. THE INTERACTION OF HOUSEHOLD AND LABOUR MARKET MODULES IN MICROSIMULATION MODELS

Jan H.M. Nelissen

Faculty of Social Sciences
Tilburg University
P.O. Box 90153
5000 LE Tilburg
The Netherlands

Abstract. This chapter examines the modelling of household and labour market transitions in dynamic microsimulation models. For this purpose, the demographic module and, to a lesser extent, the labour market module in the Dutch microsimulation model NEDYMAS is described. Next, the interaction of demographic and labour market transitions in this model and some other microsimulation models are indicated. The effect of demographic variables on labour market transitions is generally included in the more recent models. This is in accordance with the developments in labour economics during the last decades that led to a narrowing of the gap between stylized textbook models and reality. The reverse, the implementation of socio-economic variables in the household module, is at the same level as two decades ago, and also here, this partially reflects the developments within this field, to wit, the lack of empirical applications, although theory has been developed more and more. We mainly find empirical applications in this field in the United States. This is reflected in the interactions between the demographic module and the labour market module in American microsimulation models, like DYNASIM. But the application of behavioural equations rests on ad hoc models and simultaneous models have not been used. For the Netherlands, we expect a further inclusion of demographic variables in the labour market module, whereas the inclusion of socio-economic variables in the demographic module is only in its first stage.

Household Demography and Household Modeling
Edited by E. van Imhoff *et al.*, Plenum Press, New York, 1995

14.1. INTRODUCTION

Traditional demographic forecasts are limited to the events of birth, death and migration. Marriage and divorce are included in a number of very special cases only (e.g. the Netherlands, Great Britain, and Ireland). This results in a restricted disaggregation of the population by age, sex, and (sometimes) marital status. The individual's position in the household and those of the relatives are subsequently *externally* determined by static methods such as the headship rate method. Nowadays, it is generally accepted that this method is inappropriate for making a forecast. Its shortcomings are virtually resolved by macrosimulation. Macrosimulation disaggregates the population by relevant categories and the resulting groups of individuals may experience transitions from one category to another during a certain period of time (Keilman & Keyfitz 1988, p. 267). However, the main disadvantage of macrosimulation is its restricted capacity of disaggregation, owing to the limited states or categories that can be included in the analysis (e.g. Keilman 1988). This disadvantage is not really relevant if the problem is limited to household modelling, in which the required specification is mitigated, and if no linkage is made to other socio-economic categories with more than only a very limited number of additional characteristics. However, the current generation of macrosimulation models is not suited to supply us, for example, with a household distribution by exact number and by exact age of the household members or to generate the distribution of pensions benefits, when these benefits depend on labour histories, like the ABP pension scheme in the Netherlands. This type of limitations does not hold for the microsimulation approach. Microsimulation operates at the micro level, and as a consequence it produces very detailed information, such as distributions, life histories, and data on subpopulations. The method was proposed some decades ago by Orcutt (1957). For a long time, computer costs retarded the application of microsimulation. However, developments in computer hardware, especially in speed, internal memory size, and disk capacity, have reduced these costs substantially (see Hellwig 1988).

In this chapter, we look at the modelling of household and labour market transitions in dynamic microsimulation models. For this purpose, the demographic module and, to a lesser extent, the labour market module in the Dutch microsimulation model NEDYMAS (Nelissen 1994) are described. This model has the most extended demographic aspects of all known microsimulation models, and the treatment of household and labour market modules are comparable with other models. We first give a short historical overview of the development in microsimulation. Next, we describe the demographic module of NEDYMAS, how labour market transitions are treated in this model, and finally we look at the (absence of) interaction between these modules, both in NEDYMAS and some other microsimulation models.

14.2. A SHORT HISTORY

In the 1960s and early 1970s, a number of purely demographic-oriented microsimulation models were developed. Hyrenius and others developed a reproduction model at the University of Gothenburg (Hyrenius & Holmberg 1970). Their model had a demographic-physiological framework. The well-known POPSIM model was constructed at the University of North Carolina (Horvitz *et al.* 1972) and was used for the evaluation of family planning campaigns. At the University of Toronto, Howell and Lehotay (1978) constructed a microsimulation model for exploring small human populations, AMBUSH. Emphasis was on kinship ties and similar kinds of links within small communities. Other contributions are from Ridley and Sheps (1966) — who developed REPSIM to study the effects of demographic and biological effects on natality — and SOCSIM used by Wachter, Hammel, and Laslett (1978) to analyse kinship relations in small communities in the 17th century.

Microsimulators which also contain socio-economic variables were implemented, particularly in the 1960s and 1970s. The first microsimulator, SUSSEX (Orcutt *et al.* 1961), was primarily developed to prove that this kind of model was possible. It was also limited to population simulations. The model was a dynamic one, but modelling was too rough for real applications.

The first applied (static) microsimulation model was TAX, developed at the Treasury Department (Pechman 1965) and applied from 1963. This model led to a number of more complex models in the United States and Canada. Microsimulation became the "... dominant quantitative technique for forecasting the impacts of policy changes in the social welfare policy area" (Fallows 1982, p. 2) from about 1969 in the United States. Its inception was the development in 1969 of the static Reforms in Income Maintenance (RIM) model. It was aimed especially at calculating the eligibility for public assistance programmes (Wilensky 1969; McClung 1969). RIM was developed under great time pressure at the President's Commission on Income Maintenance Program. The dissolution of this Commission led to the transfer of many researchers in the field of microsimulation — among them Nelson McClung, Gail Wilensky, and Robert Harris — from the federal government to The Urban Institute. This institute has played a major role in the further development of microsimulation since 1970.

The first applicable dynamic[1] microsimulation model — which could handle more than demographic simulations only — was developed between 1969

[1] Static models ignore the time aspect in the population, whereas dynamic models take account of this time aspect. The former assume that the sample structure of the micro-units does not change and weighting factors are used to adjust the microdata base to the population in the future.

and 1976 at The Urban Institute and is called DYNASIM. Applications of DYNASIM have been taking place in the field of the effect of expected changes in the working behaviour of women and of divorce patterns, on the distribution of earnings and transfer income, in the field of forecasting the future costs and caseloads of public transfer programmes, and in the field of the distributional impact of reforms in retirement income policy.

A second large microsimulation development project started in the early 1970s in West Germany. It was initiated by the Social Policy Research Group Frankfurt-Mannheim (Sozialpolitisches Entscheidungs- und Indikatorensystem — SPES) and later on adopted by the Special Collaborative Program 3 (Sonderforschungsbereich 3 — Sfb3).[2] Both a dynamic and a static model were initiated at the SPES. The static model (Klanberg 1975) was developed to analyse proposals for tax reforms and changes in the old age pension system. A variant of this model was used to study the distributional implications of the contributions for the Sickness Insurance Act (Hamacher 1978). At the Sfb3, the model was developed further and used, for example, for the simulation of medical services (Brennecke 1981).

The dynamic model was first developed by Hecheltjen (1974). This model did not contain links between household members, which limited its possibilities. The household mobility was improved by Steger (1980; also Galler & Steger 1978), whereas the socio-economic aspects were improved first at the SPES and later at the Sfb3 (see Krupp *et al.* 1981; Galler & Wagner 1981). In addition to the cross-sectional model, a longitudinal model was developed to study the distributional aspects of pension insurance and the financing of the educational system (Helberger 1982; Helberger & Wagner 1981).

From this overview, it is clear that most microsimulation models for the household sector are developed for specific applications. Nation-wide models are only present in limited numbers, and most of these are static. However, in the dynamic models too, the number of applied behavioural assumptions is limited. They are mainly used in the determination of labour supply, income, and consumption, and are much less sophisticated or even absent in the field of demographic transitions and educational decisions.

14.3. THE DEMOGRAPHIC MODULE IN NEDYMAS

The demographic module of NEDYMAS contains 17 submodules. Its structure can be found in Table 14.1. The first step is the determination of the number of new immigrants and the number of return immigrants. The starting-

---

[2] See Galler & Wagner (1986, p. 229) for an overview of the development of the Sfb3 micromodels.

Table 14.1. Programme module sequencing for each individual in NEDYMAS, version 1991/B (demographic part only)

Order	Programme module	Eligibility
1.	Immigration	Adding persons who have not yet lived in the Netherlands to the data file
2.	Family reunification	Adding persons whose partner or parent(s) immigrated earlier to the data file
3.	Emigration	Each family and each person over 16
4.	Return immigration	Each family and each person over 16 who had emigrated before
5.	Homes for the aged	Each couple and each person over 60
6.	Nursing homes	Each person
7.	Institutions for the mentally disabled	Each person
8.	Psychiatric institutions	Each person
9.	Other institutional households	Each person
10.	Death	Each person
11.	First marriage selection	Each never-married person over 14
12.	Remarriage selection	Each previously married person
13.	Divorce	Each married couple
14.	Dehabitation[a]	Each couple living together without being married
15.	Cohabitation selection	Each person over 14 who is not living with someone, married or not married
16.	Splitting-off of children	Unmarried persons up to 35
17.	Fertility	Women between 15 and 49

[a] We use the term 'cohabitation' only for people living together without being married. If they decide to dissolve their consensual union, we speak of 'dehabitation'.

point is the total number of immigrants. These numbers are transformed to age, sex, and marital status specific numbers. The number of return immigrants has been calculated on the basis of the return immigration rates by number of years passed since the year of emigration and the number of emigrants in those years who are still living alive abroad. In this, we limited ourselves to people who emigrated between the current year and six years earlier. Subtracting the resulting number of return immigrants from the total number of immigrants, we get the number of new immigrants (by age, sex, and marital status). The second step is the addition of new immigrants to our sample. For that purpose, the number of new immigrants is multiplied by the proportion of the sample population in the real population (the sampling ratio), in which rounding occurs by lot, using the uniform [0,1] distribution. This resulting number of new immigrants is added to our sample. In this process, we distinguish six subgroups:[3]

a. Married women, with or without children, who immigrate without a partner. We assume that these are cases of family reunification. The proportion of these women in the total number of immigrating married women is calculated using the probability that family reunification takes place with immigrated married men, living in the Netherlands without their family. The number of women in this subgroup is related to the number of immigrated married men living in the Netherlands without a partner (see e.). This gives us the probability of family reunification. Using a Monte Carlo process, we determined for each married man without a partner whether or not family reunification takes place. If this occurs, one female person is added to the sample. The number of persons in the database is raised by one and she gets this number as her personal identification number. The sex of this person is of course female, the marital status married. The pointer to her husband equals the identification number of the man in question and conversely, the pointer of the man for his wife is set equal to the identification number of the female in question. Her year of immigration is the current year. The age of the wife is set equal to the male's one minus three, unless such a woman is not available in the set of immigrating married women. In the latter case, the age difference between man and woman is minimalized. The age at which she is assumed to have left her parental home is put at the year in which she became eighteen years old. If she is younger than eighteen, she leaves the parental home in the current year. It is also assumed that she marries in the year in which she left the parental home. The number of new immigrating married women of this age is decreased by one. If family reunification did take place, the number of minor children involved in the reunion is simulated. For that purpose, we use the distribution of immigrating family heads by number of minor children. We assume that immigrating women aged 64 years and older do not have minor children in their households. The number of minor children is determined by a Monte Carlo process. The age of the children is dependent on the age of the woman and on the number of children by age in our population of new immigrants, if available. If a child is added to the immigrating female, the number of persons in the database is raised by one in the same way as described before.

[3] We only describe the database adjustments for married women who immigrate without a partner. The procedure is analogous in other cases.

b. Unmarried people older than seventeen years. We assume that these persons immigrate as single persons, without children.
c. The same applies for the group of widowed and divorced men.
d. Widowed and divorced women: these groups are also added to our sample. In this case, children may be present. The presence of children is determined in the same way as under a., on the understanding that now the number of children is determined by a random draw from the distribution of female-headed immigrating households by number of children.
e. Married men who immigrate without their family and consequently leave behind their wives and children, if present, in the country of origin.
f. Married men, immigrating with their family. The probability that this will occur has been determined above. A wife is attached to him, analogously to women in category a., if present. Children are attached in a similar manner.

Now all new immigrants are inserted in our sample. The next step is the simulation of return immigration. Earlier, we fixed the number of return immigrants by year of emigration up to and including six years ago. Dividing these numbers by our sampling ratio and the number of emigrants (who have not yet re-immigrated or died) in the last seven years in our sample, gives us the probability of return migration. These probabilities are used to determine whether a person who emigrated during the last six years or the current year will return to the Netherlands. Here we assume that married women immigrate with their husbands and that minor children immigrate with their mothers, or in case of divorce or widowhood, with their fathers. Adult children are assumed to decide independently whether they will immigrate or not.

Emigration is largely simulated analogously to immigration. However, no persons are added to our sample population. Here too, the first step is the determination of return emigrants and 'new' emigrants. Subtracting the resulting number of return emigrants from the total number of emigrants, we get the number of new emigrants (by age, sex, and marital status). On the basis of the populations at risk, the probability of return emigration and the new emigration rates are determined. Again it is assumed that the probability is independent of age and household composition. Return emigration takes place as a family, with the exception of children older than seventeen years. The latter decide for themselves whether they want to emigrate or not. Given the limited data, it is also assumed that return emigration differs from the emigration pattern of new emigrants only during the first five years of their residence in the Netherlands. The new emigration probability is dependent on age, sex, and marital status. Cohabiting people are treated in the same way as married people. The whole emigration process is regulated by Monte Carlo decisions. The process for new emigrating persons is divided into four subprocesses:

a. Married and cohabiting men, who emigrate on their own, i.e. without their family.
b. Married and cohabiting men, who emigrate with their family. Again, it is assumed that children older than seventeen years decide whether or not they want to emigrate, independent of their parents' choice.
c. Not cohabiting men and women older than seventeen years. Minor children, if present, emigrate together with their father or mother.
d. Cohabiting women, whose husbands have already emigrated in the past and who emigrate for reasons of family reunification. If minor children are present, these will accompany their mother.

The next steps are the outflow out of and the inflow into institutional households. An extensive description can be found in Nelissen (1991).

Mortality is simulated by subjecting the individuals in the sample to age, sex, and marital status specific death occurrence-exposure rates. In this process, we distinguish between persons remaining in an institutional household and persons living on their own. Each year, a Monte Carlo process decides for each person whether (s)he will die. If this occurs, the individual's year of death is put equal to the current year. If the person in question is married, his or her partner becomes widowed and the year of widowhood is put equal to the current year. If the individual cohabits, the partner's marital status changes back into the status (s)he had before the start of the cohabitation with the deceased person. Also in this case, the attribute 'year of widowhood' is put equal to the current year. If the deceased person has children, then the pointer to the father or mother (depending on who died) of these children becomes zero, whereas the child's pointer to the former father (mother) is put equal to the identification number of the deceased father (mother). If the other parent had already died before, the child has become an orphan.

For each person in our database who is older than 14 years and younger than 85 years and not married, a Monte Carlo process decides each year whether or not (s)he will be a marriage candidate. If the decision is yes, the person is stored in a file of potential marriage candidates. When all persons have passed through the marriage submodule, the matching process takes place (see Figure 14.1). First, the number of marriages to be contracted is determined. It is put at half the sum of the number of male and female candidates, in which a Monte Carlo process decides in cases of rounding off. The resulting number is indicated by N. The next step in the matching process is contracting the marriages of cohabiting persons. The female is the starting point in this. Therefore, we take all women in our stock of candidates who live in a

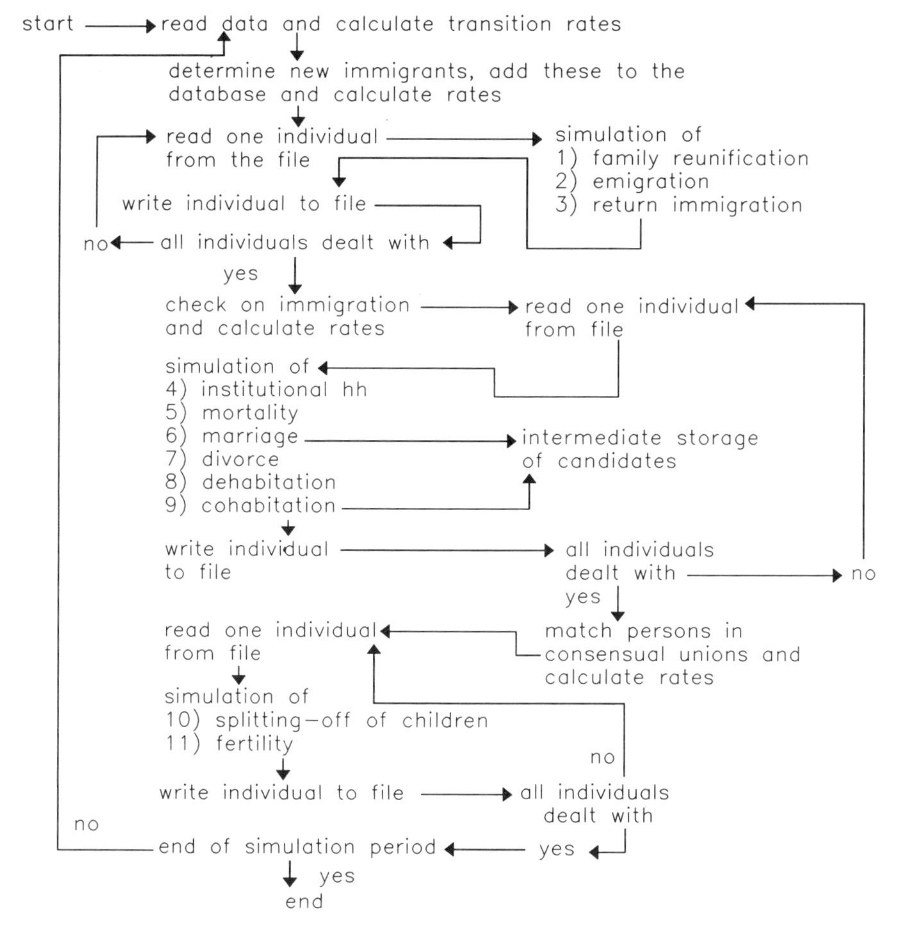

Figure 14.1. The demographic module in NEDYMAS

consensual union. They are said to marry their partner (even if this partner is not included in our stock of male candidates). These people get the marital status married and their year of marriage becomes the current year. It is possible that the wife or husband has died in the current year. In that case, the marital status of the person still alive becomes widowed. If both partners have died during the current year, nothing has changed.

If all women in our stock of candidates who lived in a consensual union have married their partners, then all men and women in this stock who cohabit are deleted. To get the intended total number of N marriages, we determine whether we have to draw extra candidates (not married, not cohabiting, aged 15 to 84 years old), or delete candidates from the stock. In both cases, the localization coefficients for marriages are used as weights. In this way, we get an equal number of men and women who have to be coupled to each other. Using the distribution of marriages by ages of both partners, the number of marriages still to be contracted after the aforementioned marriage process of cohabiting persons, and the RAS-method, we determined the number of marriages in our sample by age of the partners. This matrix is called M.

The matching process of the males and females, now left in our stock of candidates, is carried out in two steps. Again, the woman is the starting point. In the first step, a female candidate is taken from the stock and we look at whether the number of male candidates is still more than ten. If so, we draw ten male candidates using the distribution by age of the male partner, given the female's age, and under the condition that the concerning cell in the matrix M is positive. In the second step, the most suitable candidate out of these ten is determined on the basis of the female's level of education and the distribution of husbands by level of education conditional on the woman's level of education. If the remaining number of women is ten or less, then these women are coupled to a husband on the basis of the distribution by age only. When a wife has chosen her husband, the corresponding cell in the matrix M is diminished by one. Both the male's and female's year of marriage becomes the current year, their year of divorce (if positive) becomes zero, their marital status is now married. Each gets a pointer to his or her partner, which equals the partner's identification number. If one of the partners was still living with his or her parents, the year of leaving the parental home of this partner becomes the current year. If one of the partners already had children, these children are also assigned to the other partner. This means that the number of children of this last partner becomes positive, that (s)he gets pointers to these children and that the children concerned get a pointer to their new parent. If both partners already had children, these children are mutually assigned to both parents.

Also, it is possible that the wife or husband has died in the current year. In that case the marital status of the person still alive becomes widowed. If both partners died during the current year, nothing has changed.

Divorce is carried out by subjecting the married women in our sample population to an age-specific divorce occurrence-exposure rate. A Monte Carlo process decides whether or not a wife's marriage ends in divorce in the current year. If divorce occurs, the marital status of the woman and her husband are changed to divorced. The pointer to the partner becomes zero, whereas the pointer referring to the former partner is set equal to the corresponding personal identification number. For both individuals, the year of divorce becomes positive (and equals the current year). If the husband has emigrated, the number of emigrated people without a partner is diminished by one. If children live with their parents, we determined for each child whether the woman or her husband is granted custody of the child.

Dehabitation is carried out by subjecting the cohabiting women in our sample population to an age-specific dehabitation occurrence-exposure rate. The process is analogous to divorce.

The process of (unmarried) cohabitation is largely analogous to the marriage process. The most important difference is the way in which we determine the transition rate for a first cohabitation. We assume that no cohabitations took place before 1965. From Van 't Klooster-Van Wingerden (1979, p. 35-37) we find 18,291 male headed non-family households with female non-relatives, 16,671 female headed non-family households with male non-relatives, 3,126 male headed one-parent family households with other female non-relatives and 8,294 female headed one-parent family households with other male non-relatives for the year 1971. This makes 46,382 unmarried cohabitations in 1971. It is assumed that 5000 cohabitations started in 1965 and that there was an exponential growth between 1965 and 1971. For the year 1982, we find about 205,000 cohabitations on the basis of the Netherlands Fertility Survey ('Onderzoek Gezinsvorming')[4] and additional estimates for the persons older than 37 years. Between 1971 and 1982 we also interpolated exponentially. For the years 1983 and later, we assume that the percentage of people living together, married or unmarried, at every age, does not change.

Given the (expected) number of married people by age and sex, we can calculate the number of people cohabiting by age and sex. Using this number of cohabiting people, the number of people by age and sex cohabiting in the model population at the beginning of the year, the expected number of non-first cohabitations started in the current year and the expected number of dehabitations in the current year yields the number of first cohabitations in the current year. Coupling these numbers to the population at risk gives the first cohabitation rate. On the basis of the Netherlands Fertility Survey, it is also assumed that the non-first cohabitation rate equals the remarriage rate for divorced people for those 33 years or older, and it equals twice this remarriage rate for those under 25 years of age. Between the two ages, the multiplication rate is interpolated. Using these cohabitation rates, a Monte Carlo process decides whether or not a non-cohabiting person will be a cohabitation candidate. If so, (s)he is temporarily stored. After all persons have gone through the cohabitation submodule, the candidates are matched. This process is analogous to the marriage process for non-cohabiting marriage candidates.

The next submodule in the demographic part of our model is the splitting-off of children for other reasons than marriage or cohabitation. Only limited infor-mation is available on this subject. We use results from the 1971 Census and the 1977 and 1981 Housing Demand Surveys ('Woningbehoeftenonderzoek'); see Van Leeuwen (1980), Ploegmakers & Van Leeuwen (1983) and

---

[4] Personal communication with Van der Giessen of the NCBS and own calculations.

Heida & Gordijn (1985). From these data we know the percentages of unmarried men and women by age, living with their parents in the years in question. Between 1971 and 1977, and between 1977 and 1981, we interpolated these percentages. For the years 1982 and later, we assume that these percentages do not change, whereas for the years before 1971 we take those of 1971, corrected for the relative participation rate in university education by persons between 18 and 25 years (1971 = 100) squared (derived from NCBS (1984)). This last factor is considered as a proxy for the developments in splitting-off of children for reasons other than marriage or cohabitation in the 1950s and 1960s.

When marriage and cohabitation take place in our model, we can count the number of unmarried persons still living with their parents. Decreasing these numbers with the numbers resulting when applying the calculated percentages and dividing these differences by the counted numbers (being the population at risk), gives us the desired probability for splitting-off for reasons other than marriage or cohabitation. Via a Monte Carlo procedure, we determined whether an unmarried, non-cohabiting person living with his or her parents, will leave the parental home. If so, (s)he starts a one-person household. This procedure is only executed for persons younger than 40 years. If someone reaches the age of 40 and is still living with his or her parents, we decided that the person in question leaves the parental home in the current year.

Births are based on marital status specific parity progression rates for the ages 15 to 50 years. In this parity, one, two, three, and four and over, are distinguished. The first step is the determination of the number of births within and outside marriage by the mother's age and parity. These numbers are transposed into age, parity, and marital status specific birth rates. The population at risk is not known from vital statistics. Therefore, we use the simulated female population by age, marital status, and parity as population at risk. For females with four or more children, it is assumed that the parity progression rate for a fifth, a sixth, and so on child equals the parity progression rate for a fourth child. Within the category of unmarried persons, we distinguish between cohabiting and non-cohabiting women. Using these parity progression rates, we determine for each year for every female aged 15 to 49 years old, whether a child is born or not. If this occurs, the number of children of the woman is raised by one, the number of personal records is also increased by one, and this number becomes the child's identification number. The same number is used as the pointer of the mother for her newly-born child. The child's pointer for its mother is set equal to the mother's identification number. The child's sex is assigned by lot and its year of birth becomes the current year. The number of children minus one is assigned to the newly-born child as its number of brothers and sisters, and the same is done with its brothers and sisters. The same adjustments have been made for, and with reference to the partner, if present.

An outline of the variables used in the demographic module are given in Table 14.2. An overview of the personal characteristics that are used is given in Table 14.3.

Before the individuals' passage through the various demographic submodules, various occurrence-exposure rates are calculated. First, the family reunion rates and return-immigration rates are determined. After the addition of new immigrants, the return-emigration rates and the first-emigration rates are calculated. After immigration, the rates for the flows into and out of the institutional households, the death rates, marriage rates, divorce rates, dehabitation rates, and cohabitation rates are calculated. The rates for leaving the parental home and fertility are determined after the matching of persons in consensual unions has taken place. In calculating these rates, we used, among other data as generated by the simulation model, NCBS-data on the population by age, sex, and marital status for the years 1947-1985[5] and the same forecasted

Table 14.2. Review of the variables used in the demographic model

Submodule	Variables
Immigration	year of birth, age, sex, marital status, year of previous emigration if applicable, probability of immigration without family, probability of re-migration
Emigration	analogous to immigration
Death	year of birth, age, sex, marital status, yes/no in institution and if yes, type of institution
Leaving parental home	year of birth, age, sex, nuptiality rate, probability of cohabitation, participation in high education
Marriage	year of birth, age, sex, marital status, age difference of partners, level of education
Cohabitation	analogous to marriage (excluding marital status)
Divorce	year of birth, age, sex, marital status, number of children, age difference of partners
Dehabitation	analogous to marriage (excluding marital status)
Fertility	year of birth, age, marital status, living together or not, number of live-born children
Entering instutional hh	year of birth, age, sex, type of institution, and presence in another type of institutional household
Leaving instutional hh	year of birth, age, sex, and type of institution

5 These data were provided in the form of tables by Statistics Netherlands (NCBS). See also NCBS (1953 and 1970).

Table 14.3. Variables and range used in the demographic model

Variables	Range
Year of birth	1848, 1849, ... , 2059
Year of death	1947, 1948, ... , 2059
Id. no. of person in question	(1*)
Id. no. of mother	(1*)
Id. no. of father	(1*)
Year of leaving parental home	Year of birth + 15, ... , year of birth + 35
Year of emigration	Year of birth, year of birth + 1, ... , year of death
Year of immigration	Year of birth, year of birth + 1, ... , year of death
Partner emigrated	Yes, no
Number of siblings	0, 1, ... , 19
Marital status	Never married and not cohabiting, never married and cohabiting, married, divorced and not cohabiting, divorced and cohabiting, widowed and not cohabiting, widowed and cohabiting
Sex	Male, female
Number of children (n)	0, 1, ... , 20
Number of children allocated to partner when separated (p)	0, 1, ... , 20
Id. no. of children	(n*)
Id. no. of children allocated to partner	(p*)
Year of cohabitation	Year of birth + 15, ... , year of birth + 84
Year of marriage	Year of birth + 15, ... , year of birth + 84
Year of separation	Year of cohabitation (or marriage), year of cohabitation (or marriage) + 1, ... , year of birth + 85
Id. no. of partner	(1*)
Id. no. of former partner	(1*)
Year of widowhood	Year of marriage, ... , year of birth + 99
Type of institution and date of entrance	(1*)

data from the NCBS 1984 forecast.[6] From 2035, the rates are held constant at the 2034 level.

Besides, the size of a number of simulated subpopulations is determined to calculate some of the aforementioned rates. This concerns the number of new male and female immigrants and emigrants by year of immigration and emigration, respectively; the female population, aged 15 up to 50 years, by parity, marital status, and cohabiting or not; the number of dehabitations plus

[6] Also available from tables from NCBS, see also NCBS (1985).

the number of cohabitations that ended in the death of one of the partners plus the number of cohabiting men emigrating without their family; the number of married men living abroad without their family; the number of immigrated married men living in the Netherlands without their family, the number of male and female persons, aged 15 up to 40, still living with their parents, and the number of persons living in the various types of institutional households.

All demographic events are assumed to take place on June 1st of each year. Divorce and dehabitation do not occur in the year of marriage or cohabitation. And after a divorce or dehabitation, the possibility of remarrying or re-cohabiting does not exist until the next year. If a person dies, s/he is, in spite of this, subjected to the other possible demographic events. However, the probability that these events will occur is halved. In the case of emigration, the person's record is maintained. In the year of emigration, other possible demographic events also occur with a 50 per cent chance. In the next years, no demographic changes are allowed until s/he decides to return to the Netherlands, with the exception of mortality. This means that, in spite of the recursive nature of the model, competing risks are dealt with to a large extent. Only the occurrence of marriage, cohabitation, dehabitation, and divorce in the same year are excluded. The resulting error, however, is very small.

Table 14.3 gives an impression of the power of microsimulation and the large amount of differentiated simulation it provides. For example, we can generate the distribution of people by year of birth, year of death, year of leaving parental home, year of immigration, year of emigration, whether partner emigrated with or without the reference person, number of siblings, sex, marital status, number of children, number of children with former partner, year of birth of the children, year of cohabitation, year of marriage, level of education of the reference person and of his or her partner, year of separation, and year of widowhood, and all this simultaneously. If we want to generate similar information by macrosimulation we need $212 \times 100 \times 21 \times 100 \times 100 \times 2 \times 20 \times 20 \times (35 \times 20) \times 70 \times 70 \times (5 \times 5) \times 85 \times 85 = 2.2 \times 10^{24}$ states, requiring about 10^{19} megabytes, but we still do not know who is whose partner or child, whereas this kind of information is also available from microsimulation. When using a database of 100,000 persons, we need only 8 megabytes to generate the aforementioned information (excluding the transition rates). For example, the Dutch macrosimulation model LIPRO (Van Imhoff & Keilman 1991) contains only eleven household positions and deals with 100 (age) $\times$ 2 (sex) $\times$ 2 (exits) $\times$ 11 (household positions) = 2,200 states.

14.4. THE LABOUR MARKET MODULE IN NEDYMAS

In the model NEDYMAS, sixteen positions are distinguished as economic activity, namely: employee, self-employed, civil servant, retired, unemployed employee, unemployed civil servant, unemployed others, soldier, disabled employee, disabled civil servant, disabled others, student, housewife or houseman, unemployed ex-disabled employee, unemployed ex-disabled civil servant, and unemployed ex-disabled others.

The Labour Force Surveys ('Arbeidskrachtentellingen') are the starting point for the determination of the transition probabilities between the different states of economic activity. For example, the Labour Force Survey 1977 contains information on the economic activity in 1976 and 1977 for men and women of all ages between 14 and 66 years old. The states used in the Labour Force Survey are transformed to six states by us: disabled, employed, unemployed, soldier, student, and retired or working in own household (housemen and housewives). Using these data, it is determined for each year for each individual whether his or her economic activity changes. In this process, additional data are used, e.g. data that refer to unemployment.

The determination of the labour supply is an important element in this. Labour supply is determined using a labour supply equation, modelled by Van Soest, Woittiez, and Kapteyn (1990), which explicitly takes account of demand-side restrictions. The labour supply of individual household members is considered in a neo-classical framework, in which net wages, the social security and tax system, and the household composition are taken into account. The demographic variables used are LOGFS (logarithm of family size), DCH<6 (dummy children younger than 6), LAGE (logarithm of age), and L2AGE (LAGE-squared). The estimation results with respect to the preferences are shown in Table 14.4. The family characteristics appear to be insignificant for males and very significant for females. We will not deal with the possible transitions between the various states of economic activity (Nelissen 1994, chapter 6). All possible transitions between the 16 distinguished groups are simulated. When all transitions have taken place, the simulated number of employed and unemployed persons will generally differ from the actual or predicted number of employed and unemployed persons. The main reason for this is that the estimation of the parameters of the labour supply function is executed with data referring to the year 1985. It is quite debatable whether people in, for example, the 1950s show the same behaviour. Therefore we confront the simulated numbers with the actual or predicted figures. For the period 1990-2029, we use the forecasts of Departementale Werkgroep SZW (1984). For the years 2030 and further, we assume that the number of unemployed persons remains constant and that the labour force participation rates remain constant at the 2029 level.

Table 14.4. Estimation results for the preferences (standard errors in parentheses) in the extended model of Van Soest *et al.* (1990)

Parameter	Males		Females	
β (wage rate)	0.41	(0.24)	0.77	(0.24)
δ (unearned income)	- 0.0007	(0.003)	- 0.0041	(0.0013)
α_0 (constant term)	-259.	(109.)	-172.	(85.5)
α_1 (LOGFS)	0.14	(1.73)	- 14.1	(3.1)
α_2 (DCH<6)	- 0.90	(1.29)	- 9.03	(2.22)
α_3 (LAGE)	167.	(61.)	126.	(49.9)
α_4 (L2AGE)	- 23.7	(8.4)	- 19.8	(7.2)
σ_ϵ (random preference)	11.91	(0.98)	7.71	(1.96)

Source: Van Soest *et al.* (1990, p. 536).

First, the simulated and the true numbers of unemployed men and women are compared. If the simulated number is less than the true one, the simulated number is raised by giving persons (at maximum up to the moment both numbers no longer differ) who are houseman/housewife or employed, the state 'unemployed'. First labour supply is tested, if it concerns a housewife or a houseman. They are only added to the unemployed if labour supply is positive. This process continues until the simulated number of unemployed persons equals the true one. If the simulated number is larger than the true number of unemployed persons, then the simulated number is diminished by giving the corresponding number of unemployed persons in the sample, the state 'employed'. This also happens in a random way.

Next, the simulated and actual numbers of employed men and women are compared. The procedure is analogous. At last we test whether the employed population is correctly subdivided over the categories employees, civil servants, and self-employed persons. If this is not the case, economic status has been adjusted to get the right numbers.

14.5. THE INTERACTION OF THE HOUSEHOLD AND THE LABOUR MARKET MODULE IN THE MAJOR MICROSIMULATION MODELS

We see that the interaction between the household and the labour market module in NEDYMAS is one-sided. Labour market behaviour is not supposed to affect household transitions. Socio-economic variables only play a role in the matching process via the level of education of the marriage and cohabitation candidates. On the other side, demographic processes do influence labour market decisions. For the latter, we find that female labour force participation varies with the presence of young children and family size. Moreover, age and sex play a dominant role in all socio-economic transitions.

The foregoing also roughly holds for microsimulation models in other countries. For example, the labour force participation decisions of women in the German Sfb3 model are determined by variables such as family size and the age of the youngest child (Galler & Wagner 1986, p. 237). With respect to the reverse relationship, virtually the same applies as for NEDYMAS: the socio-economic part does not affect the demographic part, except for the splitting-off of children, which is affected by the beginning of a job or new school type (e.g. Ott 1986).

In the model of Canadian Statistics DEMOGEN (e.g. Wolfson 1989), first union and union dissolution depend on labour force participation, whereas labour force participation is determined by marital status and the presence of children by age. However, these relationships have not been estimated simultaneously.

The DYNASIM model (Orcutt *et al.* 1976) is somewhat more elaborated in this respect, although the labour force participation equation has not been estimated, but results from a series of cross-tabulations. The labour force participation of women depends on marital status and the presence of children younger than six. Contrary to NEDYMAS and the Sfb3 model, the demographic transitions in DYNASIM are also determined by socio-economic characteristics. However, the transitional equations are not really based on a theoretical model, but are the result of ad hoc findings. The following interactions are used. The divorce probability depends on the socio-economic status (being disabled or not, being unemployed or not) of the head of the household and the level of the wife's earnings. Birth and death rates are, among other things, determined by the level of education. First marriage selection depends on level of education, the number of hours spent in labour force participation, the wage rate, and the part of income received from the Aid to Families with Dependent Children (AFDC) program, from the food stamps program, and from unemployment compensation payments.

The foregoing makes clear that the effect of demographic variables on labour market transitions is generally included in the more recent models. This is in accordance with the developments in labour economics during the last decades that led to a narrowing of the gap between stylized textbook models and reality. The reverse, the implementation of socio-economic variables in the household module, is at the same level as two decades ago and also here, this partially reflects the developments within this field, namely the limited number of empirical applications, although theory has been developed more and more. We mainly find empirical applications in this field in the United States, and this is reflected in the interactions between the demographic module and the labour market module in American microsimulation models, like DYNASIM. But, as we saw, the application of behavioural equations rests on ad hoc models and simultaneous models have not been used.

All in all, we find that model builders are aware of the simultaneity of demographic and labour market transitions, but that putting this in practise is done in a rather opportunistic way. The problem is not the absence of theories. A broad range of (theoretical) models exists, in which labour force participation and the various demographic processes, like fertility, marriage, divorce, and leaving the parental home are treated simultaneously (see for an overview Nelissen 1990). But "(t)heory seems far ahead of measurement" (Ritzen & Van Dalen 1990, p. 178). Another problem forms the absence of data specifically gathered to estimate this type of demographic decisions. The latter results in the estimation of models in which the inclusion of socio-economic characteristics is mainly determined by their 'accidental' availability in the database to be used, instead of the theoretical model that forms the point of departure.

These problems are responsible for the absence of behavioural equations in the demographic part of NEDYMAS. Until recently, the only available results in the Netherlands with respect to the effect of socio-economic variables on demographic transitions are from Siegers (1985) and deal with fertility. However, his results are not suitable for our application. The simultaneous model is estimated using cross-sectional data, whereas the use of micro-data is limited to the Willis model, which is, however, poorly implemented. Moreover, it has only been applied for women born before 1943, so that his results are not very useful in modelling household dynamics in the 1990s.[7] At the moment, data are less and less a problem as a consequence of the availability of a large household panel survey in the Netherlands. However, it looks as if Dutch demographic researchers have had a dislike of economic-oriented research in the field of decision-making processes of demographic behaviour.[8] In spite of this, a new generation of researchers has taken up the challenge, which recently resulted in a number of useful publications (Manting 1994; Mulder & Manting 1994; Vermunt 1991). These are possibly suitable for inclusion in NEDYMAS.

---

7 We could follow the same procedure as for the labour market positions, but — in contrast with the model of Van Soest *et al.* — the explanatory variables used by Siegers cannot be considered to cover the theoretical basis.

8 We can think of the low priority that this type of research has in the research agenda of the 'NWO Prioriteitsprogramma Bevolkingsvraagstukken'.

14.6. CONCLUSIONS

From the foregoing, we can conclude that microsimulation in the field of household modelling can be very fruitful if interest goes beyond the characteristics age, marital status, sex and household position. This especially holds if a linkage is made with socio-economic variables, where the latter are affected by household characteristics (e.g. labour market transitions) or if so-called user profiles not only depend on household characteristics (e.g. eligibility for most social security schemes). Another very important advantage of microsimulation is that information is utilized optimally, whereas multidimensional methods lead to a loss of information as a consequence of aggregation, and this is reflected in the output possibilities.[9]

In this contribution, we looked at how the demographic and labour market modules have been implemented in dynamic microsimulation models, with emphasis on the Dutch model NEDYMAS. It appears that, at least for the time being, a twofold development will continue. The inclusion of demographic variables in the labour market module is already generally applied and will also be applied in the field of getting unemployed or disabled. The results of, among others, Aarts and De Jong (1990) and Van den Berg (1990) can be used for the modelling of disablement and labour mobility, respectively. On the other side, only first results with respect to the inclusion of socio-economic variables to explain demographic transitions are available.

ACKNOWLEDGEMENTS

The author is indebted to Ad Vossen, Tilburg University, for helpful comments on an earlier draft. The author would like to gratefully acknowledge the funding provided by the *Research Program for Population Studies* of the Netherlands Organization for Scientific Research (NWO, grant no. 18.051).

REFERENCES

AARTS, L.J.M., & P.R. DE JONG (1990), *Economic Aspects of Disability Behavior*, Ph.D. Dissertation, Leyden University: Leyden.

BRENNECKE, R. (1981), *Bevölkerungsentwicklung, Gesundheitsbeschwerden und Inanspruchnahmen von Ärzten*, Sfb3 Arbeitspapier No. 52, Frankfurt.

---

[9] For example, using the same data as LIPRO and applying microsimulation, it is very easy to connect children to their parents and consequently it is very simple to simulate, for example, the distribution of households by household size and age of the household members.

DEPARTEMENTALE WERKGROEP MINISTERIE VAN SZW (1984), *Demografische Ontwikkelingen in Macro-Economisch Perspectief*, Ministry of Sociale Affairs and Employment: The Hague.

FALLOWS, S.E. (1982), *Implementation by Personal Transfer: The TRIM/MATH-Model*, IPES 82.0208, Gesellschaft für Mathematik und Datenverarbeitung mbH: Bonn.

GALLER, H.P., & A. STEGER (1978), Mikroanalytische Bevölkerungssimulation als Grundlage sozialpolitischer Entscheidungen — Erste Ergebnisse. *In*: H.-J. KRUPP & W. GLATZER (eds.) *Umverteilung in Sozialstaat*, Campus Verlag: Frankfurt, pp. 237-275.

GALLER, H.P., & G. WAGNER (1981), Das Mikrosimulationsmodell. *In*: H.-J. KRUPP, H.P. GALLER, H. GROHMAN, R. HAUSER & G. WAGNER (eds.) *Alternativen der Rentenreform '84*, Campus Verlag: Frankfurt, pp. 177-211.

GALLER, H.P., & G. WAGNER (1986), The Microsimulation Model of the Sfb3 for the Analysis of Economic and Social Policies. *In*: G.H. ORCUTT, J. MERZ & H. QUINCKE (eds.) *Microanalytic Simulation Models to Support Social and Financial Policy*, Elseviers Science Publishers: Amsterdam, pp. 227-247.

HAMACHER, B. (1978), Auswirkungen gesetzlicher Beitragsregelungen der Krankenversicherung auf die ökonomische und soziale Lage von Personen und Haushalten — Ein Mikrosimulationssystem der Krankenversicherung, Frankfurt (mimeo).

HECHELTJEN, P. (1974), *Bevölkerungsentwicklung und Erwerbstätigkeit*, Westdeutscher Verlag: Opladen.

HEIDA, H., & H. GORDIJN (1985), *PRIMOS Huisoudenmodel; Analyse en Prognose van de Huishoudensontwikkeling in Nederland*, Ministerie van VROM: The Hague.

HELBERGER, C. (1982), *Auswirkungen öffentlicher Bildungsausgaben in der Bundesrepublik Deutschland auf die Einkommensverteilung der Ausbildungsgeneration*, Kohlhammer: Stuttgart.

HELBERGER, C., & G. WAGNER (1981), Beitragsäquivalenz oder interpersonelle Umverteilung in der gesetzlichen Rentenversicherung? *In*: P. HERDER-DORNEICH (ed.) *Dynamische Theorie der Sozialpolitik*, Duncker & Humblot: Berlin, pp. 331-392.

HELLWIG, O.(1988), *Micromodelling the Australian Household Sector; A Proposal*, Darmstadt University: Darmstadt.

HORVITZ, D.G., F.G. GIESBRECHT, B.V. SHAH & P.A. LACHENBRUCH (1972), *POPSIM, A Demographic Microsimulation Model*, Monograph no. 12, Carolina Population Centre, University of North Caroline: Chapel Hill.

HOWELL, N., & V.A. LEHOTAY (1978), AMBUSH: A Computer Program for Stochastic Microsimulation of Small Human Populations, *American Anthropologist*, Vol. 80, pp. 905-922.

HYRENIUS, H., & I. HOLMBERG (1970), *On the Use of Models in Demographic Research*, Demographic Institute: Gothenborg.

KEILMAN, N.W. (1988), Dynamic Household Models. *In*: N.W. KEILMAN, A.C. KUIJSTEN & A.P. VOSSEN (eds.) *Modelling Household Formation and Dissolution*, Clarendon Press: Oxford, pp. 123-138.

KEILMAN, N.W., & N. KEYFITZ (1988), Recent Issues in Dynamic Household Modelling. *In*: N.W. KEILMAN, A.C. KUIJSTEN & A.P. VOSSEN (eds.) *Modelling Household Formation and Dissolution*, Clarendon Press: Oxford, pp. 254-284.

KLANBERG, F. (1975), *Entwicklung eines Mikrosimulationsmoduls auf Mikrodatenbasis*, SPES Arbeitspapier No. 34, Frankfurt/Mannheim.

KRUPP, H.-J., H.P. GALLER, H. GROHMAN, R. HAUSER & G. WAGNER (eds.) (1981), *Alternativen der Rentenreform '84*, Campus Verlag: Frankfurt.

MANTING, D. (1994), *Dynamics in Marriage and Cohabitation*, PDOD/Thesis: Amsterdam.

McClung, N. (1969), Estimates of Income Transfer Program Direct Effects. *In*: President's Commission on Income Maintenance Programs, *Technical Studies*, US Government Printing Office: Washington D.C.

Mulder, C.H., & D. Manting (1994), Strategies of Nest-Leavers: 'Settling Down' versus Flexibility, *European Sociological Review*, Vol. 10, pp. 155-172.

NCBS (1953), *Bevolking van Nederland. Leeftijd en geslacht, 1900-1952. Leeftijd, geslacht en burgerlijke staat, 1947-1952*, De Haan NV: Utrecht.

NCBS (1970), *Bevolking van Nederland naar geslacht, leeftijd en burgerlijke staat 1930-1969*, Staatsuitgeverij: The Hague.

NCBS (1984), *Vijfentachtig Jaren Statistiek in Tijdreeksen*, Staatsuitgeverij: The Hague.

NCBS (1985), *Bevolkingsprognose voor Nederland 1984-2035*, Staatsuitgeverij: The Hague.

Nelissen, J.H.M. (1990), The Microeconomic Theory of Household Formation and Dissolution: State-of-the-Art and Research Proposals. *In*: C.A. Hazeu & G.A.B. Frinking (eds.) *Emerging Issues in Demographic Research*, Elsevier: Amsterdam, pp. 127-170.

Nelissen, J.H.M. (1991), *The Modelling of Institutional Households in the Microsimulation Model NEDYMAS*, Working Paper 63, Tilburg University: Tilburg.

Nelissen, J.H.M. (1994), *Income Redistribution and Social Security: An Application of Microsimulation*, Chapman & Hall: London.

Orcutt, G.H. (1957), A New Type of Socio-Economic System, *Review of Economics and Statistics*, Vol. 39, pp. 116-123.

Orcutt, G.H., S. Caldwell & R. Wertheimer II (1976), *Policy Exploration Through Microanalytic Simulation*, Urban Institute: Washington D.C.

Orcutt, G.H., M. Greenberger, J. Korbel & A. Rivlin (1961), *Microanalysis of Socioeconomic Systems: A Simulation Study*, Harper and Row: New York.

Ott, N. (1986), *Bevölkerungsentwicklung bis zum Jahr 2000; Modellrechnungen mit der Version 86.0 des Mikrosimulationsmodells des Sfb3*, Working Paper Sonderforschungsbereich 3 no. 212, Frankfurt University: Frankfurt.

Pechman, J.A. (1965), A New Tax Model for Revenue Estimating. *In*: A.T. Peacock & G. Hauser (eds.) *Government Finance and Economic Development*, OECD: Paris, pp. 231-244.

Ploegmakers, M.J.H., & L.Th. van Leeuwen (1983), *Veranderende Huishoudens in Nederland. Statistische Analyse over de Periode 1960-1977*, Ministerie van VROM: The Hague.

Ridley, J.C., & M.C. Sheps (1966), An Analytical Simulation Model of Human Reproduction with Demographic and Biological Components, *Population Studies*, Vol. 19, pp. 297-310.

Ritzen, J.M.M., & H.P. van Dalen (1990), A Comment. *In*: C.A. Hazeu & G.A.B. Frinking (eds.) *Emerging Issues in Demographic Research*, Elsevier: Amsterdam, pp. 175-180.

Siegers, J.J. (1985), *Arbeidsaanbod en Kindertal*, Ph.D. Dissertation, University of Groningen: Groningen.

Steger, A. (1980), *Haushalte und Familien bis zum Jahr 2000. Eine mikroanalytische Untersuchung für die Bundesrepublik Deutschland*, Campus Verlag: Frankfurt.

Van den Berg, G.J. (1990), *Structural Dynamic Analysis of Individual Labour Market Behaviour*, Ph.D. Dissertation, Tilburg University: Tilburg.

Van Imhoff, E., & N.W. Keilman (1991), *LIPRO 2.0: An Application of a Dynamic Demographic Projection Model to Household Structure in the Netherlands*, Swets & Zeitlinger: Amsterdam.

Van Leeuwen, L.Th. (1980), *Ontwikkelingsfasen van het Gezin*, Monografieën Volkstelling 1971, 9, Staatsuitgeverij: The Hague.

VAN SOEST, A., I. WOITTIEZ & A. KAPTEYN (1990), Labor Supply, Income Taxes, and Hours Restrictions in The Netherlands, *Journal of Human Resources*, Vol. 15, pp. 517-558.

VAN 'T KLOOSTER-VAN WINGERDEN, C.M. (1979), *Huishoudensvorming en Samenlevingsvormen*, Monografieën Volkstelling 1971, 11, Staatsuitgeverij: The Hague.

VERMUNT, J. (1991), Een Multivariaat Model voor de Geboorte van het Eerste Kind, *Maandstatistiek van de Bevolking*, Vol. 39, pp. 22-33.

WACHTER, K.W., E.A. HAMMEL & P. LASLETT (1978), *Statistical Studies of Historical Social Structure*, Academic Press: New York.

WILENSKY, G. (1969), An Income Transfer Computation Program. *In*: President's Commission on Income Maintenance Programs, *Technical Studies*, US Government Printing Office: Washington D.C.

WOLFSON, M.C. (1989), Divorce, Homemaker Pensions and Lifecycle Analysis, *Population Research and Policy Review*, Vol. 8, pp. 25-54.

Epilogue

15. EPILOGUE

Evert van Imhoff[1], Anton Kuijsten[2], and Leo van Wissen[1]

[1]Netherlands Interdisciplinary Demographic Institute (NIDI)
P.O. Box 11650
2502 AR The Hague
The Netherlands
[2]Department of Planning and Demography
University of Amsterdam
Nieuwe Prinsengracht 130
1018 VZ Amsterdam
The Netherlands

15.1. HOUSEHOLD DEMOGRAPHY: TEN YEARS AFTER

On the basis of the results of the 1984 Voorburg workshop mentioned in chapter 1, several years ago Keilman and Keyfitz sketched a number of elements of a research strategy "which should ultimately lead to a better performance and therefore a greater usefulness of household models" (Keilman & Keyfitz 1988). Since then, some of Keilman & Keyfitz' wishes have been fulfilled, or have at least shown considerable progress, while other components have been neglected or have until now proved to be resistant against research ingenuity.

Among the latter issues, standard definitions and typologies are still lacking, as shown by Keilman in this volume. Nevertheless, Kuijsten reports some progress in classifying countries according to their time path in going through the so-called Second Demographic Transition. Another unfulfilled wish has been the application of 'fuzzy set theory' in household demography, given the inherent uncertainties in identifying household membership. Strikingly enough, the topic was raised several times during the summer course and is also discussed in Ott's contribution. Theory in general remains a rather painful subject, and the recent history of household demography seems to reconfirm this, although in this volume Burch remains hopeful and offers several useful suggestions for future progress.

Household Demography and Household Modeling
Edited by E. van Imhoff *et al.*, Plenum Press, New York, 1995

Other parts of Keilman & Keyfitz' research agenda *have* been addressed during the past decade. Some progress has been made in solving the consistency problem in household macro-simulation models (cf. the contribution by Van Imhoff). The notion of competing risks has received ample attention, both in household models (cf. the contribution by Galler) and in the analysis of household behaviour with the upsurge of event history analysis in demography (cf. the contribution by Courgeau). Event history techniques typically require longitudinal data, the availability of which has significantly improved, although the chapter by Ott shows that there are still problems of data collection and data management.

The contributions collected in this volume can be viewed as reflecting the main themes that are currently foremost in household demography. This is not to say that all relevant topics have been covered. For instance, an explicit new economic or sociological perspective is absent in this volume, signalling the fact that at present the main focus lies elsewhere.

The strong emphasis on data and especially on models raises the question whether we really need theory to base our household models on. Bartlema and Vossen (1988) state that household models should be based on theories, but for the time being it looks like we can do without much theory in developing our multidimensional household simulation models. It is interesting to explore this situation a little further and to see what role substantive theoretical concepts play in the models presented in Part III. This issue will be pursued in the next section.

Since the emphasis in the most recent developments is on modelling, in section 15.3 we will evaluate what we call the 'mainstream' household approach: the multidimensional model and its variants, as presented in this volume. An important question raised in this respect is to what extent the present generation of multidimensional simulation models is able to cope with anticipated future developments in the field of household demography. We argue that the life course perspective is the most fruitful approach to the next generation of household models.

15.2. HOUSEHOLD DEMOGRAPHY: MODELS WITHOUT THEORY?

The first three models in Part III, namely those presented by Prinz, Nilsson, and Sellerfors, De Beer, and Van Imhoff, can be described in the words of Ryder (1992) as "formal demographic models": they are completely general, but lack any behavioural substance. All three are macro-simulation models based on the methodology of the multistate or multidimensional approach. Multidimensional models have played a prominent role in household demography during the last ten years. The quest for empirical transition probabilities is central to this modelling strategy. The data needed to support such transition models have only recently become available in some countries, and do not exist at all in most other countries. Whatever the merit of the multidimensional approach in other respects, its availability has certainly guided our thinking about the definition, organization, and collection of consistent household data, although, as Keilman shows, the results may still be very different among countries. In a number of countries, data on household transitions have started to build up since the 1980s and, within a few years, time series on these transitions will be available that should stimulate further theoretical developments in this area. So, in a way, the application of these formal demographic household models helps to pave the way for theoretical progress in household demography.

The situation is somewhat different for the two other models presented in Part III. Both models are not 'pure' demographic models, as they contain non-demographic elements as well. Heida and Hooimeijer show that relevant aspects of the household decision process can be endogenized in a multistate household model. They incorporate several aspects of housing market theory and succeed in retaining the multistate model structure, yet at the expense of increased complexity and reduced transparency. The housing market aspects act within a number of 'niches' of this overall modelling framework. What goes on within each of these niches is structured according to notions derived from housing market theory. Thus, instead of specifying an explicit time path for the probability of leaving the parental home, their model SONAR simulates the process of leaving the parental home on the basis of a time series for housing supply by relevant characteristics. Nelissen faces a similar situation when linking household demography and labour market behaviour. However, in fact the only link established in NEDYMAS is the conditioning effect of household characteristics on female labour supply. According to Nelissen, the problem is not the lack of theory but the lack of data and empirical analyses giving empirical evidence and parameter estimates on the simultaneity of the processes. Once such relationships are established empirically, his microsimulation methodology is well suited to the task of incorporating these effects.

The housing market influence on household transitions focusses on the process of leaving the parental home, whereas the demographic effects of the labour market are mainly concentrated on the choice of the number of children and leaving the parental home. It seems, therefore, that a full account of the latter process would involve both housing and labour market considerations. Whereas housing market and labour market theorists have not even come to terms with each other here, it seems unlikely that demographers will have much to say on this very complicated topic in the near future.

15.3. THE MULTIDIMENSIONAL HOUSEHOLD MODEL AND BEYOND

The multistate approach has proven to be a very powerful tool in the modelling of household structures. Indeed, without diminishing the relevance of the arguments in favour of microsimulation and recognizing the popularity of microsimulation approaches, we think that the multidimensional household model, as presented in its pure form in the chapters by Van Imhoff and by Prinz, Nilsson, and Sellerfors, is the mainstream in current household modelling. Its advantages are obvious. The model is a generalization of the traditional life table model in demography and is conceptually simple. The methodology of the standard multidimensional model is well developed and, last but not least, the model is very accessible due to the availability of excellent software.

Despite these advantages and successes, there are a number of inherent weaknesses when applying multidimensional models to household structures, which may be summarized as follows. First, the underlying Markov assumption of the process may not be valid. In particular, there are two assumptions that are very unlikely to hold in household transition processes: the model disregards effects of previously occupied states; and sojourn times are assumed to have no effect on transition probabilities. In this volume, Courgeau shows evidence that sojourn times *are* important in determining transition probabilities of changing position in the household. Second, due to the aggregated nature of the data, it is impossible to make inferences about groups in the population other than those based on the pre-defined state space. For instance, a distribution of households by size is impossible to make unless the size of the household is explicitly incorporated in the state space. A third drawback, although demographers are quite familiar with it, since almost all demographic models share this feature, is the fact that transition rates are exogenous. However, demographers have been able to live with these vast amounts of exogenously-determined parameters in the case of mortality and, although already much more difficult, fertility. If transition probabilities are stable or only change slowly over time, there is nothing wrong with this approach. In the case of changes in household position, however, the methodology breaks down, since household dynamics is a very heterogeneous process, with large fluctuations over time. Fourth, and traditionally put forward by microsimulators (see e.g. Nelissen, this volume), there is the problem of the large number of cells when the state space becomes large. Even with a moderate number of household categories, an aggregated approach will become inefficient if not infeasible from a data-handling point of view. Finally, the large number of cells in the transition matrix requires a considerable effort in estimating the required transition parameters. At present, the data available in most countries do not support the estimation of all parameters of even a moderately-sized multidimensional model.

Not all the arguments listed here are unique to the multidimensional household model. For instance, the Markov assumption underlies many microsimulation approaches as well, and exogenous parameter values are typical, not only for the multidimensional household model, but for most demographic models.

As Keilman and Keyfitz state, after listing at least eight different theoretical approaches with respect to the household (1988, p. 263), "searching for *the* universally valid theory is a futile attempt". They consider the life course perspective, also suggested by Willekens (1988, 1991), as the general framework for synthesizing various partial theoretical explanations. The life cycle of the household may be defined in terms of the individual life courses of its members, and household dynamics can be understood by reference to changes in the individual life courses (Willekens 1988, p. 88). Since the individual life course is the collection of all careers a person is engaged in, the concept is very well suited to integrate various partial theories into a consistent structure.

Indeed, the life course approach has gained much popularity recently, and offers a flexible framework for representing complex relationships among events. Galler gives a good overview of a number of important methodological issues regarding the modelling of competing risks, while Courgeau presents an excellent empirical example of life course analysis using hazard models. Although a lot of methodological problems remain to be solved, it appears that the analytical tools are there to answer many questions that arise from a theoretical household perspective. In addition, we anticipate that data availability will improve significantly in the future, due to the application of automatic registration and geo-information processing techniques. As a consequence, we expect that the testing of substantive household theories will become much more important in the future, and much more will be known about dynamic household processes in terms of causal linkages, simultaneity, timing of events, etc. The simple Markovian transition scheme used in the present models will be increasingly questioned from a theoretical perspective. Other, probably more complex, causal structures will be hypothesized and tested empirically. This will lead to more complex model structures. Heida and Hooimeijer show that at least some of these new developments can be handled in a macro-simulation context. However, in general, we anticipate that the translation of results from life course analysis into a simulation context will lead to a micro-approach.

Although the advances in computing technology have been tremendous, it is still a difficult task to implement a microsimulation model. Most existing models have been written for a mainframe computer, and even the fastest PC's still have difficulties with these large programmed micro-structures. In addition, documentation is often scarce or absent. What is therefore definitely needed for the future is a flexible 'life course microsimulation projection' program, that allows the simulation of alternative household decision structures based on sound theoretical conjectures.

15.4. THE CIRCLE GAME

It was back in the early 1970s, when household and family demography was not yet there, that Joni Mitchell recorded a song called 'The Circle Game', a song that in fact deals with man's life course, during which time seems to fly:

"And the seasons they go round and round
And the painted ponies go up and down
We're captive on the carousel of time
We can't return we can only look behind
From where we came
And go round and round and round
In the circle game."

In our Editors' Introduction we expressed the feeling that over the years, in its growth towards maturity, family and household demography seems to have managed quite well in keeping the balance between theory, empirical analysis and formal modelling. In this volume too, we have tried to do justice to all of these pillars on which scientific progress must be founded, attempting to give proper weight to issues of both past and present trends, theory, data, and modelling. Admittedly, strong emphasis has been laid on data and especially on models, if only counting space devoted to these issues. But the reader should be aware, as the editors are, of the fact that none of these issues has really pride of place when looking upon the scientific process as an everlasting cyclical movement, an eternal circle game, in which adequate data are input in models that test theories on their explanatory power for observed empirical trends, leading to new questions, collection of still more adequate data, and so on, and so forth.

In our scientific activities we are all captive on that carousel, with shining painted ponies called 'trends', 'data', 'theory', and 'models' each in their own quadrant of the circle. We sometimes think that it would be nice to ride the pony we prefer, if only temporarily, but as soon as the merry-go-round is turned on, all four ponies move with equal speed and after each full turn they are back at the place where they started. And the carousel cannot be stopped once it has started, the show must go on.

Some years after Joni Mitchell's song was released, another performer, Jim Croce, wished "If I could save time in a bottle". To some extent, saving time in a bottle is precisely what state-of-the-art publications do, and not only for the participants of the meeting on which the book is based. Even if we cannot return or stand still, time and again we must look behind, to from where we came. It is the editors' sincere hope that those who will do so in future, might find something of value in this bottle.

REFERENCES

BARTLEMA, J. & A. VOSSEN, (1988) Reflections on Household Modelling. *In*: KEILMAN, KUIJSTEN & VOSSEN (eds.).

KEILMAN, N., A. KUIJSTEN & A. VOSSEN (eds.), (1988), *Modelling Household Formation and Dissolution*, Clarendon Press: Oxford.

KEILMAN, N., & N. KEYFITZ, (1988), Recurrent Issues in Dynamic Household Modelling. *In*: KEILMAN, KUIJSTEN & VOSSEN (eds.).

RYDER, N., (1992) The Centrality of Time in the Study of the Family. *In*: E. BERQUÒ & P. XENOS (eds.), *Family Systems and Cultural Change*, Clarendon Press: Oxford.

WILLEKENS, F., (1988) A Life Course Perspective on Household Dynamics. *In*: KEILMAN, KUIJSTEN & VOSSEN (eds.).

WILLEKENS, F., (1991) Understanding the Interdependence between Parallel Careers. *In*: J. Siegers, J. de Jong Gierveld & E. van Imhoff (eds.), *Female Labour Market Behaviour and Fertility: A Rational-Choice Approach*, Springer-Verlag: Berlin.

AUTHOR INDEX

SUBJECT INDEX

ISBN 0-306-45187-5
90000
9 780306 451874